Martin Waldseemüller's 'Carta marina' of 1516

Chet Van Duzer

Martin Waldseemüller's 'Carta marina' of 1516

Study and Transcription of the Long Legends

Springer Open

Chet Van Duzer
Lazarus Project
University of Rochester
Rochester, NY, USA

ISBN 978-3-030-22705-0 ISBN 978-3-030-22703-6 (eBook)
https://doi.org/10.1007/978-3-030-22703-6

This book is an open access publication; the preparation and open access publication of "Martin Walseemüller's 'Carta marina' of 1516 – Study and Transcription of the Long Legends" were made possible thanks to generous support from the Jay I. Kislak Foundation, Inc.

This Springer imprint is published by the registered company Springer Nature Switzerland AG
The registered company address is: Gewerbestrasse 11, 6330 Cham, Switzerland

Acknowledgements

It is a pleasure to thank here those who have supported me in my work on this book. My research on the *Carta marina* was generously funded by a Kislak Fellowship at the Library of Congress from November 2011 to March 2012 and June to July 2012, and I offer my enthusiastic thanks to Jay Kislak, Arthur Dunkelman, and the Jay I. Kislak Foundation for their support. I also wish to thank Barbara Tenenbaum of the Hispanic Division of the Library of Congress, who was Curator of the Kislak Collection during my fellowship; Mary Lou Reker of the Kluge Center at the Library of Congress for facilitating my work; Ralph Ehrenberg, then Chief of the Geography and Map Division for his encouragement, and Eric Frazier of the Rare Book and Special Collections Division for his unstinting generosity. And I thank my intern in the Kluge Center, Chelsea Jimenez, for her help and good humor.

I owe particular thanks to John Hessler of the Geography and Map Division at the Library of Congress, my co-author on the book *Seeing the World Anew: The Radical Vision of Martin Waldseemüller's 1507 & 1516 World Maps* (Delray Beach, FL: Levenger, and Washington, DC: Library of Congress, 2012), which contains some of my research on the *Carta marina* that is presented here. Working on the Waldseemüller maps with John, who wrote about the 1507 map, was an exciting intellectual adventure.

And I offer my thanks again to the Jay I. Kislak Foundation and to Arthur Dukelman for generously funding the release of this book in open access, which I trust will facilitate the enjoyment and study of Waldseemüller's map by a wider audience than would otherwise be possible. Sadly, Mr. Kislak himself passed away on October 3, 2018; may he rest in peace.

Providence, RI, USA

Chet Van Duzer

Contents

Introduction to the *Carta Marina* 1

1.1 Introduction

This book is devoted to an imposing world map, printed on twelve sheets and rich in detail, that was designed by the German cartographer Martin Waldseemüller in 1516, whose only surviving exemplar is in the Jay I. Kislak Collection at the Library of Congress. This map, the *Carta marina*, has tended to live in the shadow of Waldseemüller's earlier world map, that printed in 1507, which is famous for being the first to apply the name "America" to the New World. The *Carta marina* lacks some of the striking audacity of the 1507 map, on which the cartographer not only debuts a new name for the newly discovered lands in the west, but also represents all 360 degrees of longitude at a time when the interior and the western reaches of the New World were unknown, and the vastness of the Pacific was still undiscovered by Europeans. On the *Carta marina*, by contrast, he more prudently omits as unknown everything between the eastern coast of the New World and the eastern coast of mainland Asia. Yet the *Carta marina* is the fruit of a cartographic boldness that is equally impressive: a willingness to discard almost all of the research done for the earlier map, and undertake the laborious creation of an entirely new detailed and monumental image of the world based on a new philosophy and a new projection, and using new sources. The map is a remarkable testament both to the cartographer's determination to show the true form of the world and to the dynamism of early sixteenth-century cartography.

One of the many differences between the 1507 and 1516 maps is that there is a larger number of long legends on the latter. In the long text block in the lower left corner of the map (see Legend 9.3), Waldseemüller lists many of the sources that he used in creating the map, which are also the sources of many of the long legends. He clearly viewed the textual element of his map as very important, and yet in the more than one hundred years since the rediscovery of the *Carta marina*, few of the legends have been transcribed and translated, they have never been studied together, and their correlations with the sources that Waldseemüller lists on the map have not been explored.[1] Thus an essential aspect of this important map, and of Waldseemüller's effort to convey information to the map's viewers, has remained uninvestigated.

We know little about Waldseemüller, and the general lack of scholarly attention devoted to the *Carta marina* represents not only a failure to address one of the masterpieces of the most important cartographer of the early sixteenth century, but also a lost opportunity to study the development of his cartographic thought, and thus add to our knowledge of the man. By examining how he used his sources, we can gain insight into Waldseemüller's methods and character, and by seeing how his cartographic thought evolved, we can come to appreciate his intellectual openness and flexibility.

In this introduction I offer a detailed discussion of the *Carta marina*, focusing on a comparison of that map with the 1507, and also with the maps in Waldseemüller's edition of Ptolemy's *Geography* published in 1513, in the interest of revealing all that the later map can tell us about the development of Waldseemüller's thought. Following this general discussion of the map comes a transcription, translation, and study of all of the long legends on the *Carta marina*, with particular attention devoted to the determination of their sources. My hope is that the book will be of use not only to readers with a direct interest in Waldseemüller, but also more broadly to any scholar working on early sixteenth-century cartography, and to anyone interested in seeing how an experienced cartographer of that period went about constructing a new image of the world.

[1]All of the toponyms on the *Carta marina*, but not the legends, are transcribed by Meret Petrzilka, *Die Karten des Laurent Fries von 1530 und 1531 und ihre Vorlage, die 'Carta Marina' aus dem Jahre 1516 von Martin Waldseemüller* (Zurich: Neuen Zürcher Zeitung, 1970), pp. 42–110.

C. Van Duzer, *Martin Waldseemüller's 'Carta marina' of 1516*,
https://doi.org/10.1007/978-3-030-22703-6_1

This book is accompanied by Electronic Supplementary Material, available from the page for this book on www.springer.com. These materials include high-resolution images of each sheet of the *Carta marina* and a high-resolution image of the whole map. In the book itself there is an illustration of each sheet of the map at the beginning of the section about the texts on that sheet, but the high-resolution online images of the sheets will allow the reader to zoom in and better see the details being discussed.

The ESM also includes an index PDF of the whole map that indicates with a number the location of each of the long texts on the map that are transcribed, translated, and studied below. This PDF is searchable, so that if the reader is having difficulty determining where exactly Legend 8.7 is located, a search in the PDF for "8.7" will find it.

1.2 Martin Waldseemüller and His Works

Martin Waldseemüller was born between 1470 and 1475, either in Freiburg or the nearby village of Wolfenweiler, and studied at the University of Freiburg: he is registered as a student there in 1490. In about 1505 he moved to the town of Saint-Dié in the Vosges Mountains not far from Strasbourg, and in 1513 he became a canon of the collegiate church there, though at the time he was living in Strasbourg. Aside from that stay in Strasbourg, he spent his adult life in Saint-Dié,[2] and died there in 1520.[3]

In Saint-Dié, Waldseemüller worked with a small group of humanists who sometimes called themselves the Gymnasium Vosagense. There is no evidence that this group was involved in teaching, and it seems only to have been an association of scholars. The group was led by Gualtier Ludd, secretary to Duke René II of Lorraine, and the owner of a small press in Saint-Dié[4]; the other members that we know of were Gualtier's cousin Nicholas Ludd, Matthias Ringmann, and Johannes Basinus Sendacurius.[5]

It is by following Waldseemüller's development as a cartographer through his works that we can learn the most about him. Those works are as follows[6]:

1. In 1507, Waldseemüller, in close collaboration with Matthias Ringmann, published three works that were designed to accompany each other. The first was a short book, the *Cosmographie introductio*, printed in Saint-Dié, no doubt on Ludd's press, which is an introduction to geography followed by a Latin translation of Amerigo Vespucci's account of

[2]For biographical details about Waldseemüller see Lucien Gallois, "Waldseemüller, chanoine de Saint-Dié," *Bulletin de la Société de Géographie de l'Est* 21 (1900), pp. 221–229; P. Albert, "Über die Herkunft Martin Walzenmüller's, genannt Hylacomylus," *Zeitschrift für die Geschichte des Oberrheins* 15 (1900), pp. 510–514; Hermann Flamm, "Die Herkunft des Kosmographen Martin Waldseemüller (Walzenmüller)," *Zeitschrift für die Geschichte des Oberrheins* 27.1 (1912), pp. 41–52; Franz Laubenberger, "Martin Waldseemüller. Ein kurzer Abriß über Leben und Werk des Freiburger Kartographen," *Vermessungstechnische Rundschau* 27.8 (1965), pp. 306–310; 27.9 (1965), pp. 356–358; and 27.10 (1965), pp. 384–386; and Hans Wolff, "Martin Waldseemüller: The Most Important Cosmographer in a Period of Dramatic Scientific Change," in Hans Wolff, ed., *America: Early Maps of the New World*, trans. Hugh Beyer et al. (Munich: Prestel, 1992), pp. 111–126.

[3]Chet Van Duzer and Benoît Larger, "Martin Waldseemüller's Death Date," *Imago Mundi* 63.2 (2011), pp. 217–219, and see Chet Van Duzer, "A Printer's Silent Tribute to the Passing of Martin Waldseemüller in Early Sixteenth-Century Strasbourg," *Papers of the Bibliographical Society of America* 110.3 (2016), pp. 313–333.

[4]On the printing press in Saint-Dié see Arthur Benoit, "Notes sur les commencements de l'imprimerie à Saint-Dié," *Bulletin de la Société philomatique vosgienne* 13 (1887–88), pp. 183–208; and Albert Ronsin, "L'imprimerie humaniste de Saint-Dié au XVIe siècle," in Emil van der Vekene, ed., *Refugium animae bibliotheca: Festschrift für Albert Kolb* (Wiesbaden: G. Pressler, 1969), pp. 382–425.

[5]On Waldseemüller's collaborators in the Gymnasium Vosagense see M. d'Avezac, *Martin Hylacomylus Waltzemüller, ses ouvrages et ses collaborateurs: voyage d'exploration et de découvertes à travers quelques épîtres dédicatoires, préfaces et opuscules en prose et en vers du commencement du XVI siècle: notes, causeries et digressions bibliographiques et autres* (Paris: Challamel aîné, 1867); and Lucien Gallois, "Le Gymnase Vosgien," *Bulletin de la Société de Géographie de l'Est* 21 (1900), pp. 88–94. On Ringmann see Charles Schmidt, "Mathias Ringmann (Philésius), humaniste alsacien et lorrain," *Mémoires de la Société d'Archéologie Lorraine* 25 (1875) pp. 165–233; Charles Schmidt, *Histoire littéraire de l'Alsace à la fin du XVe siècle* (Paris: Sandoz and Fischbacher, 1879–80), vol. 2, pp. 87–131; and Benoît Larger, "Le Gymnase vosgien et ses réseaux: les cardinaux: Lettre de Procuration de René II, roi de Sicile et duc de Lorraine à François Sodérini, cardinal de Sainte-Suzanne," in *Saint-Dié-des-Vosges baptise les Amériques: actes du colloque, 12 mai 2007, Centre Robert-Schuman, Saint-Dié-des-Vosges* (Saint-Dié-des-Vosges: Musée Pierre Noël, 2008), pp. 82–85. On the question of whether Waldseemüller or Ringmann should be credited with the naming of America see Franz Laubenberger (trans. Steven Rowan), "The Naming of America," *The Sixteenth Century Journal* 13.4 (1982), pp. 91–113, with some corrections in Christine R. Johnson, "Renaissance German Cosmographers and the Naming of America," *Past & Present* 191 (2006), pp. 3–43.

[6]For a good account of Waldseemüller's works see Robert W. Karrow, *Mapmakers of the Sixteenth Century and Their Maps: Bio-Bibliographies of the Cartographers of Abraham Ortelius, 1570* (Chicago: Speculum Orbis Press, 1993), pp. 568–583.

his four voyages.[7] The second was a set of woodcut gores for a terrestrial globe with a diameter of 12 cm (4.5 inches); the printer is not specified.[8] And the third was the 1507 world map, printed on twelve sheets, each approximately 45.5 × 62 cm (about 18 × 24.4 inches), which were designed to be assembled into a wall map measuring 128 × 233 cm (50.4 × 91.7 inches).[9] The map is titled *Universalis Cosmographia Secundum Ptholomaei Traditionem et Americi Vespucii Alioru[m]que Lustrationes* (A map of the whole world according to the tradition of Ptolemy and the explorations of Amerigo Vespucci and others). This map is the first to apply the name "America" to the New World, a name that Waldseemüller and Ringmann proposed in the *Cosmographiae introductio*. The printer of the map is not indicated. It survives in just one exemplar, which was owned by the astronomer, mathematician, and globemaker Johann Schöner (1477–1547) and preserved by him in a codex now called the Schöner Sammelband.[10] The codex was discovered in 1901 by Joseph Fischer in Wolfegg Castle in Baden-Württemberg, Germany; in 2003, the Library of Congress completed its acquisition of the map.[11] Elsewhere I have shown that the 1507 map, not only in terms of overall design and projection but also with regard to its long descriptive texts—but not with regard to its place names—is based closely on the large world map by Henricus Martellus at Yale, which was made c. 1491—or rather, not on the specific map at Yale, but on another, similar large map by Martellus that is now lost.[12]

[7]The full title of the book refers to the accompanying map and globe (this is the title of the first edition): *Cosmographiae introductio: cum quibusdam geometriae ac astronomiae principiis ad eam rem necessariis. Insuper quatuor Americi Vespucij nauigationes. Uniuersalis cbosmographiae [sic] descriptio tam in solido q[uam] plano eis etiam insertis qu[a]e Ptholom[a]eo ignota a nuperis reperta sunt* (Saint-Dié: [Gualtier and Nicholas Ludd], 1507). On the *Cosmographiae introductio* see Henry Harrisse, *Bibliotheca americana vetustissima: A Description of Works Relating to America, Published between the Years 1492 and 1551* (New York: G. P. Philes, 1866), #44–47, pp. 89–96; the work is reproduced in facsimile and translated into English by Joseph Fischer and Franz von Wieser in *The 'Cosmographiae introductio' of Martin Waldseemüller in Facsimile, Followed by the Four Voyages of Amerigo Vespucci, with their Translation into English* (New York: The United States Catholic Historical Society, 1907). A new translation and discussion of the *Cosmographiae introductio* by John Hessler may be found in his *The Naming of America: Martin Waldseemüller's 1507 World Map and the 'Cosmographiae introductio'* (London: D. Giles, 2008). There is a French translation by Pierre Monat in Albert Ronsin, *La fortune d'un nom: America: le baptême du Nouveau Monde à Saint-Dié-des-Vosges* (Grenoble: J. Millon, 1991), pp. 101–219. The work has also been reproduced in facsimile and translated into Spanish by Miguel León-Portilla in *Introducción a la cosmografía y las cuatro navegaciones de Américo Vespucio* (Coyoacán: Universidad Nacional Autónoma de México, 2007); the accompanying CD includes Spanish translations of all of the long legends on Waldseemüller's 1507 map. There also is a facsimile with transcription and translation into German in Martin Lehmann, *Die 'Cosmographiae Introductio' Matthias Ringmanns und die Weltkarte Martin Waldseemüllers aus dem Jahre 1507: Ein Meilenstein frühneuzeitlicher Kartographie* (Munich: Martin Meidenbauer, 2010).

[8]There are five known surviving copies of the gores: at the University of Minnesota, in the Stadtbucherei Offenburg, at Charles Frodsham and Co. Ltd. (purchased at the Christie's sale of June 8, 2005), at the Bayerische Staatsbibliothek in Munich and at the Universitätsbibliothek in Munich (discovered in 2012). Serious questions have been raised about the genuineness of the exemplar in the Universitätsbibliothek in Munich: see Michael Blanding, "Why Experts Don't Believe This Is a Rare First Map of America," *New York Times*, Dec. 10, 2017. For general discussion of the gores see Henry Harrisse, *The Discovery of North America* (London: H. Stevens, 1892; Amsterdam: N. Israel, 1961), pp. 440–442, no. 67, and 467–468, no. 82; Fischer and von Wieser in *The 'Cosmographiae introductio' of Martin Waldseemüller* (see note 7), pp. 23–30; Edward L. Stevenson, *Terrestrial and Celestial Globes: Their History and Construction* (New Haven: Yale University Press, 1921), vol. 1, pp. 70–71; *Americana vetustissima: Fifty Books, Manuscripts, & Maps Relating to America from the First Fifty Years after its Discovery (1493–1542)* (New York: H. P. Kraus, 1990), pp. 30–31; and *Cartography, Including the Waldseemüller Gores: Wednesday 8 June 2005* (London: Christie, Manson & Woods Ltd., 2005).

[9]Waldseemüller's 1507 map has been published in facsimile in Joseph Fischer and Franz Ritter von Wieser, *Die älteste Karte mit dem Namen Amerika aus dem Jahre 1507 und die Carta marina aus dem Jahre 1516 des M. Waldseemüller (Ilacomilus)* (Innsbruck: Wagner'schen Universitäts-Buchhandlung, 1903; Amsterdam: Theatrum Orbis Terrarum, 1968), with a good but brief introduction; and in John W. Hessler and Chet Van Duzer, *Seeing the World Anew: The Radical Vision of Martin Waldseemüller's 1507 & 1516 World Maps* (Washington, DC: Library of Congress and Delray Beach, FL: Levenger Press, 2012). There are two excellent high-resolution scans of the map available on the internet site of the Library of Congress. For discussion of the 1507 map, in addition to the introductions in the two facsimile editions see Charles G. Herbermann, "The Waldseemüller Map of 1507," *Historical Records and Studies* 3.2 (1904), pp. 320–342; Elizabeth Harris, "The Waldseemüller World Map: A Typographic Appraisal," *Imago Mundi* 37 (1985), pp. 30–53; and Toby Lester, *The Fourth Part of the World: The Race to the Ends of the Earth, and the Epic Story of the Map that Gave America its Name* (New York: Free Press, 2009).

[10]For a description of the Schöner Sammelband see John W. Hessler, "The Schöner *Sammelband*," in Arthur Dunkelman, ed., *The Jay I. Kislak Collection at the Library of Congress: A Catalog of the Gift of the Jay I. Kislak Foundation to the Library of Congress* (Washington, DC: Library of Congress, 2007), pp. 99–108.

[11]On the Library of Congress's acquisition of Waldseemüller's 1507 map see John Hébert, "The Map that Named America: Martin Waldseemüller's 1507 World Map," *Coordinates*, Series B, No. 4 (August, 2005), available at http://purl.oclc.org/coordinates/coordinates.htm.

[12]See Chet Van Duzer, *Henricus Martellus's World Map at Yale (c. 1491): Multispectral Imaging, Sources, Influence* (New York: Springer, 2018). Also see Chet Van Duzer, "Multispectral Imaging for the Study of Historic Maps: The Example of Henricus Martellus's World Map at Yale," *Imago Mundi* 68.1 (2016), pp. 62–66; and my talk "New Light on Henricus Martellus's World Map at Yale (c. 1491): Multispectral Imaging and Early Renaissance Cartography," delivered in the Caroline Werner Gannett Distinguished Speaker Series in Digital Humanities at the Chester F. Carlson Center for Imaging Science, Rochester Institute of Technology, March 18, 2015, with video available at https://www.cis.rit.edu/martellus.

2. In 1511, Waldseemüller produced a wall map of Europe on four sheets that measured 141 × 107 cm (about 56 × 42 inches). This map, the *Carta Itineraria Europae*,[13] is the first printed wall map of the continent, and the first map of Europe to show (as its name suggests) the most important trade routes. No copies of the 1511 printing survive, but one exemplar of the 1520 printing is extant.[14] The map was accompanied by a short book written by Ringmann that supplies a more detailed description of the regions of Europe than there is room for on the map, titled *Instructio manuductionem prestans in cartam itinerariam Martini Hilacomili* (Strasbourg: Grüninger, 1511).[15]
3. In 1513 a new edition of Ptolemy's *Geography*, on which Waldseemüller together with Matthias Ringmann and other colleagues had begun work in 1505, but which suffered various delays, was printed by Johann Schott in Strasbourg.[16] In addition to the standard twenty-seven Ptolemaic maps, this edition has a very full collection of *tabulae modernae* or modern maps based on more recent data, all but one based on information from nautical charts. This was the most important edition of Ptolemy published in the sixteenth century.[17]
4. In 1516, Waldseemüller published the *Carta marina*, which like the 1507 map is printed on twelve sheets, each approximately 45.5 × 62 cm (about 18 × 24.4 inches), which were designed to be assembled into a wall map measuring 128 × 233 cm (50.4 × 91.7 inches).[18] The printer of the map is not specified. Also like the 1507 map, the *Carta marina* survives in just one copy, which was owned by Johann Schöner (1477–1547), and which was preserved together with a copy of the 1507 map in the Schöner Sammelband.[19] Following the Library of Congress's purchase of the 1507 map, the collector Jay I. Kislak bought the Sammelband, minus the 1507 map and a star chart by Albrecht Dürer. He has donated much of his large collection of Americana, including the Sammelband with the *Carta marina*, to the Library of Congress, where the map now resides.[20] In a separate transaction in 2016 the Library of Congress acquired the Dürer star chart that had been in the Sammelband, and thus now owns all of the elements that originally comprised the Sammelband.

Towards the end of his life Waldseemüller was working on books titled *Itineraria* and *Chronica mundi*, which had been promised to the publisher Johann Grüninger, but were never completed or printed.[21] The brief references to these works have

[13]The map bears the title *Carta itineraria Europae*, "Road map of Europe," along its bottom edge; along the top and bottom borders, there is a fuller title: *Carta Europae topica neoterica civitatum fluviorum et montium [di]tancias eciam situs tam medi[d]os quam veros indicans, opus corographic[um] et geographicum Martini Ilacomili Friburgensis*, "Modern map of Europe showing the distances and locations (both measured and accurate) of cities, rivers, and mountains, a chorographic and geographical work by Martin Waldseemüller of Freiburg."

[14]The single exemplar of the 1520 printing of the *Carta itineraria Europae* is in Innsbruck, Tiroler Landesmuseum Ferdinandeum, Historische Sammlungen, Kartographie, K 9/39. For discussion of the map see Leo Bagrow, "'Carta Itineraria Europae' Martini Ilacomili, 1511," *Imago Mundi* 11 (1954), pp. 149–150; Peter H. Meurer, *Corpus der älteren Germania-Karten: ein annotierter Katalog der gedruckten Gesamtkarten des deutschen Raumes von den Anfängen bis um 1650* (Alphen aan den Rijn: Canaletto Uitgeverij, Repro-Holland, 2001), pp. 155–160; and Albert Ronsin, "Carta itineraria Europae. La première carte routière murale d'Europe, éditée à Saint-Dié en 1511," *Mémoire des Vosges* 3 (2001), pp. 6–12, and 4 (2002), pp. 10–13. The map has been reproduced in facsimile as Martin Waldseemüller, *Carta itineraria Europae* (Bonn: Kirschbaum Verlag, 1972), with an accompanying study by Karl-Heinz Meine, *Erläuterungen zur ersten gedruckten (Strassen-)Wandkarte von Europa, der Carta itineraria Evropae der Jahre 1511 bzw. 1520 von Martin Waldseemüller (um 1470 bis etwa 1521), Kostbarkeit des Tiroler Landesmuseum Ferdinandeum, Innsbruck* (Bonn-Bad Godesberg: Kirschbaum, 1971).

[15]There is a brief discussion of the *Instructio manuductionem prestans* in M. d'Avezac, *Martin Hylacomylus Waltzemüller* (see note 5), pp. 135–141.

[16]Incidentally in my article "Colored as its Creators Intended: Painted Maps in the 1513 Edition of Ptolemy's Geography," forthcoming in *Imago temporis*, I identify the workshop hand-coloring scheme for the maps in the 1513 edition of Ptolemy's *Geography*, that is, the coloring scheme intended by Waldseemüller and the other creators of the edition.

[17]For discussion of the 1513 edition of Ptolemy's *Geography* see Ruthardt Oehme, "Martin Waldseemüller und der Straßburger Ptolemäus von 1513," in Karl Friedrich Müller, ed., *Beiträge zur Sprachwissenschaft und Volkskunde: Festschrift für Ernst Ochs zum 60 Geburtstag* (Lahr: M. Schauenburg, 1951), pp. 155–167; R. A. Skelton, "Bibliographical Note," in Ptolemy, *Geographia, Strassburg, 1513* (Amsterdam: Theatrum Orbis Terrarum, 1966), pp. v–xx; Patrick Gautier Dalché, "The Reception of Ptolemy's *Geography* (End of the Fourteenth to Beginning of the Sixteenth Century)," in David Woodward, ed., *The History of Cartography*, vol. 3, *Cartography in the European Renaissance* (Chicago: University of Chicago Press, 2007), part 1, pp. 285–364, at 348–349; and Alfred Hiatt, "Mutation and Supplement: The 1513 Strasbourg Ptolemy," in Zur Shalev and Charles Burnett, eds., *Ptolemy's 'Geography' in the Renaissance* (London: Warburg Institute, and Turin: Nino Aragno Editore, 2011), pp. 143–161.

[18]The *Carta marina* was reproduced in facsimile together with the 1507 map in Joseph Fischer and von Wieser, *Die älteste Karte mit dem Namen Amerika* (see note 9), with a good brief discussion in the introduction; and in Hessler and Van Duzer, *Seeing the World Anew* (see note 9). The map is also discussed by Seymour I. Schwartz, *Putting 'America' on the Map: The Story of the Most Important Graphic Document in the History of the United States* (Amherst, NY: Prometheus Books, 2007), pp. 197–206.

[19]On the Schöner Sammelband see note 10 above.

[20]See Arthur Dunkelman, ed., *The Jay I. Kislak Collection at the Library of Congress: A Catalog of the Gift of the Jay I. Kislak Foundation to the Library of Congress* (Washington, D.C.: Library of Congress, 2007), esp. 105–107 on the *Carta marina*.

[21]On these two unfinished books of Waldseemüller see Oskar von Hase, *Die Koberger: Eine Darstellung des buchhändlerischen Geschäftsbetriebes in der Zeit des Überganges vom Mittelalter zur Neuzeit* (Leipzig: Breitkopf & Härtel, 1885), p. 138, and "Briefanhang" nos. 106, 117, and 123; and Hildegard Binder Johnson, *Carta marina: World Geography in Strassburg, 1525* (Minneapolis, MN: University of Minnesota Press, 1963), pp. 39, 41, 44, 45, 57, and 95–96. On p. 95 Johnson translates part of letter 123 in von Hase.

given rise to claims that some of the maps in the 1522 edition of Ptolemy's *Geography* published by Grüninger were made by Waldseemüller and intended for one of these books, but this is unlikely to be true, as Waldseemüller had been moving away from Ptolemy for some years before his death. It has been plausibly argued, however, that Waldseemüller's notes for these books were used by Lorenz Fries in writing his *Uslegung der mercarthen oder Charta Marina*, the booklet that accompanied Fries's 1525 edition of Waldseemüller's *Carta marina*, and was published by Grüninger in Strasbourg.[22]

In addition, the *Uslegung* contains a map showing the route that Alvise Cadamosto took on his voyage to Madeira and the Canary Islands in 1455,[23] and it is very likely that Waldseemüller made this map for either the *Chronica mundi* or the *Itineraria*.[24] Further, a copy of Francanzio de Montalboddo's *Itinerarium Portugallensium* (Milan, 1508) in the Österreichische Nationalbibliothek in Vienna (signature 394.092-C.Kar) has a set of six maps that have been added to the book, a world map and five maps of the coast of West Africa that illustrate the voyages of Cadamosto (1455/56) and Pedro de Sintra (1463). One of these maps is very similar to (but not identical with) that in the 1525 *Uslegung*, and it seems likely that they were produced as part of the preparations for the *Chronica mundi* or *Itineraria*, either by Waldseemüller or by a closely affiliated cartographer.[25] But some passages in the *Uslegung* and these maps are all that we have of Waldseemüller's final projects.

1.3 Comparing and Contrasting the 1507 and 1516 Maps

As mentioned above, Waldseemüller's *Carta marina*, like his 1507 map, is printed on twelve sheets that were designed to be assembled into a wall map measuring 128 × 233 cm (50.4 × 91.7 inches). But while they share these physical characteristics, in most other respects the two maps are very different, and the differences are reflected in their titles. The title of the 1507 map, as mentioned above, is *Universalis cosmographia secundum Ptholomaei traditionem et Americi Vespucii alioru [m]que lustrationes* ("A map of the whole world according to the tradition of Ptolemy and the explorations of Amerigo Vespucci and others). Waldseemüller's use of both Ptolemy and Vespucci as sources—of both ancient and modern authorities—is indicated in the portraits of them at the top of the 1507 map.

But the title of the 1516 map indicates a radical repudiation of ancient authorities:

> Carta marina navigatoria portugallen[siorum] navigationes atque totius cogniti orbis terre marisque formam naturamque situs et terminos nostri[s] temporibus recognitos et ab antiquorum traditione differentes eciam quorum vetusti non meminuerunt autores, hec generaliter indicat.
>
> A nautical chart that comprehensively shows the Portuguese voyages and the shape and nature of the whole known world, both land and sea, its regions, and its limits as they have been determined in our times, and how they differ from the tradition of the ancients, and also areas not mentioned by the ancients.

The change from Waldseemüller's following Ptolemy to repudiating him is dramatic, and illustrates a dichotomy of Renaissance culture: on the one hand, admiration for the methods of enquiry and systems of knowledge of the ancients, and on the other, recognition that new investigations or explorations could produce results superior to those of the ancients—for example, the discovery that the equatorial Torrid Zone, which various classical authors held to be uninhabitable and uncrossable, was a myth.[26] In a long introductory text in the lower left corner of the *Carta marina*, Waldseemüller discusses his earlier map (certainly the 1507 map),[27] and his reasons for creating a new one. He concedes that a map with ancient place

---

[22]On Fries's use of Waldseemüller's notes in the *Uslegung* see Johnson, *Carta marina* (see note 21), pp. 96–99.

[23]The map of Cadamosto's voyage is only in the 1525 edition of the *Uslegung*, not in the 1527 or 1530 editions.

[24]So Robert W. Karrow Jr., *Mapmakers of the Sixteenth Century and Their Maps* (Chicago: Speculum Orbis Press, 1993), pp. 582–583.

[25]See Peter H. Meurer, "Sechs Karten der westafrikanischen Küste aus der Waldseemüller-Schule," *Cartographica Helvetica* 45 (2012), pp. 15–26.

[26]For discussion of passages in which Renaissance authors express their doubts about the correctness of the ancients see R. Hooykaas, *Humanism and the Voyages of Discovery in 16th Century Portuguese Science and Letters* (Amsterdam and New York: Noord-Hollandsche U.M., 1979). For discussion of the Torrid Zone see Milton V. Anastos, "Pletho and Strabo on the Habitability of the Torrid Zone," *Byzantinische Zeitschrift* 44 (1951), pp. 7–10; and Angelo Cattaneo, "Réflexion sur les climats et les zones face à l'expansion des XVe et XVI siècles," *Bulletin du Comité Français de Cartographie* 199 (March, 2009), pp. 7–21.

[27]Peter W. Dickson, *The Magellan Myth: Reflections on Columbus, Vespucci, and the Waldseemueller Map of 1507* (Mount Vernon, Ohio: Printing Arts Press, 2007), questions whether the earlier map described is in fact the 1507 map, asserting that Waldseemüller says that the earlier map represents the world according to Ptolemy, while the 1507 map contains more than that. Unfortunately this doubt is based on a misinterpretation or incomplete reading of Waldseemüller's text. After Waldseemüller writes that he designed his earlier map so that "it would only have in it those customs and features that are known to have been extant or in use in Ptolemy's time," he continues: "Many things were added that were discovered and confirmed by the testimony of experience by the Venetian citizen Marco during the papacies of Clement IV and Gregory X, and by the Portuguese captains Christopher Columbus and Amerigo Vespucci.".

names, like his earlier map, is of limited utility, since it is difficult to recognize modern places according to their ancient names, and also remarks that recent explorers have detected various errors in the geographical writings of the ancients, particularly in Ptolemy's *Geography*. He then writes[28]:

> Quibus ipse permotus communi eruditorum utilitati studens hunc secundarium totius orbis typum primo adieci, ut sicut illic veterum constetit auctorum totius orbis terra marique descriptio, sic reluceat hic non noua solum ac presens totius orbis facies, sed cum hoc mediorum temporum indita rebus mortalibus consueta et naturalis permutatio pateat ut unico habeas (si ita dici iubet) contuitu quid, quales, quomodo res caduce nunc fiunt, qualesque priscis fuerint temporibus et quales aliquin future a nobis nullatenus dubitari possint. Hanc igitur iuxta Neotericorum traditionem totius orbis spetiem & descriptionem Chartam placuit appellare marinam, eo que in maris descriptionibus vulgarem fuerimus & approbatissimam nauticarum tabularum notificationes insequuti, sumus insuper in mediterranea Asie atque Aphrice descriptione Ne[o]tericorum itinerarios, particulares tabulas, chorographias, & quorundam recensiorum [for recentiorum] lustratorum relationes plerumque imitati....
>
> Moved by these considerations, and in the interest of the common utility of scholars, I have added this second image of the world to my first, so that just as in the first the image of the whold world, land and sea, agreed with that of the ancient authors, so in this one, not only may the new and present face of the world shine forth, but together with that, the customary and natural change introduced into worldly affairs in the intervening times, so that you can see (if I may say so) at a single glance why, of what kind, and how transitory things have come to be now, what they were like in former times, and how they will be in the future, without a doubt. Therefore, it seemed good to call this image and description of the whole world, made in accordance with the tradition of modern authors, a Carta marina, and for that reason, as far as the depiction of the oceans, I have followed the commonly used and the most approved nautical charts and their indications, while in the depiction of the Mediterranean, Asia and Africa I have made ample use of recent authors' travel narratives, regional maps, descriptions of countries, and the accounts of some recent explorers....

Though his 1507 map shows the New World, Waldseemüller here describes his earlier work as an image of the earth according to the ancients, no doubt to increase the attractiveness of his new map, which is based on the most recent information available.

Together with this change in his thought about what a world map should be came a closely related change in cartographic models. His 1507 map is based on Ptolemy's *Geography*—not only on Ptolemy's geographical data regarding the locations of cities and other features in Europe, Africa and Asia, but also on his system for representing geographical space, using a grid of latitude and longitude. More specifically, Waldseemüller used as the model for his 1507 map a large world map of c. 1491 by Henricus Martellus Germanus (Fig. 1.1).[29] Martellus's map uses the Ptolemaic grid of latitude and longitude, and is laid out using a modification of Ptolemy's second projection. Waldseemüller followed Martellus closely (Fig. 1.2), and used the same projection, has similar decorative wind-heads in the border of the map, arranged things so that Japan is at the eastern or right-hand edge of the map as it is on Martellus's, and borrowed many descriptive texts from Martellus.[30] Waldseemüller of course added the New World, often used different sources for place names, and depicted southern Africa very differently, but in other respects he made heavy use of Martellus, particularly for the outlines of Asia and for his long descriptive texts.

In his *Carta marina* Waldseemüller abandoned the Ptolemaic model, and instead adopted the model of nautical charts or portolan charts.[31] The origin of nautical charts is unclear, but the earliest surviving examples date to the late thirteenth

[28]For the full Latin text and English translation of this introductory paragraph see Legend 9.3.

[29]Martellus's world map of c. 1491 is 122 × 201 cm, almost as large as Waldseemüller's 1507 map (128 × 233 cm). The map is on permanent display in the Beinecke Rare Book and Manuscript Library at Yale University and has the identifying number Art Store 1980.157. For discussion of the map see Alexander O. Vietor, "A Pre-Columbian Map of the World, Circa 1489," *Imago Mundi* 17 (1963), pp. 95–96; Marcel Destombes, *Mappemondes, A.D. 1200–1500* (Amsterdam: N. Israel, 1964), pp. 229–233 with plates 37–38; Carlos Sanz, "Un mapa del mundo verdaderamente importante en la famosa Universidad de Yale," *Boletín de la Real Sociedad Geográfica* 102 (1966), pp. 7–46; Arthur Davies, "Behaim, Martellus and Columbus," *Geographical Journal* 143 (1977), pp. 450–59, esp. 456–58; and Ilaria Luzzana Caraci, "Il Planisfero di Enrico Martello della Yale University Library e i Fratelli Colombo," *Rivista Geografica Italiana* 75 (1978), pp.132–43, translated into English as "Henricus Martellus' Map in the Yale University Library and the Columbus Brothers" in the author's *The Puzzling Hero: Studies on Christopher Columbus and the Culture of His Age* (Rome: Carocci, 2002), pp. 281–291.

[30]On Waldseemüller's borrowing of the descriptive texts on his 1507 map from Martellus see note 12 above.

[31]For good general accounts of nautical charts see Tony Campbell, "Portolan Charts from the Late Thirteenth Century to 1500," in J. B. Harley and David Woodward, eds., *The History of Cartography* (Chicago: University of Chicago Press, 1987-), vol. 1, pp. 371–463; Corradino Astengo, "The Renaissance Chart Tradition in the Mediterranean," in David Woodward, ed., *The History of Cartography*, vol. 3.1: *Cartography in the European Renaissance* (Chicago and London: The University of Chicago Press, 2007), pp. 174–262; and Ramon J. Pujades i Bataller, *Les cartes portolanes: la representació medieval d'una mar solcada* (Barcelona: Institut Cartogràfic de Catalunya, 2007), with CD.

Fig. 1.1 World map made by Henricus Martellus c. 1491. Beinecke Rare Book and Manuscript Library, Art Store 1980.157. Image by Lazarus Project/MegaVision/RIT/EMEL, courtesy of the Beinecke Rare Book and Manuscript Library

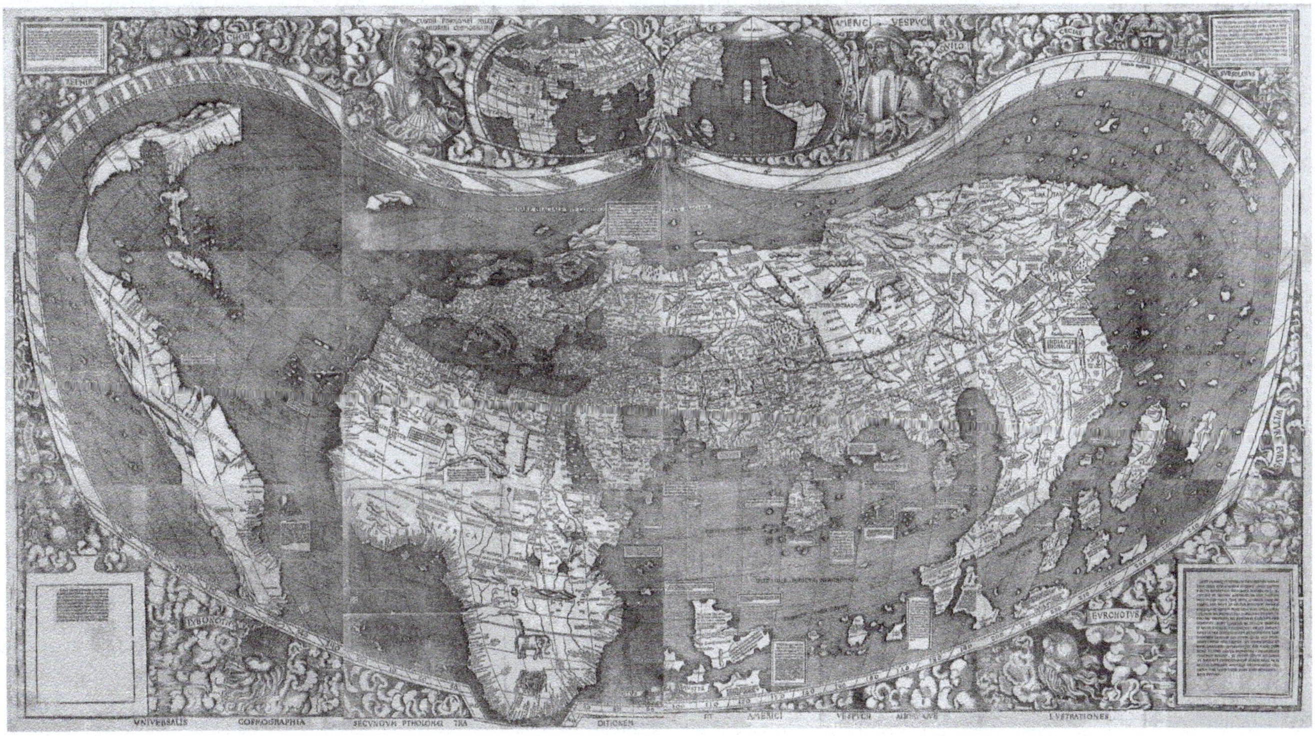

Fig. 1.2 Martin Waldseemüller's world map of 1507. Library of Congress, Geography and Map Division, G3200 1507 .W3. Courtesy of the Library of Congress

Fig. 1.3 Nautical chart by Nicolo de Caverio, c. 1503 (Paris, Bibliothèque nationale de France, Cartes et plans, SH archives 1). Courtesy of the Bibliothèque nationale de France

century. In essence they are practical tools for navigation, usually hand-drawn on parchment, with the emphasis on coastal features and place names; rather than being marked with latitude and longitude, they have a system of rhumb lines that radiate out in the standard compass directions (or directions of the traditional winds) from points organized in one or two large circles. In addition to the relatively plain nautical charts used for navigation, others were elaborately decorated with cities, kings, animals, flags, and compass roses, and had descriptive texts added to them. This was the type of map that Waldseemüller chose as the model for his 1516 map, and in fact we know the specific map he used: the nautical chart by Nicolo de Caverio of Genoa, made c. 1503 (Fig. 1.3).[32] We know this because of the close similarities of place names between Caverio's chart and Waldseemüller's *Carta marina*,[33] and the striking similarities of layout between the two maps (Fig. 1.4), including the area of the world depicted and the locations of the nodes of the systems of rhumb lines.

What caused Waldseemüller to abandon the Ptolemaic model and projection he had used in his 1507 map and adopt a nautical chart model—and to abandon the bold idea he had implemented in his 1507 map of depicting the whole circumference of the earth?[34] The former question is particularly intriguing, as we know that Waldseemüller had access to the Caverio map when he made his 1507 map.[35] Yet he still chose to use Ptolemy's system of cartography rather than the nautical chart system, and also to use Ptolemy's information for the shape of North Africa, for example, while the shape of

[32]The Caverio chart is in Paris, Bibliothèque nationale de France, Cartes et plans, SH archives 1, and measures 115 × 225 cm, similar in size to Waldseemüller's maps of 1507 and 1516. Edward L. Stevenson has studied the map in *Marine World Chart of Nicolo de Caneiro Januensis 1502 (circa)* (New York: American Geographical Society and the Hispanic Society of America, 1908). The longer legends on the map are transcribed by Armando Cortesão, *Cartografia e cartógrafos portugueses dos séculos XV e XVI* (Lisbon: Seara Nova, 1935), vol. 1, pp. 154–155. There is a color facsimile, which is smaller, however, than the original: *Planisphère nautique sur vélin du Génois Nicolao de Caverio* (Paris: Bibliothèque nationale, 1992). For discussion of the map in addition to that in Stevenson's work see L. Gallois, "Le Portulan de Nicolas de Canerio," *Bulletin de la Société de Géographie de Lyon* 9 (1890), pp. 97–119, reprinted in *Acta Cartographica* 9 (1970), pp. 76–98.

[33]For tables comparing the place names on Caverio's map and Waldseemüller's *Carta marina* for the New World and the coast of Africa see Stevenson, *Marine World Chart* (see note 32), pp. 84–110.

[34]Waldseemüller's boldness in depicting all 360° of the earth's circumference on the 1507 map is emphasized by the accompaniment of that map with a small globe based on the same geography as the map.

[35]For a list of some of the place names that are similar in the Caverio chart and Waldseemüller's 1507 map—including the copying of errors—see Joseph Fischer and Franz Ritter von Wieser, *Die älteste Karte mit dem Namen Amerika* (see note 9), pp. 26–29.

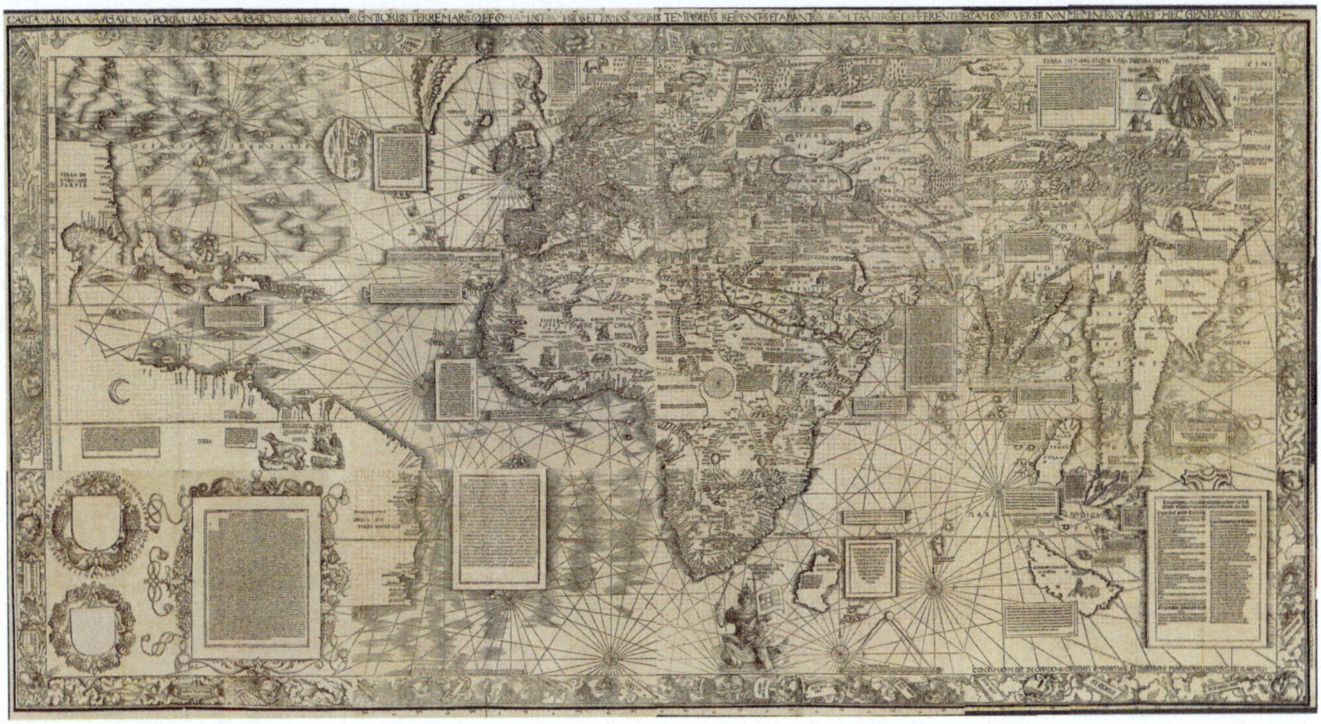

Fig. 1.4 Martin Waldseemüller's *Carta marina* of 1516. Library of Congress, Geography and Map Division, G1015 .S43 1517 Vault. Courtesy of the Library of Congress

that same region is markedly different (and more accurate) on Caverio's chart. Evidently at some point between 1507 and 1516, perhaps while he was involved in the production of the 1513 edition of Ptolemy's *Geography*, he became convinced of the superiority of the more recent geographical data available in nautical charts. I will explore this question, and the development of Waldseemüller's cartographic thought, in more detail below, but certainly one factor in his decision to follow the nautical chart model was that the best data available was already in that format.

With regard to the latter question, namely why Waldseemüller chose not to depict the whole circumference of the earth in his *Carta marina*, although Waldseemüller clearly decided to be less venturous in depicting little-known regions, the answer cannot be simply that he did not have good data about the parts of the world he does not depict on the *Carta marina*. One notable difference between Caverio's map and the *Carta marina* is that Waldseemüller depicts less of the eastern part of the world than his model: Caverio shows substantial portions of the northeastern coast of continental Asia and of the ocean we now call the Pacific that Waldseemüller chose not to copy (compare Figs. 1.3 and 1.4). Waldseemüller had reasonably good information about the location of Japan from reading Marco Polo's account of his travels, who placed Japan 1500 miles east of mainland China,[36] and Waldseemüller depicted it on his 1507 map, but not on his *Carta marina*. The answer seems to be that Waldseemüller designed the *Carta marina* to be more practical than his 1507 map: it shows only the parts of the world where Europeans had traveled, and where trade was known to occur, and it shows those parts using a fundamentally practical cartographic format, one developed for use on ships.

In addition to omitting some 128 degrees of longitude from the *Carta marina*, Waldseemüller depicts much less of the northern polar regions: his 1507 map runs all the way to the North Pole, but the *Carta marina* only to a bit more than 70° N. The *Carta marina* does include several more degrees of latitude in the southern ocean than the 1507 map, but overall, Waldseemüller's exclusion of large parts of the earth's surface from the *Carta marina*, together with the *Carta marina* being almost exactly the same physical size as the 1507 map, and its border being much narrower, meant that Waldseemüller was able to show far more detail, including both texts and images, in the areas that he does depict than he could on the 1507

[36]For the distance of Cipangu or Japan from mainland Asia see Marco Polo, *Marka Pavlova z Benátek, Milion: Dle jediného rukopisu spolu s prilusnym zakladem latinskym*, ed. Justin Václav Prásek (Prague: Nákl. Ceské akademie císare Frantiska Iozefa, 1902), p. 153; and Marco Polo, *The Book of Ser Marco Polo*, ed. and trans. Henry Yule (New York: C. Scribner's Sons, 1903), Book 3, Chap. 2, vol. 2, p. 253.

map. In comparison with the 1507 map, the *Carta marina* offers a "zoomed in" view of the known parts of the world. Thus, for example, in Arabia on sheet 6 of the 1507 map, there is room only for place names from Ptolemy and indications of mountains and rivers, but on the *Carta marina* there are images of Mecca and Medina as well as long legends describing the cities and features of the region (see Legends 7.6, 7.8, and 7.9).

One of the most striking differences between Waldseemüller's *Carta marina* and its principal model, the Caverio chart, is in the interiors of the continents, particularly in Africa and Asia. On Caverio's chart the emphasis (as is common on nautical charts) is on the coastlines, and he provides very few geographical details of the interior. In Africa there are images of two mountain ranges, three cities and three animals, some banners indicating the names of regions, and a decorative circular world map in place of a compass rose. Asia is largely empty, aside from some compass roses and a few banners with place names. The situation on the *Carta marina* is entirely different: both Africa and Asia are full of rivers, mountains, images of cities, sovereigns, peoples and animals, as well as descriptive texts. Waldseemüller also takes advantage of the open spaces in the unknown interior of South America and in the southern ocean to supply long texts, one the long introduction to the map quoted from earlier, another describing South America, and in the southeastern corner of the map (Legend 12.11), a list of the sources and prices of the spices in the great trading center of Calicut (now Kozhikode, India).

The abundance of geographical information and texts on the *Carta marina* should be considered from a few different perspectives. First, the advent of printed maps represented a great democratization of cartography, so that the information in a very expensive manuscript map like Caverio's could be made available to many people through the printing press at a much lower cost.[37] What Waldseemüller chose to democratize, however, was not just Caverio's chart, but a richer, more detailed, and more edifying version of the chart, with many more decorative elements and much more textual information. A number of manuscript nautical charts have a similar high level of expensive optional elements, including images and descriptive texts, such as the Catalan Atlas of 1375[38] and Mecia de Viladestes's nautical chart of 1413,[39] but Waldseemüller's *Carta marina* is the first large printed nautical chart, and it matches or exceeds these particularly elaborate nautical charts in the amount of information it offers.

It is also possible that the large amount of text on the *Carta marina*, particularly the long introduction in the lower left corner, was intended to render a booklet to accompany the map unnecessary: Waldseemüller and his colleague Matthias Ringmann had accompanied the 1507 world map with the booklet titled *Cosmographiae introductio*, and the 1511 wall map of Europe with the booklet *Instructio manuductionem prestans in cartam itinerariam Martini Hilacomili.*[40] We cannot be certain about this surmise, however, as Lorenz Fries's later German-language version of the *Carta marina* was accompanied by a booklet titled *Uslegung der mercarthen oder Charta marina* (Explanation of the Sea Map or *Carta marina*), which was probably written in part from Waldseemüller's notes for his unfinished *Chronica mundi* or *Itineraria*.[41]

1.4 Waldseemüller's Textual Sources on the *Carta Marina*

Waldseemüller's 1507 map, like his *Carta marina*, has a large number of descriptive texts, particularly in Africa and Asia, but one of the most remarkable things about the texts on the *Carta marina* is that the overwhelming majority of them are different from the ones on the earlier map. Waldseemüller abandoned not only his earlier cartographic model (i.e. Ptolemy, by way of Martellus), but also most of his earlier textual sources, in order to create an entirely new and modern image of the world. This must have been exciting but also time-consuming, carefully studying various texts looking for just the right passages to explain different regions or cities or peoples, and also for clues about the relative locations of those places.

---

[37]We do not know the sale price of either of Waldseemüller's large world maps, but according to a letter dated 26 February 1525, from Hans Grüninger to the Nurnberg printer and publisher Anton Koberger, a copy of Lorenz Fries's 1525 version of Waldseemüller's *Carta marina* was worth 5 florins. See Oskar von Hase, *Die Koberger: Eine Darstellung des buchhändlerischen Geschäftsbetriebes in der Zeit des Überganges vom Mittelalter zur Neuzeit* (Leipzig: Breitkopf & Härtel, 1885), p. cxxxviii, document 116.

[38]The Catalan Atlas is in Paris, Bibliothèque nationale de France, MS Espagnol 30; the map has been reproduced in facsimile a few times, including *Mapamundi del año 1375* (Barcelona: S.A. Ebrisa, 1983), and more recently *El món i els dies: L'Atles Català* (Barcelona: Enciclopèdia Catalana, 2005). The atlas is also reproduced in Pujades, *Les cartes portolanes* (see note 31), on the accompanying CD, number C16.

[39]The chart by Mecia de Viladestes is in Paris, Bibliothèque nationale de France, Rés. Ge AA 566; it is illustrated in Michel Mollat and Monique de la Roncière, *Sea Charts of the Early Explorers: 13th to 17th Century*, trans. L. le R. Dethan (New York: Thames and Hudson, 1984), chart 12; and Pujades, *Les cartes portolanes* (see note 31), pp. 202–203, and on the accompanying CD, number C30.

[40]On the *Cosmographiae introductio* see note 7 above, and on the *Instructio manuductionem prestans* see note 15.

[41]On these unfinished works by Waldseemüller see note 21 above.

Waldseemüller lists his textual sources in the long legend in the lower left corner of the map (Legend 9.3); here follows the list, rearranged chronologically by the dates of the authors:

- Giovanni da Pian del Carpine, or John of Plano Carpini (c. 1182–1252), who as papal legate traveled via a northern route to the Great Khan in China, by way of Russia and Mongolia. Manuscripts of Carpini's travel narrative, which exists in two redactions, and are titled either *Ystoria Mongalorum* or *Liber Tartarorum*, are rare, and the whole of the *Historia Mongalorum* was not published until the nineteenth century.[42] Waldseemüller almost certainly consulted his travel narrative via the excerpts that Vincent of Beauvais (c. 1190–c. 1264) incorporated into Book 32 of his *Speculum historiale*, which exists in many manuscript copies and was first printed in 1473.[43]
- Friar Ascelinus, who was part of a group of Dominicans who visited the encampment of the Mongol prince Baiju in 1247. Simon of Saint Quentin, who traveled with Ascelinus, wrote an account of this mission,[44] and excerpts of his narrative (like the excerpts from Plano Carpini's account) were included in Book 32 of Beauvais's *Speculum historiale*.[45]
- Marco Polo (c. 1254–1324), the famous Venetian traveler who spent twenty-four years in the East, and whose account of his travels includes descriptions of places in the Middle East, Central Asia, China, and the Indian Ocean. Marco Polo's text exists in many manuscripts, translations, and editions; the first Latin edition was printed by Gheraert Leeu in approximately 1484.[46] Waldseemüller made heavy use of Marco Polo in his 1507 map, mostly through borrowings from a large world map by Henricus Martellus. He makes dramatically less use of the Venetian author's work in his *Carta marina*: instead of copying descriptive texts from Martellus's map, Waldseemüller searched out his own descriptive texts in a variety of geographical texts and travel narratives.
- Odorico of Pordenone (c. 1286–1331), an Italian missionary and diplomat who traveled from Venice across the Middle East to India, visited some islands in the Indian Ocean, and spent three years in China. His account of his journey was first published in Pesaro in 1513, under the title *Odorichus de rebus incognitis* (Odoric on Unknown Things) (despite the Latin title, the text is in Italian),[47] but was also available in manuscripts.[48]

---

[42]See M. A. P. d'Avezac, ed., "Johannis de Plano Carpini Antivariensis Archiepiscopi Historia Mongalorum quos nos Tartaros appellamus," *Recueil de Voyages et de Mémoires* 4 (1839), pp. 603–779. The standard edition is Giovanni da Pian del Carpine, *Storia dei Mongoli*, ed. Enrico Menestò and trans. Maria Cristiana Lungarotti (Spoleto: Centro italiano di studi sull'alto Medioevo, 1989), which includes an Italian translation and a map of Plano Carpini's travels in plate 2 following p. 224.

[43]See C. Raymond Beazley, ed., *The Texts and Versions of John de Plano Carpini and William de Rubruquis* (London: The Hakluyt Society, 1903); and Gregory G. Guzman, "The Encyclopedist Vincent of Beauvais and His Mongol Extracts from John of Plano Carpini and Simon of Saint-Quentin," *Speculum* 49.2 (1974), pp. 287–307. The 1473 edition of the *Speculum historiale* was printed in Strasbourg, probably by Adolf Rusch, in four volumes. In the edition *Bibliotheca mvndi. Vincentii burgvndi ... Specvlvm qvadrvplex, natvrale, doctrinale, morale, historiale* (Douai: Baltazaris Belleri, 1624), which is readily available through its facsimile reprinting with the title *Speculum quadruplex; sive, Speculum maius* (Graz: Akademische Druck-u. Verlaganstalt, 1964–66), the material in question appears in Book 31 of the *Speculum historiale* (which is vol. 4 of the set), rather than in Book 32.

[44]The Latin text of Simon's account has been published as Simon de Saint-Quentin, *Histoire des Tartares*, ed. Jean Richard (Paris: P. Geuthner, 1965); it was translated into French in Pierre Bergeron, ed., *Voyages de Benjamin de Tudelle autour du monde, commencé l'an 1173, de Jean du Plan Carpin en Tartarie, du frère Ascelin et de ses compagnons vers la Tartarie, de Guillaume de Rubruquin en Tartarie et en Chine en 1253, suivis des additions de Vincent de Beauvais et de l'histoire de Guillaume de Nangis pour l'éclaircissement des précédents voyages* (Paris: Imprimé aux frais du gouvernement pour procurer du travail aux ouvriers typographes, 1830), pp. 215–235.

[45]On Ascelinus see Igor de Rachewiltz, *Papal Envoys to the Great Khans* (Stanford, Calif.: Stanford University Press, 1971), pp. 115–119. A more detailed but diffuse account is in Paul Pelliot, "Les Mongols et la Papauté," *Revue de l'Orient chrétien* 23 (1922–23), pp. 3–30; 24 (1924), pp. 225–335; and 28 (1931–32), pp. 3–84, esp. 24 (1924), pp. 262–335. On Simon of Saint Quentin's account see Gregory G. Guzman, "Simon of Saint-Quentin and the Dominican Mission to the Mongols, 1245–1248," Ph.D. Dissertation, University of Cincinnati, 1968; and the same author's "Simon of Saint-Quentin and the Dominican Mission to the Mongol Baiju: A Reappraisal," *Speculum* 46.2 (1971), pp. 232–249.

[46]On this first Latin edition of Marco Polo see Lotte Hellinga, "Marco Polo's Description of the Far East and the Edition Printed by Gheraert Leeu," in Elly Cockx-Indestege, ed., *E codicibus impressisque: opstellen over het boek in de Lage Landen voor Elly Cockx-Indestege* (Louvain: Peeters, 2004), vol. 1, pp. 309–328. For discussion of the Latin manuscripts of Polo see Consuelo Wager Dutschke, "Francesco Pipino and the Manuscripts of Marco Polo's Travels," Ph.D. Dissertation, University of California at Los Angeles, 1993. For an account of manuscripts and editions (including those not in Latin) of Polo see Marco Polo, *The Description of the World*, trans. and ed. A. C. Moule and Paul Pelliot (London: G. Routledge, 1938), vol. 1, pp. 509–519.

[47]The Latin and Italian text of Odoric and an English translation are provided by Henry Yule, ed. and trans., *Cathay and the Way Thither: Being a Collection of Medieval Notices of China*, revised by Henri Cordier (London: The Hakluyt Society, 1913–16), vol. 2.

[48]For a bibliography on manuscripts of Odoric's narrative see Marianne O'Doherty, "The *Viaggio in Inghilterra* of a *Viaggio in Oriente*: Odorico da Pordenone's *Itinerarium* from Italy to England," *Italian Studies* 64.2 (2009), pp. 198–220, esp. 200.

- Pierre d'Ailly (1351–1420), a French cardinal, theologian and cosmographer whose *Imago mundi* (Image of the World) is well known for having influenced Christopher Columbus's geographical thought, specifically his conception of the width of the Atlantic[49]: Columbus owned a copy of the book which he heavily annotated.[50] The work survives in several manuscripts,[51] and it was first published c. 1480–1483.[52]
- Alvise Cadamosto (c. 1432–1483), a Venetian merchant and navigator who explored the western coast of Africa for Portugal in 1455 and 1456.[53] Cadamosto's *Navigazioni* were first published as Chaps. 1–47 of *Paesi novamente retrovati* (Newly Discovered Countries), an important collection of travel narratives first published in 1507 that was quickly translated into Latin and German.[54] The *Paesi* was in fact the second most frequently printed early account of the discoveries in the New World, after Vespucci's *Mundus novus*.[55] The interest of Waldseemüller or his associates in the voyage of Cadamosto is indicated by the existence of proof sheets of maps from Waldseemüller's workshop that show the course Cadamosto took down the coast of Africa.[56]
- Caspar the Jew of India, also called Gaspar de Gama (1444–c. 1510–1520), a Jewish merchant who met Vasco da Gama in India and acted as an interpreter for da Gama and other Portuguese navigators.[57] Material from Caspar was transmitted

---

[49]On d'Ailly's *Imago mundi* see Jeannine Quillet, "L'*Imago Mundi* de Pierre d'Ailly," in Danielle Buschinger and Wolfgang Spiewok, eds., *Nouveaux mondes et mondes nouveaux au Moyen Age: Actes du Colloque du Centre d'Etudes Médiévales de l'Université de Picardie Jules Verne, Amiens, mars 1992* (Greifswald: Reineke Verlag, 1994), pp. 107–114; for discussion of his other works as well see Jean-Patrice Boudet, "Un prélat et son équipe de travail à la fin du Moyen Âge: remarques sur l'oeuvre scientifique de Pierre d'Ailly," in Didier Marcotte, ed., *Humanisme et culture géographique à l'époque du Concile de Constance: autour de Guillaume Fillastre: Actes du colloque de l'Université de Reims, 18–19 novembre 1999* (Turnhout: Brepols, 2002), pp. 127–150.

[50]The copy of d'Ailly's *Imago mundi* annotated by Columbus, which is in the Biblioteca Colombina in Seville, has been reproduced in facsimile as Pierre d'Ailly, *Imago mundi* (Madrid: Testimonio Compañía Editorial, 1990), with a volume of commentary by Juan Pérez de Tudela, and a translation of d'Ailly into Spanish by Antonio Ramírez de Verger. Also see George E. Nunn, "The Imago Mundi and Columbus," *American Historical Review* 40.4 (1935), pp. 646–661; and Elisabetta Sarmati, "Le postille di Colombo all'*Imago Mundi* di Pierre d'Ailly," *Columbeis* 4 (1990), pp. 23–42.

[51]For manuscripts of d'Ailly's work see Destombes, *Mappemondes* (see note 29), pp. 161–163.

[52]The standard modern edition of d'Ailly's work, which includes the Latin text and a French translation, is Pierre d'Ailly, *Ymago mundi*, ed. Edmond Buron (Paris: Maisonneuve frères, 1930).

[53]Gerald R. Crone, ed. and trans., *The Voyages of Cadamosto and Other Documents on Western Africa in the Second Half of the Fifteenth Century* (London: Printed for the Hakluyt Society, 1937).

[54]The full title of the book is *Paesi nouamente retrouati et Nouo Mondo da Alberico Vesputio Florentino intitulato* (Vicenza: Henrico Vicentino, 1507). For a brief discussion of the 1507 edition of the *Paesi novamente retrovati* see Henry Harrisse, *Bibliotheca Americana Vetustissima: A Description of Works Relating to America, Published Between the Years 1492 and 1551* (New York: G. P. Philes, 1866), no. 48, pp. 96d-99; and António Alberto Banha de Andrade, *Mundos novos do mundo: panorama da difusão, pela Europa, de notícias dos descobrimentos geográficos portugueses* (Lisbon: Junta de Investigações do Ultramar, 1972), vol. 1, pp. 527–532. The most detailed account of the *Paesi* in English is in Antony Vallavanthara, *India in 1500 A.D.: The Narratives of Joseph the Indian* (Mannanam: Research Institute for Studies in History, 1984), pp. 11–24 and 295–311. Vallavanthara discusses the Latin translation of the *Paesi*, the *Itinerarium Portugallensium*, on pp. 24–41 and 312–315. There is a detailed discussion of the work's contents in Norbert Ankenbauer, *'Das ich mochte meer newer dyng erfaren': die Versprachlichung des Neuen in den 'Paesi novamente retrovati' (Vicenza, 1507) und in ihrer deutschen Übersetzung (Nürnberg, 1508)* (Berlin: Frank & Timme, 2010), esp. pp. 51–123. The most readily available printed edition of the *Paesi* is a facsimile of the 1508 edition, published in *Vespucci Reprints, Texts and Studies* (Princeton, N.J.: Princeton University Press, 1916), vol. 6, under the title *Paesi nouamente retrovati & Novo mondo da Alberico Vesputio Florentino intitulato*. Better still, Norbert Ankenbauer has made a digital edition of the 1507 edition and also of the 1508 German translation of the book, available at http://diglib.hab.de/edoc/ed000145/start.htm. The manuscript that served as the basis of the printed edition of the *Paesi novamente retrovati*, known as the Trevisan Manuscript, resides at the Library of Congress with the shelfmark MSS Med. & Ren., 26; for discussion see Eric Dursteler, "Reverberations of the Voyages of Discovery in Venice, ca. 1501: The Trevisan Manuscript in the Library of Congress," *Mediterranean Studies* 9 (2000), pp. 43–64. The documents associated with the manuscript at the Library of Congress include a copy of the following thesis, which in vol. 2 has a complete transcription of the manuscript: Fredi Chiappelli, "La relazione di Angelo Trevisan sui primi viaggi di Cristoforo Colombo," Mémoire de Licence, Université de Lausanne, 1968.

[55]See Rudolf Hirsch, "Printed Reports on the Early Discoveries and Their Reception," in Fredi Chiappelli, Michael J. B. Allen, and Robert Louis Benson, eds., *First Images of America: The Impact of the New World on the Old* (Berkeley: University of California Press, 1976), vol. 2, pp. 537–551, at 546.

[56]See Peter H. Meurer, "Sechs Karten der westafrikanischen Küste aus der Waldseemüller-Schule," *Cartographica Helvetica* 45 (2012), pp. 15–26; the maps are in a copy of the *Itinerarium Portugallensium* in Vienna, Österreichische Nationalbibliothek, shelfmark 394.092-C.Kar. There is an earlier discussion of the same maps in Erich Woldan, "Die altesten gedruckten modernen Karten Afrikas," *Anzeiger der Philosophisch-Historischer Klasse der Osterreichischen Akademie der Wissenschaften* 118 (1981), pp. 252–257.

[57]On Gaspar see E. G. Ravenstein, ed., *A Journal of the First Voyage of Vasco da Gama, 1497–1499* (London: Printed for the Hakluyt Society, 1898), pp. 179–180; Franz Hümmerich, "Studien zum 'Roteiro' der Entdeckungsfahrt Vascos da Gama 1497–1499," *Revista da Universidade de Coimbra* 10 (1927), pp. 53–302, esp. 93–137; Elias Lipiner, *Gaspar da Gama: um converso na frota de Cabral* (Rio de Janeiro, RJ: Editora Nova Fronteira, 1987); and Stefania Elena Carnemolla, "Un certo Gaspar da Gama: sfuggente figura di 'interprete' dei viaggiatori portoghesi del Cinquecento," in Maria José de Lancastre, Silvano Peloso, and Ugo Serani, eds., *E vós, tágides minhas: miscellanea in onore di Luciana Stegagno Picchio* (Viareggio, Lucca: M. Baroni, 1999), pp. 229–240; reprinted in *L'Erasmo* 16 (2003), pp. 59–74.

in a letter by Girolamo Sernigi included in *Paesi novamente retrovati*, Chaps. 60–62.[58] This book was among Waldseemüller's most important sources, but Caspar is not identified by name in those passages, and Waldseemüller seems not to have made use of those chapters, as he had a superior version of Caspar's account. In the text block on sheet 9 of the *Carta marina* (see Legend 9.3) Waldseemüller says that he had access to a travel narrative by Caspar that was sent to the King of Portugal, but that document does not appear to have survived.

- Francisco de Almeida (c. 1450–1510),[59] a Portuguese nobleman, soldier and explorer who was essential in the establishment of Portuguese power in the Indian Ocean.[60] On Almeida's voyage of 1505 from Portugal to India, one of the passengers was Balthasar Springer, or Sprenger, the representative of a trading company in Augsburg, Germany. Springer wrote an account of the voyage, which is the work that Waldseemüller is really citing.[61] The first edition of Springer's narrative, which was published in German in 1509,[62] was illustrated by Hans Burgkmair.[63] Waldseemüller made relatively little use of it, just for some toponyms in India and an image of an Indian man and an African man.
- Christopher Columbus (1451–1506), the famous Genoese explorer. Peter Martyr d'Anghiera began writing an account of the discovery of the New World in 1494,[64] and in 1504 some chapters from his work that give an account of Columbus's first three voyages were translated into Italian and published in the now very rare *Libretto de tutta la nauigatione de Re de Spagna de le isole et terreni nouamente trouati*[65]; this material was incorporated into the *Paesi novamente retrovati* a few years later in 1507 (Chaps. 84–108), which is where Waldseemüller may have found it.
- Pedro Álvares Cabral (c. 1467–c. 1520), the Portuguese explorer who in 1500–1501 discovered Brazil, sailed on to India, and then returned to Portugal.[66] One of the earliest and most complete accounts of Cabral's voyage was written by an

[58]The letter with material from Gaspar is translated into English in Ravenstein, *A Journal of the First Voyage* (see note 57), pp. 137–141, with an introduction on pp. 119–123.

[59]Banha de Andrade, *Mundos novos do mundo* (see note 54), vol. 2, p. 580, suggests that the authority Waldseemüller is citing is actually Francisco de Albuquerque, and that Waldseemüller was using manuscript letters of his, but this seems doubtful: there are really no similarities between the narrative about to be cited and Waldseemüller's legends. For discussion of Francisco de Albuquerque see Jean Aubin, "Francisco de Albuquerque, un juif castillan au service de l'Inde Portugaise (1510–15)," *Arquivos do Centro Cultural Português* 7 (1974), pp. 175–188.

[60]For an account of Almeida's life see Joaquim Candeias Silva, *O Fundador do 'Estado Português da Índia' D. Francisco de Almeida, 1457(?)–1510* (Lisbon: Imprensa Nacional, 1996). There is a portrait of Franciso de Almeida in New York, Pierpont Morgan Library, MS M. 525, f. 4r, the *Livro de Lisuarte de Abreu* of c. 1558, which has been reproduced in facsimile as *Livro de Lisuarte de Abreu* (Lisbon: Comissão Nacional para as Comemorações dos Descobrimentos Portugueses, 1992).

[61]For discussion of the Almeida/ Springer voyage see Franz Hümmerich, "Quellen und Untersuchungen zur Fahrt der ersten Deutschen nach dem portugiesischen Indien 1505/6," *Abhandlungen der Königlich Bayerischen Akademie der Wissenschaften, Philosophisch-philologische und historische Klasse* 30.3 (1918), pp. 1–153; and Thomas Horst, "The Voyage of the Bavarian Explorer Balthasar Sprenger to India (1505/1506) at the Turning Point between the Middle Ages and Early Modern Times: His Travelogue and the Contemporary Cartography as Historical Sources," in Philipp Billion, Nathanael Busch, Dagmar Schlüter, and Zenia Stolzenburg, eds., *Weltbilder im Mittelalter = Perceptions of the World in the Middle Ages* (Bonn: Bernstein-Verlag, 2009), pp. 167–197.

[62]Balthasar Springer, *Die Merfart vn[d] erfarung nüwer Schiffung vnd Wege zu viln onerkanten Jnseln vnd Künigreichen von dem großmechtigen Portugalische[n] Kunig Emanuel Erforscht funden bestritten vnnd Jngenomen* ([Oppenheim]: [Köbel], 1509). For an English translation see Balthasar Springer, *The Voyage from Lisbon to India, 1505–6: Being an Account and Journal*, ed. C. H. Coote (London: B. F. Stevens, 1894). Note that the work is misattributed to Vespucci in this edition.

[63]On Burgkmair's illustrations of Springer's narrative see Jean Michel Massing, "Hans Burgkmair's Depiction of Native Africans," *RES: Anthropology and Aesthetics* 27 (1995), pp. 39–51. As Massing notes on p. 39, "Burgkmair's woodcuts... were, for decades, the only representations of Africans and Indians based on visual information rather than mere literary testimony." Also see Stephanie Leitch, "Burgkmair's Peoples of Africa and India (1508) and the Origins of Ethnography in Print," *The Art Bulletin* 91.2 (2009), pp. 134–159; and Sandra Young. "Envisioning the Peoples of 'New' Worlds: Early Modern Woodcut Images and the Inscription of Human Difference," *English Studies in Africa* 57.1 (2014), pp. 33–54.

[64]On Peter Martyr's writings about the New World and Columbus see John Boyd Thacher, *Christopher Columbus: His Life, His Work, His Remains as Revealed by Original Printed and Manuscript Records* (New York and London: G.P. Putnam's Sons: The Knickerbocker Press, 1903–1904), vol.1.1, pp. 3–110; David Beers Quinn, "New Geographical Horizons: Literature," in Fredi Chiappelli, Michael J. B. Allen, and Robert Louis Benson, eds., *First Images of America: The Impact of the New World on the Old* (Berkeley: University of California Press, 1976), vol. 2, pp. 635–658, esp. 647–651; and Juan Gil, "Pedro Mártir de Anglería, intérprete de la cosmografía colombina," *Anuario de Estudios Americanos* 39 (1982), pp. 487–502.

[65]Chapters 22 to 25 of the *Libretto*, which contain the description of Columbus's First Voyage, are transcribed in Guglielmo Berchet, *Fonti italiane per la storia della scoperta del Nuovo mondo*, in *Raccolta di documenti e studi pubblicati dalla R. Commissione colombiana* (Rome: Ministero della pubblica istruzione, 1892–1896), part 3, vols. 1–2, in vol. 2, pp. 173–177. For an introduction to the *Libretto*, facsimile, and English translation see Thacher, *Christopher Columbus* (see note 64), vol. 2, pp. 438–456, 457–485, and 486–514, respectively. There is also a later facsimile of the copy in the John Carter Brown Library: Pietro Martire d'Anghiera, *Libretto de tutta la nauigatione de re de Spagna de la isole et terreni nouamente trouati, Venice, 1504*, ed. Lawrence Wroth (Paris: H. Champion, 1929); and a Spanish translation in Marisa Vannini de Gerulewicz, *El Mar de los descubridores: documentos y relatos inéditos o poco conocidos sobre el descubrimiento y la exploración de los mares, islas y tierras del Nuevo Mundo (siglos XV–XVI)* (Caracas: Comisión Organizadora de la III Conferencia de las Naciones Unidas sobre Derecho del Mar, 1974), pp. 111–158.

[66]On Cabral's voyage see Max Justo Guedes, ed., *A viagem de Pedro Alvares Cabral e o descobrimento do Brasil, 1500–1501* (Lisbon: Academia de Marinha, 2003), and the references cited in the following note.

unnamed member of the fleet and is known as the "Anonymous Narrative"; it was published in the *Paesi novamente retrovati*, Chaps. 63–83.[67]

- Ludovico de Varthema (c. 1470–1517), an Italian adventurer and keen observer who wrote an account of his travels to Egypt, the Middle East, India and the islands of the Indian Ocean, though there is some dispute about whether in fact he traveled anywhere east of Cairo.[68] His narrative was published soon after his return to Europe in 1508, first in Italian (1510), then in Latin (1511), and then in an illustrated edition in German (1515).[69]
- Joseph the Indian, or Priest Joseph (fl. 1490–1518), a Christian priest from Cranganore, India, who shipped with Cabral on his return to Portugal so that he could visit Rome and Jerusalem. During the voyage, and also during his stay in Portugal, he supplied detailed information about southwestern India that may have been published in 1505, and was certainly printed in 1507 as the final chapters of the *Paesi novamente retrovati*.[70]

For Waldseemüller, the majority of these sources were recent: Varthema's book, of which Waldseemüller made heavy use, was printed just a few years before Waldseemüller created the *Carta marina.* A number of the other sources he cites were published in the *Paesi novamente retrovati* in 1507, about a decade before he made the *Carta marina.*

There are a couple of interesting omissions from Waldseemüller's list. The first is the *Travels* of Sir John Mandeville,[71] who claimed to have traveled widely in Asia and Africa. The book was written in the fourteenth century and circulated very widely both in manuscript and print, with incunable editions in ten different languages,[72] including multiple editions published in Strasbourg (near Waldseemüller) with woodcut illustrations.[73] The second is the narrative of the travels of Arnold von Harff to the Holy Land, Egypt, and the Indian Ocean in 1496–1499. The work was not published until 1860,[74] but circulated in

[67]An English translation of the "Anonymous Narrative" is in William Brooks Greenlee, *The Voyage of Pedro Álvares Cabral to Brazil and India from Contemporary Documents and Narratives* (London: The Hakluyt Society, 1938), pp. 53–94. The earliest surviving version of the Anonymous Narrative is in the Trevisan Manuscript in the Library of Congress: Dursteler, "Reverberations of the Voyages of Discovery" (see note 54), p. 53. There is a Portuguese translation of the Anonymous Narrative in T. O. Marcondes de Sousa, "Relação do Pilôto Anônimo," *Revista do Instituto Histórico e Geográfico de São Paulo* 45 (1945/1950), pp. 82–108; and in Guedes, *A viagem de Pedro Alvares Cabral*, pp. 205–226.

[68]For a good discussion of Varthema's travels see Joan-Pau Rubiés, "Ludovico de Varthema: The Curious Traveller at the Time of Vasco da Gama and Columbus," in his *Travel and Ethnology in the Renaissance: South India through European eyes, 1250–1625* (Cambridge and New York: Cambridge University Press, 2000), pp. 125–163; for an English translation of Varthema see Lodovico de Varthema, *The Travels of Ludovico di Varthema in Egypt, Syria, Arabia Deserta and Arabia Felix, in Persia, India, and Ethiopia, A.D. 1503 to 1508, translated from the original Italian edition of 1510... by John Winter Jones and edited, with notes and an introduction, by George Percy Badger* (London: The Hakluyt Society, 1863). Notes on the different editions and translations of the work are on pp. iii–xvi.

[69]The illustrated edition in question is Lodovico de Varthema, *Die Ritterlich und lobwirdig rayß des gestrengen und über all ander weyt erfarnen ritters und Lantfarers herren Ludowico vartomans von Bolonia* (Augsburg: H. Miller, 1515). On the re-use of the illustrations from this edition by authors other than Waldseemüller see Lisa Voigt and Elio Brancaforte, "The Traveling Illustrations of Sixteenth-Century Travel Narratives," *PMLA* 129.3 (2014), pp. 365–398.

[70]For discussion of the information Joseph conveyed see Antony Vallavanthara, *India in 1500 A.D.: The Narratives of Joseph the Indian* (Mannanam: Research Institute for Studies in History, 1984); and Antony Vallavanthara, "The Indian Coasts and the Malabar Society as Brought to the Notice of the Europeans by Joseph the Indian," in K. S. Mathew, Teotónio R. de Souza, Pius Malekandathil, eds., *The Portuguese and the Socio-Cultural Changes in India, 1500–1800* (Tellicherry, Kerala: Institute for Research in Social Sciences and Humanities, MESHAR, 2001), pp. 93–119. For an English translation of Joseph's account see William Brooks Greenlee, "The Account of Priest Joseph," in *The Voyage of Pedro Álvares Cabral to Brazil and India from Contemporary Documents and Narratives* (London: The Hakluyt Society, 1938), pp. 95–113.

[71]John Larner suggests that "John Mandeville" was the pen name of Jan de Langhe in "Plucking Hairs from the Great Cham's Beard: Marco Polo, Jan de Langhe, and Sir John Mandeville," in Suzanne Conklin Akbari and Amilcare Iannucci, eds., *Marco Polo and the Encounter of East and West* (Toronto: University of Toronto Press, 2008), pp. 133–155.

[72]On the diffusion of Mandeville's work see Iain Higgins, "Imagining Christendom from Jerusalem to Paradise: Asia in Mandeville's Travels," in Scott Westrem, ed., *Discovering New Worlds: Essays on Medieval Exploration and Imagination* (New York: Garland, 1991), pp. 91–114, esp. 95–96. There is a good preliminary list of manuscripts of Mandeville in Josephine Waters Bennett, *The Rediscovery of Sir John Mandeville* (New York: Modern Language Association of America, 1954), Appendix 1, "The Manuscripts," pp. 265–334.

[73]Johann Prüss in Strasbourg published editions of Mandeville in 1483, 1484, and 1488; Bartholomäus Kistler in 1499; and Johann Knobloch in 1507—all of these editions illustrated. There were many other non-Strasbourg editions in both Latin and German to which Waldseemüller might have had access.

[74]Arnold von Harff, *Die Pilgerfahrt des ritters Arnold von Harff von Cöln durch Italien, Syrien, Aegypten, Arabien, Aethiopien, Nubien, Palästina, die Türkei, Frankreich und Spanien, wie er sie in den jahren 1496 bis 1499 vollendet, beschrieben und durch zeichnungen erläutert hat*, ed. Eberhard von Groote (Cologne: J. M. Heberle, 1860). The modern edition in German is Helmut Brall-Tuchel and Folker Reichert, eds., *Rom—Jerusalem—Santiago: das Pilgertagebuch des Ritters Arnold von Harff (1496–1498)* (Cologne: Böhlau, 2007), and for an English translation see Arnold von Harff, *The Pilgrimage of Arnold von Harff, Knight, from Cologne, through Italy, Syria, Egypt, Arabia, Ethiopia, Nubia, Palestine, Turkey, France and Spain, Which He Accomplished in the Years 1496 to 1499*, trans. Malcolm Letts (London: Hakluyt Society, 1946). For bibliography on his narrative see Werner Paravicini, ed., *Europäische Reiseberichte des späten Mittelalters: eine analytische Bibliographie* (Frankfurt am Main and New York: Peter Lang, 1994–), vol. 1, pp. 274–281.

manuscript, a number of which were illustrated.[75] Given Waldseemüller's wide knowledge of recent travel literature, it is difficult to imagine that he was not familiar with von Harff's book, and in fact it would be at least somewhat ironic, as his 1507 map has data that also appears in von Harff, probably by way of a map by Martellus.[76] Despite the extravagance of Mandeville's narrative, he was accepted as an authority by some other Renaissance cartographers and geographers,[77] but it seems likely that Waldseemüller chose not to use Mandeville and von Harff because he considered them unreliable.

If we look at Waldseemüller's list again in the light of his 1507 map, one of the authors included in this list and another who is omitted from it are surprising. The 1507 map proclaims Amerigo Vespucci as the discoverer of the New World: Vespucci's portrait is at the top of the map, and the southern part of the New World bears the name "America," which Waldseemüller and Ringmann created from "Amerigo."[78] Moreover, their book *Cosmographiae introductio* includes Vespucci's accounts of his four voyages. So it is surprising that Waldseemüller does not include Vespucci in his list of sources for the *Carta marina*, but does include Columbus.[79] Indeed, the *Carta marina* makes it clear that Waldseemüller had realized, probably through the account of Columbus's 1492 voyage in the *Paesi novamente retrovati*, the precedence of Columbus as discoverer: the name "America" does not appear on the map, and a legend in the South Atlantic explicitly names Columbus as the first discoverer of the New World, Cabral as the second, and Vespucci as the third (Legend 10.2).

Thus in making his *Carta marina* Waldseemüller not only abandoned the Ptolemaic cartographic model in favor of the nautical chart model; he also abandoned Vespucci as principal discoverer of the New World in favor of Columbus. The fact that the two figureheads, Ptolemy and Vespucci, displayed so prominently at the top of the 1507 map, had both fallen by the wayside in 1516 is a powerful testament to the rapid development of Waldseemüller's cartographic thought and his willingness to change his ideas in light of new information, as well as to the dynamism of early sixteenth-century cartography in general. Waldseemüller's willingness to discard all of the work he had invested in the 1507 map is all the more impressive given that the map was evidently well received.[80] It was not the demands of customers, but rather his own

[75]On the manuscripts of von Harff's narrative see Patrick de Never and Volker Honemann, "Zur Überlieferung der Reisebeschreibung Arnolds von Harff," *Zeitschrift für deutsches Altertum und deutsche Literatur* 107.2 (1978), pp. 165–178; and Peter A. Jorgensen and Barbara M. Ferré, "Die handschriftlichen Verhältnisse der spätmittelalterlichen Pilgerfahrt des Arnold von Harff," *Zeitschrift für deutsche Philologie* 110.3 (1991), pp. 406–421. The illustrated manuscripts of von Harff which are in more or less public collections are Bonn, Universitätsbibliothek, Cod. S 447; Schloß Burgsteinfurt in Steinfurt (Westf.), Fürstl. Bentheim-Steinfurtische Schloßbibl., Hs. 4; Darmstadt, Landes- und Hochschulbibliothek, Hs. 138; Schloß Erpenburg (bei Büren), Archiv der Freihernn von und zu Brenken, Cod. HX. 100; Gießen, Universitätsbibliothek, Hs. 163, ff. 5r–155r; Cologne, Historisches Archiv der Stadt, Cod. W* 382; Munich, Staatsbibliothek, Cgm 2213/32, ff. 451r–615v; Trier, Stadtbibliothek, Hs. 1938/1469 8°, and Hs. 2424/2387 2°; and Wolfenbüttel, Herzog August Bibliothek, Cod. 177 Helmst, ff. 207r–258v.

[76]I discussed the commonalities between Waldseemüller and von Harff in my lecture "Evidence for a Lost Map Used by Waldseemüller in his Depiction of Eastern Africa and the Indian Ocean," delivered May 15, 2009, at the conference "Exploring Waldseemüller's World," May 14 and 15, 2009, at the Library of Congress, Washington, DC. Video of the talk is available at http://www.loc.gov/today/cyberlc/feature_wdesc.php?rec=4569 (third talk in panel).

[77]See C. W. R. D. Moseley, "Behaim's Globe and 'Mandeville's Travels,'" *Imago Mundi*, 33.1 (1981), pp. 89–91; and Christiane Deluz, "Le Livre Jehan de Mandeville, autorité géographique à la Renaissance," in Jean Céard and J.-Cl. Margolin, eds., *Voyager à la Renaissance: Actes du colloque de Tours 1983* (Paris: Maisonneuve et Larose, 1987), pp. 205–220. Gerard Mercator also cites Mandeville as a source about Java and the southern hemisphere on his world map of 1569, but Mercator ends by saying that Mandeville is "an author unbelievable in other respects"—see Nicolas Crane, *Mercator: The Man Who Mapped the Planet* (New York: Henry Holt, 2002), p. 226. Also see "Text and Translation of the Legends of the Original Chart of the World by Gerhard Mercator, Issued in 1569," *Hydrographic Review* 9.2 (1932), pp. 7–45, at 44–45.

[78]The relevant passage in the *Cosmographiae introductio* is in Chap. 7, on signature a iii^r in the 1507 edition. In Fischer and von Wieser, *The 'Cosmographiae introductio' of Martin Waldseemüller* (see note 7), the passage is on p. xxv (Latin) and 63 (English); and in Hessler, *The Naming of America* (see note 7), it is on p. 94. For bibliography on the naming of America see note 7 above.

[79]Waldseemüller and Ringmann's favoring of Vespucci over Columbus as discoverer of the New World in the *Cosmographiae introductio* and the 1507 map is to be seen in the context of the greater interest in and popularity of accounts of Vespucci's voyages rather than Columbus's: see Rudolf Hirsch, "Printed Reports on the Early Discoveries and Their Reception," in Fredi Chiappelli, Michael J. B. Allen, and Robert Louis Benson, eds., *First Images of America: The Impact of the New World on the Old* (Berkeley: University of California Press, 1976), vol. 2, pp. 537–551, esp. 546–547.

[80]Waldseemüller mentions the positive reception of his 1507 map in a dedicatory letter addressed to his colleague Matthias Ringmann in Waldseemüller's *Architecturae et perspectivae rudimenta*, which was published with Gregor Reisch's *Margarita philosophica nova* (Strasbourg: Grüninger, 1508). Waldseemüller wrote: *Cosmographiam universalem tam solidam quam planam non sine gloria et laude per orbem disseminatam nuper composuimus, depinximus, et impressimus*: "We recently composed, drew, and printed a world map and globe which have been disseminated across the world, not without fame and praise." This letter is quoted by M. d'Avezac, *Martin Hylacomylus Waltzemüller* (see note 5), pp. 109–110. Johnson, *Carta marina* (see note 21), p. 125, note 22, also cites part of the passage. Waldseemüller also records René II's enthusiastic reception of his 1507 map in the dedicatory letter in Ringmann's *Instructio manuductionem prestans in Cartam itinerariam* (Strasbourg: Grüninger, 1511): *Neque enim obliti sumus qua aurium clementia: quam hilari vultu et quam grato animo generalem orbis descriptionem: ac alia etiam litterarij laboris nostri monimenta sibi oblata a nobis susceperit*, "For we have not forgotten with what indulgent hearing, with what a happy face, and with what a grateful spirit he received our general map of the world, and other samples of our literary works that we presented to him." This passage is quoted and translated into French by M. d'Avezac, *Martin Hylacomylus Waltzemüller* (see note 5), pp. 136–137; and into English by Toby Lester, *The Fourth Part of the World: The Race to the Ends of the Earth, and the Epic Story of the Map that Gave America its Name* (New York: Free Press, 2009), p. 373.

determination to find the best method for representing the world that led him to undertake the creation of an entirely new world map in the *Carta marina.*

In addition to recognizing Columbus's primacy as discoverer of the New World, in the *Carta marina* Waldseemüller adopted a Columbian conception of the New World. This can be seen particularly clearly in the different indications of what is west of the New World on the two maps. One of the most striking and oft-discussed aspects of Waldseemüller's 1507 map is his depiction of an ocean west of the New World before the European discovery of the Pacific by Vasco Nuñez de Balboa in September 1513. He also explicitly stated in the *Cosmographiae introductio* that the New World was an island.[81] This depiction and statement have generated claims of an earlier, pre-Balboa discovery of the Pacific by a European voyage of which no other record survives. But there is a much simpler explanation for Waldseemüller's depiction. Marco Polo had said that Japan was 1500 miles east of mainland Asia[82]; Polo clearly stated that Japan was an island; so there must be water east of Japan, and thus between Japan and the New World. Quite probably on the basis of this reasoning, Waldseemüller shows water separating the New World from Asia on the 1507 map: they are two distinct regions.

Columbus had been seeking a route to Asia by sailing west, and during all four of his voyages and to the end of his life believed that he had been in Asia, albeit in some previously unknown outlying reaches of the continent.[83] This is the view of the New World that Waldseemüller adopts in the *Carta marina*. It is particularly clear in the legend on North America, on sheet 1, which reads TERRA DE CVBA • ASIE PARTIS, "The land of Cuba, part of Asia" (Legend 1.1)[84] Other evidence for this view is in the legend on sheet 5 describing Hispaniola in the Caribbean, which begins *Spagnolla que et Offira dicitur*, "Hispaniola, which is also called Ophir" (Legend 5.1)—identifying the island with the region mentioned in the Bible from which gold and other riches were brought to King Solomon.[85] Columbus had shown great interest in the location of Ophir, and had himself asserted that Hispaniola was to be identified with that region.[86] Thus in the *Carta marina* Waldseemüller has adopted a Columbian view of the New World. His decisions to show less of the ocean east of Asia than Caverio on his chart, and to exclude Japan, should be seen as part of this same new perspective.

In attempting to update his information about and depiction of the New World in the *Carta marina*, Waldseemüller inadvertently took a step backwards, since Columbus's belief that the New World was Asia was incorrect.[87] But in other parts of the map, particularly in Asia and the Indian Ocean, the updating is breathtaking, and in examining the changes one appreciates Waldseemüller's assertion that the 1507 map shows the world according to old authors, while the *Carta marina* depicts the world according to the very latest information.

On the 1507 map, in western Asia Waldseemüller follows Ptolemy, while in eastern Asia—which was unknown to Ptolemy—he follows Marco Polo. Thus in western Asia his source was more than a thousand years old, while in eastern Asia it was some two hundred years old. On the 1516 *Carta marina*, Waldseemüller uses information from John of Plano Carpini,

---

[81]In Chap. 9 of the *Cosmographiae introductio*, Waldseemüller writes of the four parts of the world, "the first three parts [i.e. Europe, Africa and Asia] are continents, and the fourth [America] is an island, since it is seen to be completely surrounded by water."

[82]On the distance of Japan from mainland Asia see note 36 above.

[83]On Columbus's belief that his discoveries were part of Asia see George E. Nunn, "Did Columbus Believe that he Reached Asia on his Fourth Voyage?" in *The Geographical Conceptions of Columbus: A Critical Consideration of Four Problems* (New York: American Geographical Society, 1924); expanded edition with an essay titled "The Test of Time" by Clinton R. Edwards (Milwaukee: American Geographical Society Collection of the Golda Meir Library, University of Wisconsin-Milwaukee; and New York: American Geographical Society, 1992), pp. 54–90; E. G. R. Taylor, "Idée Fixe: The Mind of Christopher Columbus," *The Hispanic American Historical Review* 11.3 (1931), pp. 289–301; John H. Parry, "Asia-in-the-West," *Terrae Incognitae* 8 (1976), pp. 59–72; Folker Reichert, "Columbus und Marco Polo—Asien in Amerika. Zur Literaturgeschichte der Entdeckungen," *Zeitschrift für historische Forschung* 15 (1988), pp. 1–63, and Chet Van Duzer, "Geography," in John Hessler, Daniel De Simone, and Chet Van Duzer, *Christopher Columbus Book of Privileges: 1502, the Claiming of a New World* (Delray Beach, FL: Levenger, and Washington, DC: Library of Congress, 2014), pp. 1–26, at 8–11.

[84]For discussion of the background of Waldseemüller's depiction of the "Terra de Cuba," see Donald L. McGuirk Jr., "The Depiction of Cuba on the Ruysch World Map," *Terrae Incognitae* 20 (1988), pp. 89–97.

[85]On the riches from Ophir see 1 Kings 9:28, 10:11, and 22:48; and 2 Chronicles 8:18 of the Bible. For discussion of claims that Ophir was located in the New World see James Romm, "Biblical History and the Americas: The Legend of Solomon's Ophir, 1492–1591," in Paolo Bernardini and Norman Fiering, eds., *The Jews and the Expansion of Europe to the West, 1450 to 1800* (New York: Berghahn Books, 2001), pp. 27–46; Gerald Arthur Ward, "Columbus, Jerusalem, and Ophir: A Voyage to the Ends of Earth and Time," in his "Restoring the Shattered World: The Apocalyptic Mercantilism of Samuel Purchas and the Revelation of World Trade," Ph.D. Dissertation, Boston University, 2003, pp. 227–248; and Jorge Magasich-Airola and Jean-Marc de Beer, "King Solomon's Mines in America," in *America Magica: When Renaissance Europe Thought it Had Conquered Paradise*, trans. Monica Sandor (London and New York: Anthem Press, 2007), pp. 53–67.

[86]Columbus asserts that Hispaniola is to be identified with Ophir in his account of his Third Voyage. See Christopher Columbus, *The Four Voyages of Columbus: A History in Eight Documents*, ed. and trans. Cecil Jane (New York: Dover Publications, 1988), vol. 2, p. 6.

[87]On the development of ideas about the New World see David Beers Quinn, "New Geographical Horizons: Literature," in Fredi Chiappelli, Michael J. B. Allen, and Robert Louis Benson, eds., *First Images of America: The Impact of the New World on the Old* (Berkeley: University of California Press, 1976), vol. 2, pp. 635–658.

Simon of Saint Quentin, and an unidentified source to describe Russia (the upper left part of sheet 3), using medieval names of Grand Duke's domain (*Russia*) and the Principality of Moscow (*Moscovia Regalis*); the inhabitants are said to follow the "Greek Rite," meaning that they belong to the Orthodox Church (see Legends 3.2 and 3.3). The contrast with the 1507 map is stark: there the information comes from Ptolemy, who wrote long before the East-West Schism of 1054 and the founding of Moscow in 1147. In addition to much of the information being more recent on the 1516 map, it is far more detailed, with particulars about the religion and political relationships of the inhabitants and sovereigns.

In the Middle East on the 1507 map, Persia is merely a collection of place names from Ptolemy, while on the *Carta marina*, in addition to there being modern names for the cities, there is a legend describing the region that mixes information from Marco Polo and a recent account from Varthema (see Legend 3.35):

> persia prouincia nobilis destructa multum per tartaros sed nunc sub ditione victoriosssimi [sic] regis Sophi reparata est enim diuisa in octo regna sunt Macometani et homines fallaces

> The noble country of Persia was largely destroyed by the Tartars, but now, under the control of the unstoppable king Sophi, it has been restored and divided into eight realms. The people are followers of Mohammed and are deceitful.

"Sophi" is Shah Isma'il es-Sufi (1487–1524), the founder of the Safavid dynasty, who gained control over Persia and Khorasan (now Iran and adjoining territories to the east) around the year 1500.[88] This information was very recent indeed, compared with that on the 1507 map, and much more detailed, as it does not merely list place names, but also reveals the current political situation.

Northern India (the lower left hand part of sheet 4) is another area where Waldseemüller was following the most recent sources, but since those sources recycled traditional information, his depiction of the area is not, in fact, particularly modern. Both John of Plano Carpini and Pierre d'Ailly describe monstrous races of men in India—men with the heads of dogs, Cyclopes, men whose faces are in their chests (elsewhere called *blemmyae*), pygmies, and so on. In listing these races, Plano Carpini and d'Ailly are availing themselves of the traditional view of India as a land of marvels and monsters, a perception that goes back to ancient Greece.[89] Waldseemüller is the first cartographer to depict several of the monstrous races of India together on a map, but the source material, although it appears in relatively recent books, is old.[90]

In the northeastern corner of the map (sheet 4) is a large image of the Great Khan in his tent, and to the left of him, a long legend describing Tartaria (northern and central Asia, from the Caspian Sea to the Pacific)—the terrain, the customs, and the Khan's great power (Legend 4.16). Marco Polo gives detailed descriptions of the Khan and his realm, and Waldseemüller made heavy use of Polo in his 1507 map, but curiously, he says almost nothing about the Khan either on that map or in the *Cosmographiae introductio*.[91] He indicates some lands under the Khan's control with little escutcheons, or shield-shaped

---

[88]For bibliography on Sophi see Legend 3.35. Sophi is also illustrated in some of the manuscript maps by Battista Agnese, for example on his atlas of nautical charts in New Haven, at Yale's Beinecke Rare Book and Manuscript Library, MS 560, map 3, an image of which is available via the Beinecke's Digital Library. Sophi also appears on the unsigned chart Munich, Bayerische Staatsbibliothek, Cod. Icon. 131, which is discussed (but misdated to c. 1505) and well-illustrated in Ivan Kupčík, *Münchner Portolankarten: Kunstmann I–XIII und zehn weitere Portolankarten = Munich Portolan Charts: Kunstmann I–XIII and Ten Other Portolan Charts* (Munich: Deutscher Kunstverlag, 2000), pp. 115–119; and on an undated and previously unattributed chart in the Nordenskiöld Collection in Helsinki University Library, illustrated in black and white in A. E. Nordenskiöld, *Periplus: An Essay on the Early History of Charts and Sailing-Directions*, trans. Francis A. Bather (Stockholm: P. A. Norstedt & Söner, 1897; New York: B. Franklin, 1967), plate 23. Sophi is also mentioned on the unsigned chart attributable to Agnese which is Budapest, National Széchényi Library, Manuscript Collection, Cod. Lat. 353. And this sovereign also appears in a nautical atlas by Aloisio Cesani, Parma, Biblioteca Palatina, MS Parm. 1616, from 1574. This atlas is briefly discussed and some maps from it are reproduced in *Carte per navigare: la raccolta di portolani della Biblioteca Palatina di Parma* (Parma: MUP, Monte Università Parma, 2009), pp. 80–97; the map with Sophi is reproduced on pp. 86–87, and there is a detail of him on p. 88.

[89]For discussion of India as a locus of marvels and monsters see Rudolf Wittkower, "Marvels of the East: A Study in the History of Monsters," *Journal of the Warburg and Courtauld Institutes* 5 (1942), pp. 159–197, reprinted in his *Allegory and Migration of Symbols* (Boulder, CO: Westview Press, 1977), pp. 45–74; James Romm, "Belief and Other Worlds: Ktesias and the Founding of the 'Indian Wonders'," in George E. Slusser and Eric S. Rabkin, eds., *Mindscapes: The Geography of Imagined Worlds* (Carbondale: Southern Illinois University Press, 1989), pp. 121–135, esp. 121–122; Andrea Rossi-Reder, "Wonders of the Beast: India in Classical and Medieval Literature," in Timothy S. Jones and David A. Sprunger, eds., *Marvels, Monsters, and Miracles: Studies in the Medieval and Early Modern Imaginations* (Kalamazoo: Medieval Institute Publications, 2002), pp. 53–66; and Chet Van Duzer, "*Hic sunt dracones*: The Geography and Cartography of Monsters," in Asa Mittman and Peter Dendle, eds., *The Ashgate Research Companion to Monsters and the Monstrous* (Farnham, England, and Burlington, VT: Ashgate Variorum, 2012), pp. 387–435, esp. 402–409.

[90]For discussion of the monstrous races in India on the *Carta marina* see Chet Van Duzer, "A Northern Refuge of the Monstrous Races: Asia on Waldseemüller's 1516 *Carta marina*," *Imago Mundi* 62.2 (2010), pp. 221–231.

[91]The apparent lack of interest in the Great Khan on the 1507 map is in line with earlier humanist geography: for example, the Mongols are absent from Aeneas Silvius Piccolomini's *Asia*, which was completed around 1461: see Margaret Meserve, "From Samarkand to Scythia: Reinventions of Asia in Renaissance Geography and Political Thought," in Zweder von Martels and Arjo Vanderjagt, eds., *Pius II 'el più expeditivo pontifice': Selected Studies on Aeneas Silvius Piccolomini (1405–1464)* (Leiden and Boston: Brill, 2003), pp. 13–39.

emblems, bearing the Khan's symbol (the anchor) in western Asia, but strangely, not in eastern Asia. The great emphasis on the Khan on the *Carta marina* should be read as part of Waldseemüller's new emphasis on practical matters: the Khan represented a threat to Europe, so Waldseemüller provides information about him, and a large image to emphasize his importance. As for the legend describing Tartaria, this is a case where Waldseemüller used an older source rather than a newer one, for his information comes primarily from John of Plano Carpini, who traveled to Asia a couple of decades before Marco Polo. This is no doubt an indication that Waldseemüller thought Plano Carpini more reliable than Polo.

The Indian Ocean is one of the regions where Waldseemüller's updating of his image of the world is the most dramatic. On the 1507 map, his information about the Indian Ocean comes from Ptolemy and Marco Polo, and his legends about sea monsters come from an illustrated encyclopedia titled *Hortus sanitatis*, first published in 1491,[92] by way of a large world map by Henricus Martellus.[93] There are only small bits of information from the recent Portuguese explorations in the Indian Ocean, which followed Vasco da Gama's first voyage from Portugal to India and back in 1497–99: for instance, there is a brief legend about the important trading center of Calicut, which da Gama had reached.[94]

On his *Carta marina*, almost everything about the Indian Ocean has changed. On the 1507 map, the depiction of Taprobana (modern Sri Lanka) is straight out of Ptolemy. On the 1516 map, in a legend in the upper left corner of sheet 12, Waldseemüller disputes the equatorial position that Ptolemy assigned to the island, siding instead with the Roman author Solinus and evidence from recent Portuguese voyages that place it further south (Legend 12.1); and on sheet 8, he discusses whether Taprobana is to be identified with Sumatra (Legend 8.10). Ptolemy had said there were 1,378 islands near Taprobana, and Waldseemüller quotes him to that effect on the 1507 map; on the 1516 map he instead quotes Varthema's statement—around 1300 years more recent—that there were 8,000 islands near Sumatra (Legend 12.2). Ptolemy's various islands of cannibals, together with the magnetic islands that pull the nails from ships, are simply gone, though there are still cannibals in the area, now on the island of Java, and the information about them now comes from Varthema (Legend 12.3). Here again, Waldseemüller has set aside the information about Java on his 1507 map that came from Marco Polo.

On the 1507 map a short legend describes the trading center of Calicut,[95] while on the 1516 map, at the right-hand edge of sheet 7 (just west of Calicut, which is at the left-hand edge of sheet 8), a long legend describes the merchandise available in that city (Legend 7.18). It also gives an account of the king and his many wives as well as the unusual sexual and religious practices in the region, and includes a few words about what the people drink and eat (rice, fruit, butter, sugar, and some fish, but no meat)—all of this from Varthema. Then in the lower right-hand corner of the map, a long legend (Legend 12.11) describes the systems of weights and money at Calicut, the regions from which the various spices were brought to that city, and the price of each of them in the Calicut markets,[96] all of which information comes from Chaps. 82–83 of the 1507 *Paesi novamente retrovati*.[97] Again, Waldseemüller is providing an abundance of current practical information about trade, navigation, and local customs.

---

[92]The *Hortus sanitatis* "major," which is the work that interests us here, is to be distinguished from the *Hortus sanitatis* "minor," which is a Latin translation of the German herbal often titled *Gart der Gesundheit*, first published by P. Schoeffer, Mainz, 1485. The herbal published in 1485 has 435 Chapters, while the *Hortus sanitatis* "major" of 1491 has 1,066 chapters. Details and discussion of the early editions of the *Hortus sanitatis* are provided by Arnold C. Klebs, "Herbals of 15th Century," *Papers of the Bibliographical Society of America* 11 (1917), pp. 75–92; and 12 (1918), pp. 41–57, esp. 48–51 and 54–57. A more detailed discussion is in Joseph Frank Payne, "On the 'Herbarius' and 'Hortus sanitatis'," *Transactions of the Bibliographical Society* 6.1 (1901), pp. 63–126, esp. 105–24. The first edition of the work was published in Mainz by Jacob Meydenbach, 23 June 1491.

[93]See Chet Van Duzer, *Sea Monsters on Medieval and Renaissance Maps* (London: British Library, 2013), pp. 71–76.

[94]The legend about Calicut on Waldseemüller's 1507 map derives from that on Caverio's chart. Translated into English, it says: "The noble province of Calicut: in it there are many minerals, pepper, and other goods of trade that come from many regions: canella (cinnamon), ginger, cloves, sandalwood, and all types of spices. It was found by the King of Portugal." For good accounts of Calicut see Richard M. Eaton, "Multiple Lenses: Differing Perspectives of Fifteenth-Century Calicut," in Laurie J. Sears, ed., *Autonomous Histories, Particular Truths: Essays in Honor of John R. W. Smail* (Madison, WI: University of Wisconsin, Center for Southeast Asian Studies, 1993), pp. 71–86; reprinted in Eaton's *Essays on Islam and Indian History* (New Delhi: Oxford University Press, 2001), pp. 76–93; and Geneviève Bouchon, "Un microcosme: Calicut au 16e siècle," in Jean Aubin and Denys Lombard, eds., *Marchands et hommes d'affaires asiatiques dans l'Océan Indien et la Mer de Chine 13e-20e siècles* (Paris: Editions de l'École des Hautes Études en Sciences Sociales, 1988), pp. 50–57; translated as "A Microcosm: Calicut in the Sixteenth Century," in Denys Lombard and Jean Aubin, eds., *Asian Merchants and Businessmen in the Indian Ocean and the China Sea* (New Delhi and New York: Oxford University Press, 2000), pp. 40–49.

[95]For a translation of the legend about Calicut on the 1507 map see note 94 above.

[96]It seems likely that Waldseemüller's long legend about the spice trade on the *Carta marina* was inspired by a long legend on the spice trade on Martin Behaim's globe of 1492. For a transcription of Behaim's legend on the spice trade see E. G. Ravenstein, *Martin Behaim, His Life and His Globe* (London: G. Philip & Son, Ltd., 1908), pp. 89–90. Behaim describes the alleged stops that cargoes of spices make on their journey from islands near Java Major, to Java Major itself, to Ceylon, and so on to Europe. Behaim seems to ascribe considerably too many stages to the journey, and the subject of his legend (the route the spices take) is different from Waldseemüller's (the sources and prices of the spices), but nonetheless Behaim's legend probably prompted Waldseemüller to think about a more detailed and informative legend about spices, with updated information.

[97]The Chapters of the *Paesi* on the systems of weights and money used in Calicut and the sources and prices of the spices are translated into English in Greenlee, *The Voyage of Pedro Álvares Cabral* (see note 70), pp. 91–94.

Waldseemüller updated his portrayal of the Indian Ocean in other ways, among them his treatment of the sea monsters. On his 1507 map, several legends describe sea monsters in the Indian Ocean, such as this one just north of Madagascar:

> Hic cernitur orcha mirabile monstrum mari[n]um ad modum [s]olis cum reverberat cuius figura vix describi potest nisi quod est pelle mollis et carne in mensa.
>
> Here is seen the orca, an extraordinary sea monster, like the sun when it glitters, whose form can hardly be described, except that its skin is soft and its body huge.

Although this information ultimately comes from the illustrated encyclopedia *Hortus sanitatis*, Waldseemüller's source for his legends about Indian Ocean sea monsters on the 1507 map was Henricus Martellus's large world map discussed earlier.[98] On the 1516 map, these sea monsters are gone, and the cartographer presents just one image of a sea monster in the southern ocean (sheet 11), south of the southern tip of Africa. In this image King Manuel of Portugal rides a sea monster through the waves, holding aloft a scepter and the banner of Portugal, proclaiming his nation's mastery of the ocean, particularly of the passage to India around the Cape of Good Hope. The image alludes to a new title that Manuel had adopted following Vasco da Gama's return from his first voyage to India, *Senhor da conquista e da navegação e comércio de Etiópia, Arábia, Pérsia e da Índia*, "Lord of the conquest, and navigation, and commerce of Ethiopia, Arabia, Persia and India"—the adoption of which title is reported in the *Paesi novamente retrovati*, Chap. 62.[99]

The differences between the sea monsters on the two maps reflect a radical reconceptualization of the Indian Ocean. Most of the sea monsters on the 1507 map—all of which derive from Martellus—are dangerous, and thus would discourage navigation, while the image of King Manuel riding a sea monster on the *Carta marina* boldly proclaims human control over the dangers of the sea, and by extension, dominion over the oceans themselves.[100] The riches on the distant shores of the Indian Ocean are no longer mere abstractions, things told of in tales; they are now commodities that are weighed out and sold in markets at specific prices (which are listed in the lower right corner of the map). And those markets can be reached by ship along well-established routes[101] that are evidently untroubled by sea monsters. In the short space of nine years, Waldseemüller had set aside an essentially medieval view of the ocean and adopted a much more modern conception.

1.5 The *Carta Marina*'s Iconographical Program, and Its Sources

As mentioned earlier, in creating the *Carta marina*, Waldseemüller used Caverio's chart as a basis, but added so many features as to create something essentially new: it has much more geographical detail in the hinterlands, far more textual information, and a more elaborate artistic decoration. This increased level of decoration sets the map apart not only from Caverio's chart but also from the 1507 map. The borders of the 1507 map are decorated with finely depicted wind-heads and the portraits of Ptolemy and Vespucci, but the map proper is artistically quite plain: there are renderings of mountains and small trees, flags and some small coats of arms, one city in Asia, one elephant and a few people in Africa, one ship in the South Atlantic and one parrot in South America, but little more. The *Carta marina*, on the other hand, boasts a rich and ambitious iconographical program, particularly in Asia.

---

[98] This legend about the orca comes from the *Hortus sanitatis*, "De piscibus," Chap. 64. I discuss the dependence of the sea monster legends on Waldseemüller's 1507 map on those on the Yale Martellus map in Chet Van Duzer, *Henricus Martellus's World Map at Yale (c. 1491): Multispectral Imaging, Sources, Influence* (New York: Springer, 2018), pp. 64–66. Also see my *Sea Monsters on Medieval and Renaissance Maps* (London: British Library, 2013), pp. 71–76.

[99] The passage about Manuel's title appears in the so-called Second Letter of Girolamo Sernigi, and is translated into English in Ravenstein, *A Journal of the First Voyage* (see note 57), pp. 137–141, at p. 141. Manuel's title also appears in another document that Waldseemüller had (see Legend 12.5), namely the *Epistola potentissimi ac inuictissimi Emanuelis Regis Portugaliae & Algarbiorum &c. De victoriis habitis in India & Malacha: ad S. in Christo Patrem & D[omi]n[u]m nostrum D[omi]n[u]m Leonem X. Pont. Maximum* (Rome: Impressa per Iacobum Mazochium, 1513). For an excellent discussion of Manuel's title see Luís Filipe F. R. Thomaz, "L'idée imperiale manueline," in Jean Aubin, ed., *La découverte, le Portugal et l'Europe: actes du colloque, Paris, les 26, 27 et 28 mai 1988* (Paris: Fondation Calouste Gulbenkian, Centre Culturel Portugais, 1990), pp. 35–103, at 41–47; also see Paulo Pereira, "'Armes divines': la propagande royale, l'architecture manuéline et l'iconologie du pouvoir," *Revue de l'Art* 133 (2001), pp. 47–56, esp. 49.

[100] See Chet Van Duzer, *Sea Monsters on Medieval and Renaissance Maps* (London: British Library, 2013), pp. 75–76; and "Sea Monsters on Maps—The Transition from the Middle Ages to the Renaissance," *BIMCC Newsletter* 46 (May, 2013), pp. 18–19.

[101] For discussion of the commercial opportunities offered by the new discoveries see Christine R. Johnson, *The German Discovery of the World: Renaissance Encounters with the Strange and Marvelous* (Charlottesville and London: University of Virginia Press, 2008), esp. pp. 88–122. On changing conceptions of the ocean during this period see Hildegard Binder Johnson, "New Geographical Horizons: Concepts," in Fredi Chiappelli, ed., *First Images of America: The Impact of the New World on the Old* (Berkeley: University of California Press, 1976), vol. 2, pp. 615–633.

On traditional manuscript nautical charts, many of the decorative elements were optional: the person commissioning the chart could choose to have various elements added to a basic chart, including images of cities, animals, trees, ships, and sovereigns (Fig. 1.5). On sumptuous nautical charts made in the fourteenth century, the sovereigns depicted are in North Africa, but on later charts, sovereigns in Asia and sometimes Europe appear as well. The Caverio chart has just one image of a sovereign, the Magnus Tartarus, or Great Khan. Waldseemüller included many images of sovereigns on the *Carta marina* —a far larger number than on any surviving manuscript nautical chart. The abundance of sovereigns can be interpreted as reflecting Waldseemüller's interest in the world's politics—that is, in adding practical information to the *Carta marina*. Moreover, Waldseemüller made use of a simple graphical convention that, although common in other media in ancient, medieval and Renaissance art, had essentially not been employed in the depictions of sovereigns on nautical charts[102]: he used size to indicate the relative importance of the different sovereigns. Waldseemüller's decision to make many of the sovereigns quite small allowed him to include the large number that appear on the map, and also meant that most of the sovereigns are artistically rather simple, and thus took less time to design and to cut into the woodblocks.

Two particularly large images of sovereigns appear on the map, one of the Great Khan in the northeast corner of sheet 4, and the other of King Manuel of Portugal riding the sea monster on sheet 11. It seems likely that Waldseemüller intended the viewer to compare and contrast the greatest power in the East with the greatest power in the West, one powerful on land, the other on the oceans. On Caverio's map, the Great Khan is pudgy and unimposing, while on the *Carta marina* he is large, stern, and warlike (Fig. 1.6). The image is finely executed, and was probably made by a special artist rather than by Waldseemüller himself (we have no evidence that Waldseemüller had any woodcutting skills).[103] This likelihood is increased by the fact that some details of the image do not agree as well with Waldseemüller's textual sources as we might expect. The Khan's facial features do not agree with Plano Carpini's description of typical Tartar features, for example, and while the Khan's braided hair accords with Plano Carpini's description, he also is quite clear that most Tartars do not have beards, but on the *Carta marina* the Khan does have one.[104]

While the model of Waldseemüller's image of the Khan is unknown, in the case of the image of King Manuel riding the sea monster (Fig. 1.7), which also seems to be the work of a specialized artist, we can identify the likely iconographical sources, as there are few surviving earlier Renaissance images of humans riding sea monsters. The Italian painter Andrea Mantegna produced a print in about 1485–88 known as the *Battle of the Sea Gods*, in which one of the gods rides a sea monster much as King Manuel does on the *Carta marina* (Fig. 1.8).[105] But Waldseemüller's direct source was more likely Jacopo de' Barbari's monumental six-sheet view of Venice of c. 1500, which was itself no doubt influenced by Mantegna: in front of the city, right in front of St. Mark's Square, de' Barbari has an image of Neptune riding a sea monster and holding aloft on his trident a sign that reads AEQVORA TVENS PORTV RESIDEO HIC NEPTVNVS ("I, Neptune, reside here, watching over the seas at this port") (Fig. 1.9).[106] This is a powerful image of the protection that Venice enjoyed from the

[102]The only earlier chart I am aware of where greater size is used to indicate the greater importance of a sovereign is on the Catalan Atlas of 1375: Antichrist in the northeastern corner of the map is much larger than the other sovereigns. There are sovereigns of different sizes on the first map in an atlas of nautical charts by Vesconte Maggiolo (Parma, Biblioteca Palatina, MS parm. 1614), but it is not clear that these differences in size always reflect differences in importance: for illustrations see *Carte per navigare: la raccolta di portolani della Biblioteca Palatina di Parma* (Parma: MUP, Monte Università Parma, 2009), 56–57 and 64. But Waldseemüller did not have to look far afield for the idea of using size to indicate importance, for this convention was commonly used in depictions of cities on nautical charts.

[103]The sophistication of the *Carta marina*'s image of the Great Khan can be appreciated by comparing it with earlier images of that sovereign, in addition to that on the Caverio map. For example, Paris, Bibliothèque nationale de France, MS 2810, f. 2v, in a famous fifteenth-century manuscript that includes Marco Polo's travels, which shows the Polos paying homage to the Khan, and may be viewed via www.mandragore.bnf.fr ; and Madrid, Biblioteca Nacional, MS 9267, f. 70r, in a manuscript of Louis de Langle's *Tractatus de figura seu imagine mundi* made c. 1460; the image is reproduced in Marc-Édouard Gautier, ed., *Splendeur de l'enluminure: le roi René et les livres* (Angers: Ville d'Angers, and Arles: Actes Sud, 2009), p. 235.

[104]For Plano Carpini's description of the Tartars' physical features see Giovanni da Pian di Carpine, *Storia dei Mongoli*, ed. Enrico Menestò and trans. Maria Cristiana Lungarotti (Spoleto: Centro italiano di studi sull'alto Medioevo, 1989), pp. 232–233 (Latin) and 340 (Italian); and Christopher Dawson, ed., *The Mongol Mission: Narratives and Letters of the Franciscan Missionaries in Mongolia and China in the Thirteenth and Fourteenth Centuries* (New York: Sheed and Ward, 1955), pp. 6–7.

[105]For discussion of Mantegna's image see Michael A. Jacobsen, "The Meaning of Mantegna's *Battle of Sea Monsters*," *The Art Bulletin* 64.4 (1982), pp. 623–629.

[106]On de' Barbari's image of the city see Juergen Schulz, "Jacopo de' Barbari's View of Venice: Map Making, City Views, and Moralized Geography before the Year 1500," *The Art Bulletin* 60.3 (1978), pp. 425–474, esp. 468 on Neptune; and Deborah Howard, "Venice as a Dolphin: Further Investigations into Jacopo de' Barbari's View," *Artibus et Historiae* 18.35 (1997), pp. 101–111, esp. 106 (from which I borrow the translation of Neptune's sign).

Fig. 1.5 Detail of an elaborately decorated nautical chart made by Matteo Prunes in 1559. Library of Congress, Geography and Map Division, G5672.M4P5 1559 .P7 Vault: Vellum 7. Courtesy of the Library of Congress

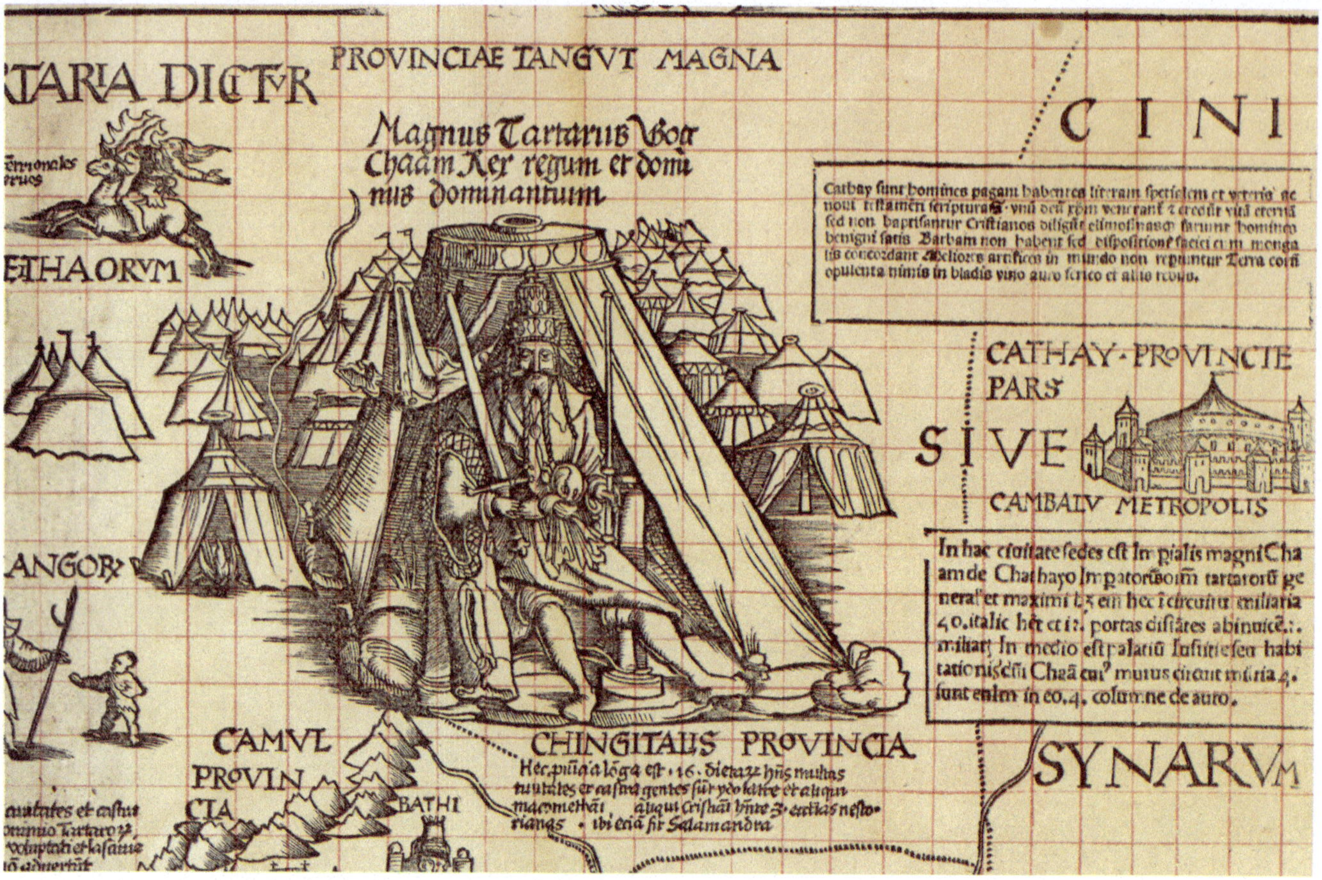

Fig. 1.6 Detail of the Great Khan on Waldseemüller's *Carta marina* (sheet 4). Courtesy of the Library of Congress

god of the sea, but Waldseemüller's recasting of this image is still more powerful: he has replaced the classical god Neptune with a contemporary king, thus almost effecting an apotheosis of Manuel[107]; and rather than indicating the protection passively enjoyed by Venice, the new image illustrates Portugal's active control of the lanes of navigation to India.[108]

Waldseemüller also made use of recent sources for other images on his map, as part of his effort to create an entirely fresh and modern image of the world. In South America Waldseemüller has an image of an opossum (sheet 5), and this is the earliest surviving European depiction of that animal. Vicente Yáñez Pinzón was the first European to see an opossum in 1499; in fact it was the first marsupial that Europeans had ever seen, and was regarded as a marvel. Pinzón brought an opossum back to Spain and left this description of the creature[109]:

> Between these Trees he saw a strange Monster, the foremost part resembling a Fox, the hinder a Monkey, the feet were like a Mans, with Ears like an Owl; under whose Belly hung a great Bag, in which it carry'd the Young, which they drop not, nor forsake till they can feed themselves.

---

[107]Waldseemüller had included an image of Neptune riding a sea monster on his 1511 *Carta itineraria Europae*, of which only one exemplar of the 1520 printing survives, in Innsbruck, in the Tiroler Landesmuseum Ferdinandeum. For bibliography on the map see note 14.

[108]On Manuel's status as King of the Ocean see Luís Filipe F. R. Thomaz, "L'idée imperiale manueline," in Jean Aubin, ed., *La découverte, le Portugal et l'Europe: actes du colloque, Paris, les 26, 27 et 28 mai 1988* (Paris: Fondation Calouste Gulbenkian, Centre Culturel Portugais, 1990), pp. 35–103, esp. 37 and 41–47. On the politicization of the ocean reflected in the image on Waldseemüller's *Carta marina* see Elizabeth Mancke, "Early Modern Expansion and the Politicization of Ocean Space," *Geographical Review* 89.2 (1999), pp. 225–236; and for a brief discussion of the image of Manuel on the sea monster see Gaetano Ferro, Luisa Faldini, Marica Milanesi, and Gianni Eugenio Viola, *Columbian Iconography*, trans. Luciano F. Farina and Carla Onorato Wysokinski (Rome: Istituto poligrafico e Zecca dello Stato, Libreria dello Stato, 1996), pp. 450–451.

[109]See Charles R. Eastman, "Early Portrayals of the Opossum," *The American Naturalist* 49.586 (1915), pp. 585–594; and Susan Scott Parrish, "The Female Opossum and the Nature of the New World," *The William and Mary Quarterly* 54.3 (1997), pp. 475–514, esp. 485. The translation quoted here is from John Ogilby's *America* (1671), as cited by Parrish.

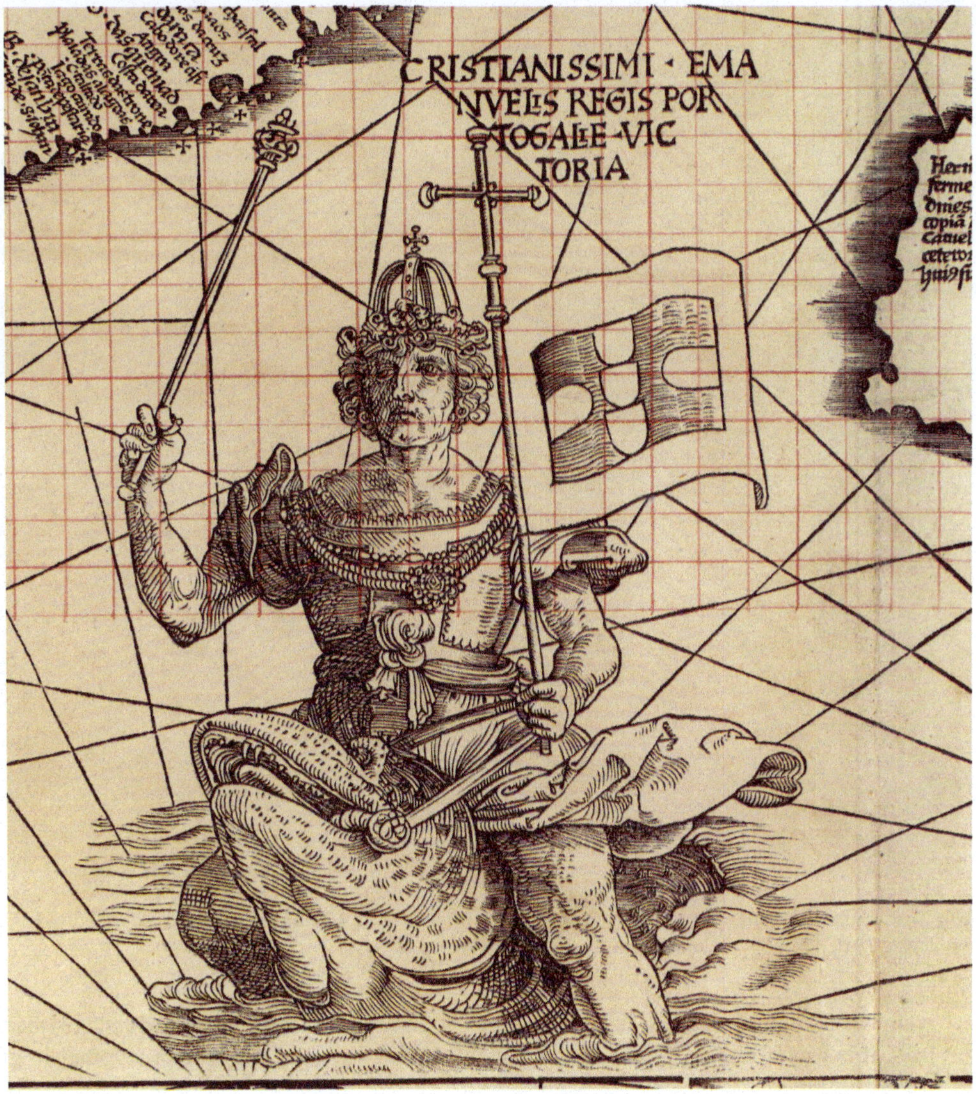

Fig. 1.7 Detail of King Manuel of Portugal riding a sea monster on Waldseemüller's *Carta marina* (sheet 11). Courtesy of the Library of Congress

As Waldseemüller's image is quite detailed, it seems likely that it was taken from a contemporary illustration of Pinzón's opossum that no longer survives.[110] Waldseemüller's image was copied both on later maps and in other media, mostly by way of the reproduction of Waldseemüller's image in Lorenz Fries's *Carta marina* of 1525, 1530, and 1531.[111]

In Scandinavia, near the northern edge of the map (sheet 2), the animal that looks like an elephant is intended to be a walrus (Fig. 1.10). The accompanying legend reads (see on Legend 2.3):

> Morsus animal ingens quantitate Elephantis huius dentes longos duos et quadrangulares carens quibus iuncturis in pedibus. Reperitur in promontoriis septentrionalibus Norbegie incedit gregatim agmine ducentorum animalium.The

[110]Waldseemüller's legend about and image of the opossum are discussed in Gaetano Ferro et al., *Columbian Iconography* (see note 108), pp. 546–547. Waldseemüller could have read about the opossum in the *Libretto de tutta la navigatione*, Chap. 31 see Thacher, *Christopher Columbus* (see note 64), vol. 2, pp. 483 (Italian) and 511–512 (English); and in the *Paesi*, Chap. 113.

[111]On Fries's *Carta marina* there is a baby opossum suckling from its mother, and this is clearly the image that Sebastian Münster copied in his *Cosmographiae uniuersalis Lib*[*ri*] *VI* (Basel: apud Henrichum Petri, 1552), Book 5, in the section titled "Pinzonus socis Admirantis quaerit nouas insulas," p. 1108; this is also the image copied by Gerard Mercator in his world map of 1569, which is reproduced in facsimile in *Gerard Mercator's Map of the World (1569) in the Form of an Atlas in the Maritiem Museum 'Prins Hendrik' at Rotterdam* (Rotterdam, 1961); and now as *Atlas van der Wereld: De wereldkaart va Gerard Mercator uit 1569* (Zutphen: Walberg Pers, 2011). It should be remarked that the image of the opossum in the 1522 (Strasbourg), 1525 (Strasbourg), 1535 (Lyon), and 1541 (Vienna) editions of Ptolemy's *Geography* is essentially identical to that on the *Carta marina*, so images of opossums that seem similar to Waldseemüller's may have derived from one of these editions, rather than from the *Carta marina*.

Fig. 1.8 Detail showing a god riding a sea monster from Andrea Mantegna's *Battle of the Sea Gods*, c. 1485–88. Washington DC, National Gallery of Art, 1984.53.1. Courtesy of the National Gallery of Art

walrus is a huge animal, the size of an elephant, and it has two long teeth which are quadrangular, and lacks joints in its legs. It is found in the northern promontories of Norway, and they travel together in groups of two hundred animals.

There is at least one discussion of a walrus by a medieval author, namely Albertus Magnus in his *On Animals*,[112] but Waldseemüller did not make use of Albertus.[113] The word Waldseemüller uses for the walrus, *morsus*, is usually held to have entered European literature in 1517, in Maciej of Miechów's *Tractatus de duabus Sarmatiis*.[114] In fact, William Caxton used it in 1480 in his *Chronicles of England*[115]—but certainly Waldseemüller's is among the earliest uses of it. On the one hand, it is clear that Waldseemüller (or his unknown iconographical source, if there was one) made the image from a vague description that emphasized the creature's elephant-like tusks and size, rather than on the basis of seeing a walrus. On the other hand, Waldseemüller's text about the animal is surely based on information derived from an examination of a walrus, as no other known early document mentions the walrus's quadrangular tusks.[116] The image and text pertaining to the walrus thus contain a mixture of good information and extrapolation, a common characteristic of attempts to interpret incomplete reports about unfamiliar things.

---

[112]See Albertus Magnus, *De animalibus libri XXVI, nach der Cölner Urschrift*, ed. Hermann Stadler (Münster: Aschendorff, 1916), vol. 2, p. 1525; and Albertus Magnus, *On Animals: A Medieval Summa Zoologica*, trans. Kenneth F. Kitchell Jr., and Irven Michael Resnick (Baltimore: Johns Hopkins University Press, 1999), vol. 2, p. 1671.

[113]There is also an interesting allusion to the walrus in Pletho's treatise *Correction of Certain Errors made by Strabo*; the Greek text is edited by Aubrey Diller, "A Geographical Treatise by Georgius Gemistus Pletho," *Isis* 27.3 (1937), pp. 441–451; and it is translated in Milton V. Anastos, "Studies in Pletho," Ph.D. Dissertation, Harvard University, 1940, pp. 68–163, with the passage about the walrus on p. 152.

[114]Valentin Kiparsky, "L'Histoire du morse," *Annales Academiae Scientiarum Fennicae*, Ser. B, 73.3 (1952), pp. 1–53, at p. 6.

[115]William Caxton, *Chronicles of England* (Westminster: William Caxton, 1480), Chap. 257 on the year 1456. See also Caxton's edition of Ranulf Higden's *Polycronicon* (Westminster: W. Caxton, 1482), f. 423, sig. 55 2 recto.

[116]I am indebted to Klaus Barthelmess (d. 2011) for his discussions of Waldseemüller's image of the walrus and its possible sources.

Fig. 1.9 Detail of Neptune riding a sea monster on Jacopo de' Barbari's six-sheet view of Venice of c. 1500. Chicago, Newberry Library, Novacco 8f 7. Courtesy of the Newberry Library

Fig. 1.10 Detail of the elephant-like walrus on Waldseemüller's *Carta marina* (sheet 2). Courtesy of the Library of Congress

Fig. 1.11 Detail of the rhinoceros in West Africa on Waldseemüller's *Carta marina* (sheet 6). Courtesy of the Library of Congress

Fig. 1.12 Albrecht Dürer's print of the rhinoceros, 1515. Washington DC, National Gallery of Art, Rosenwald Collection 1964.8.697. Courtesy of the National Gallery of Art, Washington DC

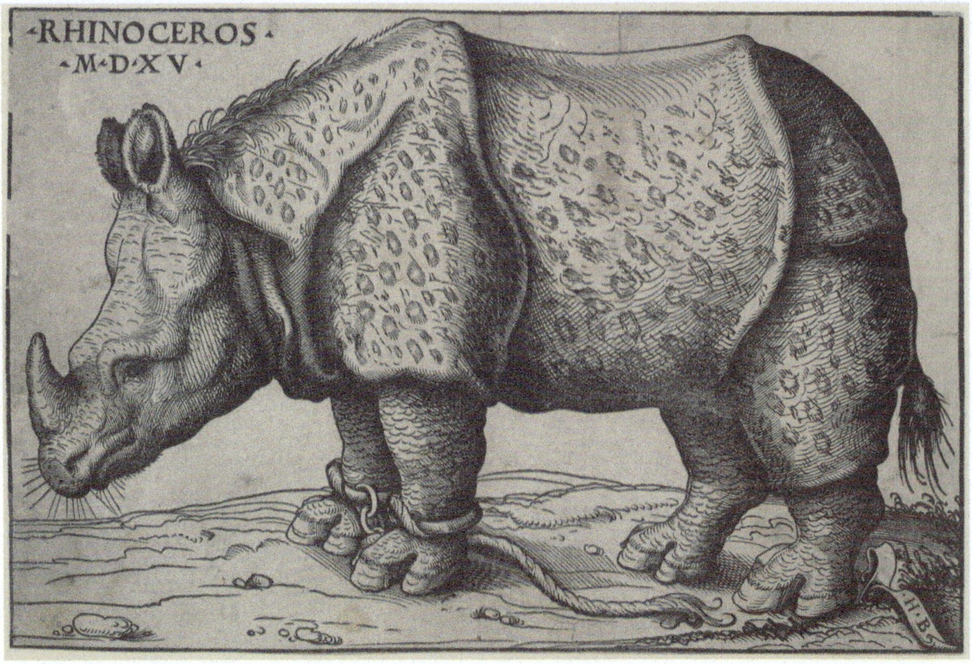

Fig. 1.13 Hans Burgkmair's print of the rhinoceros, 1515. Vienna, Grafische Sammlung Albertina, Inv. DG1934/123. Courtesy of the Albertina Museum, Vienna

In West Africa (sheet 6), Waldseemüller shows a small image of a rhinoceros (Fig. 1.11), and this is another instance where he was using the most recent iconographical sources available. In 1514 Sultan Muzafar II of Gujarat had presented a rhinoceros to Afonso de Albuquerque, the governor of Portuguese India. Albuquerque sent the rhinoceros to King Manuel of Portugal,[117] and Albrecht Dürer made an influential print depicting the animal in 1515 (Fig. 1.12).[118] Waldseemüller used as his model not Dürer's print, however, but a different one, also made in 1515, by Hans Burgkmair, that survives in only one copy (Fig. 1.13).[119] The images of the rhinoceroses are similar in the two artists' prints, but there are differences, and those differences indicate that Waldseemüller used Burgkmair's print: in particular, Dürer shows hard plating and a small ancillary horn on the crest of the creature's neck where Burgkmair places hair—and it is hair that we see in Waldseemüller's image. It has been suggested that Dürer might have been involved in the engraving of the *Carta marina*,[120] but given that Waldseemüller used Burgkmair's image of the rhinoceros rather than Dürer's, this seems unlikely.

In northeastern Asia, specifically in northern India at the bottom of sheet 4, is an image of *sati* or *suttee*, the Hindu practice whereby a widow burned herself to death on the funeral pyre of her husband. This custom very much surprised Western visitors to India, and there are legends on the subject on earlier maps, for example the metal Borgia *mappamundi* from the first half of the fifteenth century,[121] and Andreas Walsperger's *mappamundi* of 1448.[122] The image on the *Carta*

---

[117]On the history of this rhinoceros see A. Fontoura da Costa, *Les déambulations du rhinocéros de Modofar, roi de Cambaye, de 1514 à 1516* (Lisbon: Division de Publications et Bibliotheque, Agence Générale des Colonies, 1937); and Silvio A. Bedini, "The Ill-Fated Rhinoceros," in *The Pope's Elephant* (Nashville: J. S. Sanders & Company, 1998), pp. 111–136.

[118]On Dürer's print of the rhinoceros see F. J. Cole, "The History of Albrecht Dürer's Rhinoceros in Zoological Literature," in A. E. Underwood, ed., *Science, Medicine, and History* (London: Oxford University Press, 1953), pp. 337–356; Donald Lach, *Asia in the Making of Europe*, vol. 2, bk. 1, *A Century of Wonder*: *The Visual Arts* (Chicago: University of Chicago Press, 1970), pp. 158–172; and T. H. Clarke, "The First Lisbon or 'Dürer' Rhinoceros of 1515," in his *The Rhinoceros from Dürer to Stubbs, 1515–1799* (London: Sotheby's Publications, 1986), pp. 16–27.

[119]The surviving copy of Burgkmair's print is in Vienna, Graphische Sammlung Albertina, Inv. DG1934/123. For discussion of it see Lach, *Asia in the Making of Europe* (see note 118), vol. 2, bk. 1, p. 164 and plate 120; Clarke, "The First Lisbon or 'Dürer' Rhinoceros" (see note 118), pp. 24–25; and Jim Monson, "The Source for the Rhinoceros," *Print Quarterly* 21.1 (2004), pp. 50–53.

[120]For the suggestion that Dürer was involved in the production of the *Carta marina* see Joseph Fischer and Franz Ritter von Wieser, *Die älteste Karte mit dem Namen Amerika* (see note 9), p. 19; and Johnson, *Carta marina* (see note 21), p. 27.

[121]The Borgia *mappamundi* is in the Biblioteca Apostolica Vaticana, Borgia XVI, and is described and reproduced in Destombes, *Mappemondes* (see note 29), pp. 239–241 and plate 29. For the legend about *sati* see A. E. Nordenskiöld, "Om ett aftryck från XV:de seklet af den i metall graverade världskarta, som förvarats i kardinal Stephan Borgias museum i Velletri, Med 1 facsimile," *Ymer* 11 (1891), pp. 83–92, at 90; the map is reproduced between 130 and 131.

[122]Walsperger's map is in the Biblioteca Apostolica Vaticana, MS Pal. Lat. 1362 B. For the legend about *sati* see Konrad Kretschmer, "Eine neue mittelalterliche Weltkarte der vatikanischen Bibliothek," *Zeitschrift der Gesellschaft für Erdkunde* 26 (1891), pp. 371–406, esp. 398. The article is reprinted in *Acta Cartographica* 6 (1969), pp. 237–272.

Fig. 1.14 The image of *suttee* or *sati* on Waldseemüller's *Carta marina* (sheet 4). Courtesy of the Library of Congress

Fig. 1.15 Jörg Breu's woodcut image of *suttee* or *sati* from the 1515 edition of Varthema's travels. Munich, Bayerische Staatsbibliothek, Rar. 894, f. 51r. Courtesy of the Bayerische Staatsbibliothek

marina (Fig. 1.14) shows a woman who has leapt into a fire pit, with a man standing over her and about to strike her with something in his hand, no doubt so that she will die sooner; on the right a horned devil stands looking on. No legend accompanies the scene, but it illustrates a passage in Varthema about the practice of *sati* in the city of Tarnassari. Varthema describes the fire pit, the beating of the woman with sticks and balls of pitch, and the participation of men clothed like devils.[123] Waldseemüller's scene here was clearly inspired by that in the first illustrated edition of Varthema, which has

[123]See Lodovico de Varthema, *The Travels of Ludovico di Varthema in Egypt, Syria, Arabia Deserta and Arabia Felix, in Persia, India, and Ethiopia, A.D. 1503 to 1508, Translated from the Original Italian Edition of 1510… by John Winter Jones and Edited, with Notes and an Introduction, by George Percy Badger* (London: The Hakluyt Society, 1863), pp. 206–208.

Fig. 1.16 Jörg Breu's woodcut image of Medina from the 1515 edition of Varthema's travels. Munich, Bayerische Staatsbibliothek, Rar. 894, f. 9v. Courtesy of the Bayerische Staatsbibliothek

woodcuts by Jörg Breu[124] and was published in 1515,[125] just one year before the *Carta marina* (Fig. 1.15): in this scene two men instead of one are beating down the woman, and a king and noble are present, but in other respects the scenes are quite similar, particularly the poses of the women and the depiction of the flames. Once again, Waldseemüller was using a recent iconographical source, and the evidence from these images gives us a window into Waldseemüller's workshop, for we now know that the cartographer had a copy of the 1515 edition of Varthema on his bookshelf.

Waldseemüller used this same edition of Varthema as the source for two other images on his map. On f. 9v of the book there is an image of the mosque of Medina, with Varthema himself on the far left, and his guide beside him (Fig. 1.16). The guide points to the flames emanating from the top of the building, which he claims indicate the presence of the prophet's body.[126] This image of Medina, with its buttresses, windows, and other features, is used by Waldseemüller with very few changes to depict Mecca on the *Carta marina* (Fig. 1.17). The cartographer removes the flames issuing from the building's roof, and replaces them, in effect, with an Islamic crescent which he may have borrowed from the building to the left in the image in Varthema. Waldseemüller's choice to use an image of Medina for Mecca was perhaps guided by a desire to use the example he had of supposedly Islamic architecture for the more famous city; but certainly his use of this image demonstrates again his interest in giving his map a rich and accurate iconographical program from recent sources.

The important trading center of Calicut in India, mentioned above, is the only city in eastern Asia to bear a flag, and the flag is a curious one: it is of a black devil or demon (Fig. 1.18). The allusion is to the worship of devils that Varthema says

---

[124]For discussion of Breu's work see Robert Stiassny, "Jörg Breu von Augsburg," *Zeitschrift für Christliche Kunst* 7 (1894), pp. 101–120; Campbell Dodgson, "Beiträge zur Kenntnis des Holtzschnittwerks Jörg Breus," *Jahrbuch der Preussischen Kuntsammlungen* 21 (1900), pp. 192–214; Campbell Dodgson, "Jörg Breu als Illustrator des Ratdoltschen Offizin: Nachtrag," *Jahrbuch der Preussischen Kunstsammlungen* 24 (1903), pp. 335–337; and H. Roettinger, "Zum Holzschnittwerke Jörg Breus des Älteren," *Repertorium für Kunstwissenschaft* 31 (1908), pp. 48–62. The woodcuts for the Varthema cycle are reproduced in Max Geisberg, *Die deutsche Buchillustration in der ersten Hälfte des XVI Jahrhunderts* (Munich: Hugo Schmidt Verlag, 1930–32), vol. 1.

[125]The illustrated edition in question is Varthema, *Die Ritterlich und lobwirdig rayß...* (see note 69), with the illustration of *sati* on f. 51r. For discussion of the program of illustration see Stephanie Leitch, "Recuperating the Eyewitness: Jörg Breu's Images of Islamic and Hindu Culture in Ludovico de Varthema's *Travels* (Augsburg: 1515)," in her *Mapping Ethnography in Early Modern Germany: New Worlds in Print Culture* (Basingstoke and New York: Palgrave Macmillan, 2010), pp. 101–145, esp. 128–131 on the image of *sati*.

[126]Varthema was one of first Europeans to realize the falsity of the common medieval myth that Mohammad's sarcophagus was at Mecca, when in fact he was buried in Medina: for discussion of this myth see Sandra Sáenz-López Pérez, "La peregrinación a La Meca en la Edad Media a través de la cartografía occidental," *Revista de poética medieval* 19 (2007), pp. 177–218. The scene of Varthema in Medina in the 1515 edition of Varthema's travels is discussed by Leitch, "Recuperating the Eyewitness" (see note 125), p. 108.

Fig. 1.17 Detail of Mecca on Waldseemüller's *Carta marina* (sheet 7). Courtesy of the Library of Congress

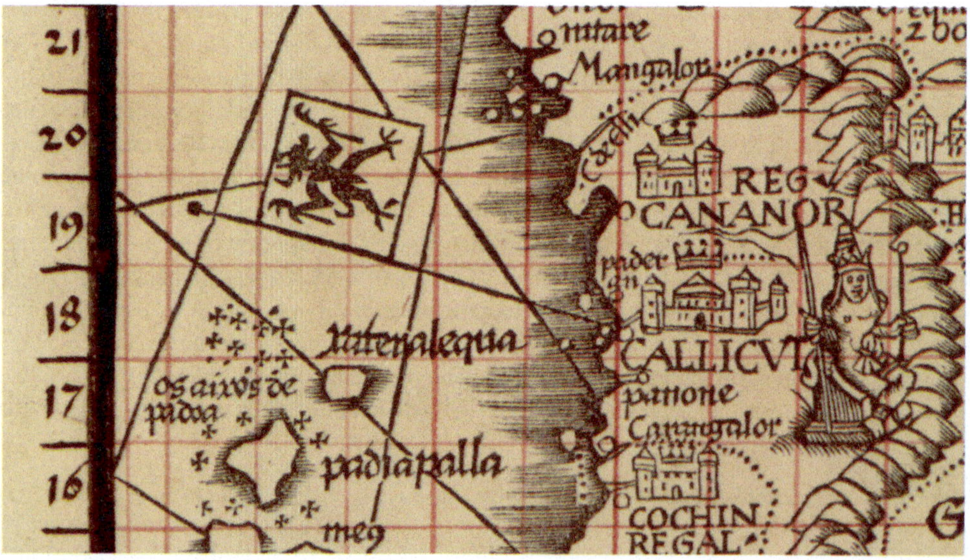

Fig. 1.18 Image of Calicut, India, with its flag displaying a demon on Waldseemüller's *Carta marina* (sheet 8). Courtesy of the Library of Congress

Fig. 1.19 Jörg Breu's woodcut image of the devil of Calicut, India, from the 1515 edition of Varthema's travels. Munich, Bayerische Staatsbibliothek, Rar. 894, f. 35r. Courtesy of the Bayerische Staatsbibliothek

took place in Calicut,[127] though this is merely a misunderstanding of Hindu art,[128] and the image on the flag is a simplification of an image of the devil supposedly worshipped in Calicut in the same 1515 edition of Varthema (Fig. 1.19).

On sheet 8, in southwestern India, there is a curious image of an Indian king with a crown on his turban who is standing rather than sitting on a throne (Fig. 1.20). Waldseemüller borrowed this image from another recent source, the 1509 edition of Balthasar Springer's account of Francisco de Almeida's voyage to India, which was illustrated by Hans Burgkmair (Fig. 1.21). Waldseemüller has rotated the man to the left, added the crown to his turban to make him a king, and shortened the handle of his weapon, but there can be no doubt where he obtained the image. Waldseemüller used Burgkmair's image of this Indian man a second time on the map, in West Africa (sheet 6), to illustrate a king in the delta of the Senegal River (Fig. 1.22). The king stands almost with his back to us, and Waldseemüller seems to have borrowed this stance from that of a figure in another of Burgkmair's illustrations of Springer's narrative, namely the man shading the king in the fold-out *Triumphus Regis Gosci sive Gutschmin*, or *Triumph of the King of Cochin* (Kochi, India) (Fig. 1.23).[129] It is interesting that Waldseemüller did not hesitate to use an image of an Indian man to illustrate an African, particularly as the 1509 edition of Springer's book has images of Africans that Waldseemüller could have copied.[130] One reason might have been that the

---

[127]On the alleged worship of devils in Calicut see *The Travels of Ludovico di Varthema*, trans. Jones (see note 123), pp. 136–138.

[128]For analysis of this misperception of Hindu art see Partha Mitter, *Much Maligned Monsters: History of European Reactions to Indian Art* (Oxford: Clarendon Press, 1977), pp. 16–20; Joan-Pau Rubiés, "Ludovico de Varthema: The Curious Traveller at the Time of Vasco da Gama and Columbus," in his *Travel and Ethnology in the Renaissance: South India through European Eyes, 1250–1625* (Cambridge and New York: Cambridge University Press, 2000), pp. 125–163, esp. 155–156; Leitch, "Recuperating the Eyewitness" (see note 125), pp. 132–135; and Jennifer Spinks, "The Southern Indian 'Devil in Calicut' in Early Modern Northern Europe: Images, Texts and Objects in Motion," *Journal of Early Modern History* 18.1-2 (2014), pp. 15–48.

[129]For discussion of the Triumph scene see Götz Pochat, "*Triumphus Regis Gosci sive Gutschmin.* Exoticism in French Renaissance Tapestry," *Gazette des Beaux-Arts*, 6th ser., 82, 105 (1973), pp. 305–310; reprinted in Götz Pochat, *Kunst, Kultur, Ästhetik: gesammelte Aufsätze* (Vienna, Berlin, and Münster: LIT, 2007), pp. 279–286.

[130]For discussion of the images of Africans in the 1509 edition of Springer's narrative see Jean Michel Massing, "Hans Burgkmair's Depiction of Native Africans," *RES: Anthropology and Aesthetics* 27 (1995), pp. 39–51, who notes (p. 51) that "Burgkmair's woodcuts… were, for decades, the only representations of Africans and Indians based on visual information rather than mere literary testimony." Also see Ronald Singer and Werner Jopp, "The Earliest Illustration of Hottentots: 1508," *The South African Archaeological Bulletin* 22.85 (1967), pp. 15–19; and Renate Kleinschmid, "Balthasar Springer. Eine quellenkritische Untersuchung," *Mitteilungen der Anthropologischen Gesellschaft in Wien* 96–97 (1967), pp. 147–190.

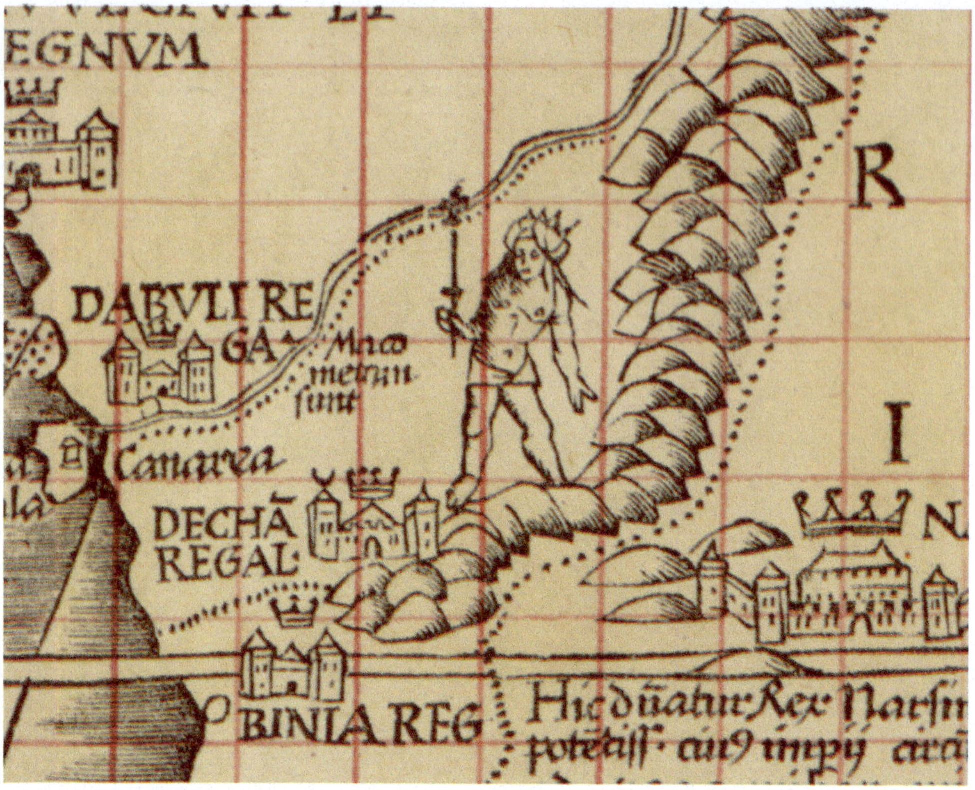

Fig. 1.20 Detail of a king in southwestern India on Waldseemüller's *Carta marina* (sheet 8). Courtesy of the Library of Congress

Africans in Springer's book are naked, and Waldseemüller's colleague Matthias Ringmann, who had worked closely with him on the *Cosmographiae introductio*, the 1511 *Carta itineraria Europae* with its accompanying booklet, and the 1513 edition of Ptolemy,[131] had written against the illustration of full-frontal nudity in 1509.[132] On the other hand, the natives in southern Africa on the 1507 map are naked, and it would not have been difficult to add clothes to Burgkmair's naked figures, so the matter is not clear.

We know that Waldseemüller had the Caverio chart in his workshop, but he twice alludes to having more than one nautical chart at his disposal. In the *Cosmographiae introductio* he says that he has followed nautical charts (plural) particularly with regard to the newly discovered lands,[133] and in the long text block on sheet 9 of the *Carta marina* he says that *eo que in maris descriptionibus vulgarem fuerimus & approbatissimam nauticarum tabularum notificationes insequuti*, "as far as the depiction of the oceans, I have followed the commonly used and the most approved nautical charts and their indications" (plural). It seems very likely that some of the illustrations on the *Carta marina* were inspired by nautical chart legends or illustrations—legends or illustrations that do not appear on the Caverio chart—and an examination of these images can give us information about the other chart or charts that Waldseemüller had.

[131]Franz Laubenberger (trans. Steven Rowan), "The Naming of America," *The Sixteenth Century Journal* 13.4 (1982), pp. 91–113, with some corrections in Christine R. Johnson, "Renaissance German Cosmographers and the Naming of America," *Past & Present* 191 (2006), pp. 3–43.

[132]See Matthias Ringmann, *Grammatica figurata* (Saint-Dié: s.n., 1509), ff. 12v–13r; there is a facsimile edition of the work, Franz Ritter von Wieser, ed., *Grammatica figurata des Mathias Ringmann (Philesius Vogesigena) in Faksimiledruck* (Strassburg: Heitz, 1905). For discussion of the book see Jean-Claude Margolin and Diana Wormuth, "Mathias Ringmann's *Grammatica figurata*, or, Grammar as a Card Game," *Yale French Studies* 47 (1972), pp. 33–46.

[133]See Joseph Fischer and Franz von Wieser in *The 'Cosmographiae introductio' of Martin Waldseemüller*, Chap. 9, p. xxxvii (Latin), and p. 78 (English); and Hessler, *The Naming of America* (see note 7), p. 106. The passage is quoted below on p. 39.

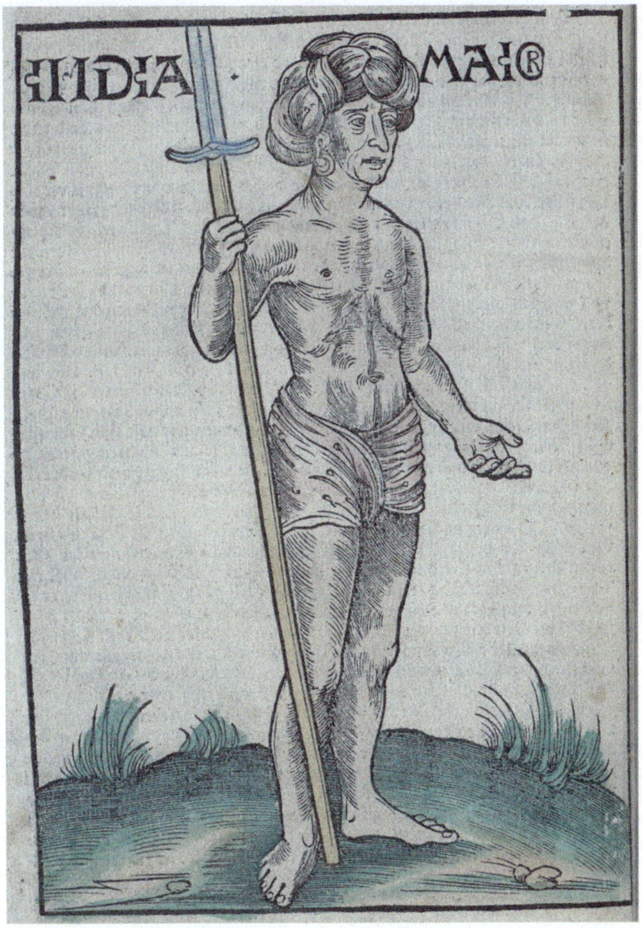

Fig. 1.21 Hans Burgkmair's woodcut illustration of an Indian man in the 1509 edition of Balthasar Springer's account of Francisco de Almeida's voyage to India. Munich, Bayerische Staatsbibliothek, Rar. 470, f. 12v. urn:nbn:de:bvb:12-bsb00045403-6. Courtesy of the Bayerische Staatsbibliothek

Fig. 1.22 Detail of a man in Senegal on Waldseemüller's *Carta marina* (sheet 6). Courtesy of the Library of Congress

Fig. 1.23 Detail of the fold-out *Triumphus Regis Gosci* in the 1509 edition of Balthasar Springer's account of Francisco de Almeida's voyage to India. Munich, Bayerische Staatsbibliothek, Rar. 470, ff. 2v–3r. urn:nbn:de:bvb:12-bsb00045403-6. Courtesy of the Bayerische Staatsbibliothek

At the top of sheet 7 of the *Carta marina* there is a small image of Moses kneeling before Mount Sinai and receiving from God the two tablets with the commandments written on them (Fig. 1.24). I do not know of an earlier nautical chart that has a similar image of Moses, but many nautical charts have a legend that probably inspired Waldseemüller to include this scene. The Pizzigani chart of 1367,[134] for example, has a legend that reads (in very idiosyncratic Latin) *Mons Sinay quo dominus jesus a de moyss instrudebat et ey legem cunferebat propter populum*, "Mount Sinai where Lord Jesus or God instructed Moses and gave him the law for the people."[135] The legend on the Catalan Atlas of 1375[136] is similar, *Mont de Sinai en lo*

[134]The 1367 Pizzigani chart is in Parma, Biblioteca Palatina, Carta nautica no. 1612, and there is a good hand-drawn facsimile of the chart in Edme-François Jomard, *Les monuments de la géographie* (Paris: Duprat, 1842–1862), nos. 44–49; and a photographic reproduction in Guglielmo Cavallo, ed., *Cristoforo Colombo e l'apertura degli spazi: mostra storico-cartografica* (Rome: Istituto Poligrafico e Zecca dello Stato, Libreria dello stato, 1992), vol. 1, pp. 432–433. There is a good digital reproduction of the chart in Pujades, *Les cartes portolanes* (see note 31), on the accompanying CD, number C13. The Library of Congress has a hand-painted copy of the map made by Agostinho Sardi in Parma in 1802, which is very briefly described in Walter W. Ristow and R. A. Skelton, *Nautical Charts on Vellum in the Library of Congress* (Washington, DC: The Library, 1977), p. 2. There is also a hand-painted copy of the map made in 1827, which is in Vienna, Österreichische Nationalbibliothek, Cod. Ser. Nov. 4676.
[135]Transcribed by Mario Longhena, "La carta dei Pizigano del 1367 (posseduta dalla Biblioteca Palatina di Parma)," *Archivio storico per le province Parmensi*, series 4, vol. 5 (1953), pp. 25–130, at 106.
[136]For references on the Catalan Atlas see note 38.

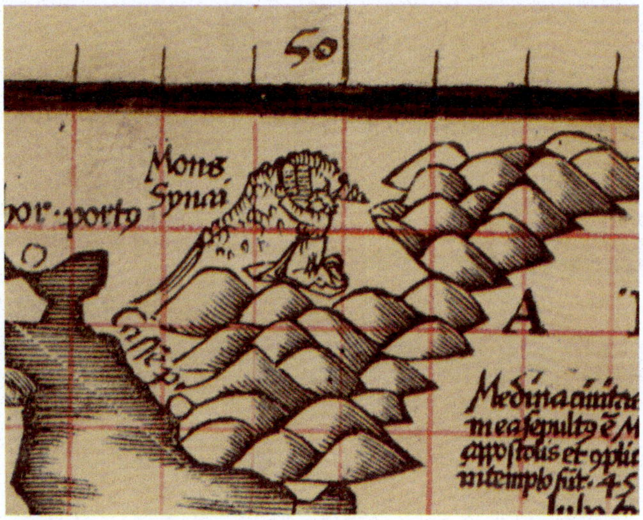

Fig. 1.24 Detail of Moses kneeling before Mount Sinai and receiving from God the two Tablets of the Law, from Waldseemüller's *Carta marina* (sheet 7). Courtesy of the Library of Congress

qual Déu dona la Ley a Moyssés, "Mount Sinai where God gave the law to Moses."[137] Similar legends appear on many other nautical charts, including some sixteenth-century works by Ottomano Freducci, such as London, British Library, Add. MS 11548, made in 1529.[138] It seems likely, then, that such a legend inspired Waldseemüller to add an image of Moses to his map; Waldseemüller may have drawn iconographic inspiration from the scene of Moses receiving the laws in Hartmann Schedel's *Liber chronicarum* (Nuremberg: Anton Koberger, 1493), f. 30v (Fig. 1.25), as in both cases Moses is receiving the tablets with his hands gripping them from the sides,[139] but there are so many representations of this scene that it is impossible to be certain that the *Liber chronicarum* was the source.[140]

On sheet 3 of the *Carta marina* just west of the Caspian (which is labeled *Mare Abacuc*...) there is a short legend that reads *Arach mons super quam requieuit Archa noe*, "Mount Ararat, upon which Noah's Ark rested," above which there is a small image of a ship on the mountains (see Fig. 1.26 and Legend 3.12). Isidore, *Etymologiae* 14.3.35, Marco Polo, and Pierre d'Ailly mention that Noah's Ark can be found on some mountains in Armenia, but they do not give the mountains' name.[141] The case is much the same with Odoric of Pordenone,[142] and Varthema does not mention Noah's Ark. But there is a nautical chart tradition

---

[137]The legend on the Catalan Atlas is transcribed in *Mapamundi del año 1375* (Barcelona: S.A. Ebrisa, 1983), p. 45.

[138]See Chet Van Duzer, "Nautical Charts, Texts, and Transmission: The Case of Conte di Ottomano Freducci and Fra Mauro," *Electronic British Library Journal*, article 6 (2017), pp. 1–65, at 44–45. The article is available at http://www.bl.uk/eblj/2017articles/article6.html.

[139]The likelihood that Waldseemüller used Schedel's *Liber chronicarum* as a source for the image of Moses is increased somewhat by the fact that he used the same book as a source for his image of a cynocephalus in India: see Van Duzer, "A Northern Refuge of the Monstrous Races" (see note 90), p. 226.

[140]For example, there is an illustration of Moses receiving the tablet with both hands in John Mandeville, *Johannes Montevilla der wyffaren de Ritter* (Strasburg: Mathias Hupfuff, 1501), fol. D 2 r, and is reproduced in *La Gravure d'illustration en Alsace au XVIe siècle* (Strasbourg: Presses universitaires de Strasbourg, 1992–2000), vol. 2, p. 176. Incidentally there are representations of Moses receiving the tablets from God in some of the manuscript atlases by Battista Agnese: see Henry R. Wagner, "The Manuscript Atlases of Battista Agnese," *Papers of the Bibliographical Society of America* 25 (1931), pp. 1–110, at 35. One of these atlases is Venice, Museo Correr, Port. 1, which is Wagner's #55, pp. 91–93; this atlas has been reproduced in facsimile as *Atlante nautico di Battista Agnese, 1553* (Venice: Marsilio, 1990), where the map in question is reproduced on plate 27. Another such atlas is New Haven, Beinecke Rare Book and Manuscript Library, MS 560, map 21: images of all of the maps in this atlas are available through the Beinecke's Digital Library at http://beinecke.library.yale.edu/digitallibrary/.

[141]Evidence from Waldseemüller's 1507 map indicates that he was not using the Latin edition of Polo published c. 1484, *De consuetudinibus et condicionibus Orientalium regionum* (Gouda: Gerard Leeu, c. 1483–1485), but rather a manuscript similar to Naples, Biblioteca Nazionale Vittorio Emanuele III, Vind. lat. 50. The text of this manuscript has been published in Marco Polo, *Marka Pavlova z Benátek, Milion: Dle jediného rukopisu spolu s prilusnym zakladem latinskym*, ed. Justin Václav Prásek (Prague: Nákl. Ceské akademie císare Frantiska Iozefa, 1902), and the passage on Noah's Ark is in Book 1, Chap. 13, pp. 17–18. For the passage in Yule's translation see Marco Polo, *The Book of Ser Marco Polo*, Book 1, Chap. 4, vol. 1, p. 46. For the passage in Pierre d'Ailly see *Ymago mundi*, ed. Edmond Buron (Paris: Maisonneuve frères, 1930), vol. 1, pp. 302–03.

[142]For the passage in Odoric see Henry Yule, ed. and trans., *Cathay and the Way Thither: Being a Collection of Medieval Notices of China*, revised by Henri Cordier (London: The Hakluyt society, 1913–16), vol. 2, pp. 101–102 (English), 280 (Latin), and 338 (Italian).

Fig. 1.25 Moses receiving from God the two Tablets of the Law, from the Hartmann Schedel's Buch der Croniken (Nuremberg, 1493), f. 30v. Library of Congress, Rare Book and Special Collection Division, Rosenwald Collection 166. Courtesy of the Library of Congress

of illustrating Noah's ark on a pair of mountains together with a brief text identifying the ship and the mountains, very much as we have on Waldseemüller's *Carta marina*.[143] The earliest surviving nautical chart that has a representation of Noah's Ark is that of Angelino Dulcert of 1339,[144] where the illustration is accompanied by the text *Archa de Noe. Mons Ararat in quo permansit Archa Noe post diluuium*, "Noah's Ark. Mount Ararat on which Noah's Ark remained after the Flood."[145] Very similar texts appeared on later luxury nautical charts including the Pizzigani chart of 1367,[146] Catalan Atlas of 1375,[147] Mecia de

[143]Noah's Ark is also illustrated on some *mappaemundi*, such as the Psalter map, the Ebstorf map, and the Hereford map, but the legend accompanying the image on the *Carta marina* is closer to what we find on nautical charts, and Waldseemüller shows no other signs of having used a large *mappamundi*. For the legend on the Ebstorf map see Hartmut Kugler, *Die Ebstorfer Weltkarte* (Berlin: Akademie Verlag, 2007), vol. 1, p. 76, 24.A1, and vol. 2, p. 135, 24/1; and for the legend on the Hereford map see Scott D. Westrem, *The Hereford Map: A Transcription and Translation of the Legends with Commentary* (Turnhout: Brepols, 2001), pp. 106–107, #224. For other appearances of Noah's Ark on *mappaemundi* see Konrad Miller, *Mappaemundi: Die ältesten Welkarten* (Stuttgart, 1895–98), vol. 3, subtitled "Die kleineren Weltkarten," pp. 6, 32, 93, 119, 145, and 148; there is also an image of the Ark, but without descriptive text, on Fra Mauro's *mappamundi* of c. 1455: see Piero Falchetta, *Fra Mauro's World Map* (Turnhout: Brepols, 2006), pp. 498–499, *1679.

[144]The 1339 Dulcert chart is in Paris, Bibliothèque nationale de France, Ge B 696 Rés. It is reproduced in Cavallo, *Cristoforo Colombo e l'apertura degli spazi* (see note 134), vol. 1, pp. 164–165, with descriptive text on pp. 162–163; and on a larger scale in Gabriel Marcel, *Choix de cartes et de mappemondes des XIVe et XVe siècles* (Paris: E. Leroux, 1896); and in Pujades, *Les cartes portolanes* (see note 31), pp. 120–121 and on the accompanying CD, number C8.

[145]The illustrations of and texts about Noah's Ark on nautical charts are surveyed by Sandra Sáenz-López Pérez, "Imagen y conocimiento del mundo en la Edad Media a través de la cartografía Hispana," Ph.D. Dissertation, Universidad Complutense de Madrid, 2007, vol. 1, pp. 475–478.

[146]On the 1367 Pizzigani chart see note 134.

[147]On the Catalan Atlas see note 38.

Viladestes's chart of 1413,[148] the chart of Joan de Viladestes of 1428,[149] and the Catalan-Estense map of c. 1460[150]; there is also an image of Noah's Ark on an elaborately decorated chart made by Grazioso Benincasa in 1482, but without the brief explanatory text.[151] When the descriptive text does appear, it is quite similar to that on Waldseemüller's *Carta marina.* On nautical charts the Ark is represented either as a chest, in a curious pyramidal shape,[152] or as a building, whereas Waldseemüller's image is distinctly a ship,[153] but it would be quite natural for Waldseemüller to change the image he found in a nautical chart to something more shiplike.[154] Thus we can be quite certain that in addition to the Caverio chart, Waldseemüller had a heavily illustrated luxury nautical chart.

In northeastern Asia on the *Carta marina* Waldseemüller has an image of a man riding a deer, certainly a reindeer, and a brief legend that says *magis Septentrionales equitant ceruos*, "In the far north they ride deer" (see Legend 4.17). On his 1507 map in the same area Waldseemüller has a legend about Balor Regio that derives from Marco Polo, and mentions that the inhabitants ride deer.[155] The image and legend on the *Carta marina* are in essentially the same location as the legend on the 1507 map, but the image comes from a nautical chart illustration of an inhabitant of Scandinavia riding a reindeer—that is, Waldseemüller has transplanted a nautical chart image relating to Scandinavia to a location that accords with what Marco Polo says about northeastern Asia. There are just a few nautical charts that have an illustration of a Scandinavian man riding a reindeer: Mecia de Viladestes's chart of 1413[156]; the Vatican Borgia XVI metal *mappamundi* from the first half of the fifteenth century, which uses nautical chart data[157]; the anonymous nautical chart which is Florence, Biblioteca Nazionale

[148]The 1413 Mecia de Viladestes chart is in Paris, Bibliothèque nationale de France, Rés. Ge AA 566; it is illustrated in Michel Mollat and Monique de la Roncière, *Sea Charts of the Early Explorers: 13th to 17th Century*, trans. L. le R. Dethan (New York: Thames and Hudson, 1984), chart 12; and in Pujades, *Les cartes portolanes* (see note 31), pp. 202–203, and on the accompanying CD, number C30.

[149]Joan de Viladestes's chart of 1428 is in Istanbul, Topkapi Sarayi, H. 1826, and is reproduced in Youssouf Kamal, *Monumenta cartographica Africae et Aegypti* (Cairo, 1926–51), vol. 4, fasc. 4, ff. 1456v–1457.

[150]The Catalan-Estense map is in Modena, Biblioteca Estense Universitaria, C. G. A. 1, and has been reproduced in facsimile, with transcription and commentary, in Ernesto Milano and Annalisa Battini, *Mapamundi Catalán Estense, escuela cartográfica mallorquina* (Barcelona: M. Moleiro, 1996); there is a high-resolution digital image of the map on the CD-ROM titled *Antichi planisferi e portolani: Modena, Biblioteca Estense Universitaria* (Modena: Il Bulino; and Milan: Y. Press, 2004), and a good study of it in Konrad Kretschmer, "Die katalanische Weltkarte der Biblioteca Estense zu Modena," *Zeitschrift der Gesellschaft für Erdkunde zu Berlin* 32 (1897), pp. 65–111 and 191–218.

[151]The 1482 chart by Grazioso Benincasa is Bologna, Biblioteca Universitaria, Rot. 3, and is reproduced in Cavallo, *Cristoforo Colombo e l'apertura degli spazi* (see note 134), vol. 1, pp. 356–357, with descriptive text on pp. 353 and 358.

[152]For discussion of the depiction of Noah's Ark as a chest and with a pyramidal shape in non-cartographic contexts see Marianne Besseyre, "L'iconographie de l'arche de Noé, du IIIe au XVe siècle: du texte aux images," Thèse de l'École Nationale des Chartes, 1997, Chap. 2; the thesis is summarized in Marianne Besseyre, "L'iconographie de l'Arche de Noé du IIIe au XVe siècle: du texte aux images," in *Positions des thèses soutenues par les élèves de la promotion de 1997 pour obtenir le diplôme d'archiviste-paléographe* (Paris: École Nationale des Chartes, 1997), pp. 53–58.

[153]On the development of the iconography of Noah's Ark depicted as a ship see Maria Teresa Lezzi, "L'arche de Noé en forme de bateau: naissance d'une tradition iconographique," *Cahiers de civilisation médiévale* 37 (1994), pp. 301–324; and Andreina Contessa, "Noah's Ark on the Two Mountains of Ararat: The Iconography of the Cycle of Noah in the Ripoll and Roda Bibles," *Word & Image* 20.4 (2004), pp. 257–270. Incidentally Noah's Ark is depicted as a ship on the fifteenth-century Zeitz *mappamundi*, which is Zeitz, Stiftsbibliothek MS Hist. Fol. 497, f. 48r, and is reproduced in color in Egon Klemp, *Africa on Maps Dating from the Twelfth to the Eighteenth Century*, trans. Margaret Stone and Jeffrey C. Stone (New York: McGraw-Hill Book Co., 1970), plate 6. For discussion and a black-and-white illustration of the map see Heinrich Winter, "A Circular Map in a Ptolemaic MS," *Imago Mundi* 10 (1953), pp. 15–22. There is a good discussion of depictions of Noah's Ark on maps in René Tebel, *Das Schiff im Kartenbild des Mittelalters und der Frühen Neuzeit: kartographische Zeugnisse aus sieben Jahrhunderten als maritimhistorische Bildquellen* (Wiefelstede: Oceanum Verlag, 2012), pp. 77–84.

[154]There is a boat-like image of Noah's Ark on a mountain in a manuscript of Jean Mansel's *La Fleur des Histoires* that was made c. 1455, namely Brussels, Bibliothèque Royale, MS 9231, f. 281v. This map is widely reproduced: see Destombes, *Mappemondes* (see note 29), p. 179, #51.1 and plate 20; Monique Pelletier, ed., *Couleurs de la terre: des mappemondes médiévales aux images satellitales* (Paris: Bibliothèque nationale de France, 1998), p. 34 (large and in color); and Peter Barber, ed., *The Map Book* (New York: Walker & Co.; and Delray Beach, FL: Levenger Press, 2005), p. 73 (large and in color).

[155]The legend on Waldseemüller's 1507 map reads: *Balor regio. Incole istius regionis habitant in montibus sunt siluestres carent vino et blada utuntur carnibus ceruorum et equitant ceruos domesticos*. For the passage in Marco Polo see Marco Polo, *Marka Pavlova z Benátek, Milion: Dle jediného rukopisu spolu s prilusnym zakladem latinskym*, ed. Justin Václav Prásek (Prague: Nákl. Ceské akademie císare Frantiska Iozefa, 1902), Book 1, Chap. 62, p. 62. For the passage in Yule's translation see Marco Polo, *The Book of Ser Marco Polo*, Book 1, Chap. 61, vol. 1, p. 269 with 271–272. Johann Schöner drew on this legend from Waldseemüller's 1507 map for a legend on his manuscript globe of 1520.

[156]On Mecia de Viladestes's chart of 1413 see note 148 above.

[157]The Borgia map is described and reproduced in Destombes, *Mappemondes* (see note 29), pp. 239–241 and plate 29; also see A. E. Nordenskiöld, "Om ett aftryck från XV:de seklet af den i metall graverade världskarta, som förvarats i kardinal Stephan Borgias museum i Velletri, Med 1 facsimile," *Ymer* 11 (1891), pp. 83–92, with the reproduction of the map between pp. 130 and 131. For a good study of the map see John Hamer, "The Borgia Map: Europe's Rise and the Re-Definition of the World," MA Thesis, University of Michigan, Ann Arbor, 1995.

Fig. 1.26 Detail of Noah's Ark on Waldseemüller's *Carta marina* (sheet 3). Courtesy of the Library of Congress

Centrale, Portolano 16 (ca. 1439–1460)[158]; and the Catalan-Estense map (c. 1460).[159] In addition, a mid fifteenth-century nautical chart which is now lost, but whose legends are preserved in a manuscript in Genoa has a text that says that in the provinces *de Scachia et de Gotia.... Sunt magni venatores et equitant ceruos*, "of Scachia and Gothia... they are great hunters and ride deer."[160]

In addition to the evidence of images, below in my discussion of the names Waldseemüller gives to the Caspian Sea (see Legend 3.25) I will show that those names come from a nautical chart, and are most similar to the names assigned to the sea on the Catalan Atlas of 1375 and the Catalan-Estense *mappamundi* of c. 1460.

The images of Noah's Ark and the man riding the reindeer, and also the names of the Caspian Sea, are of particular value in shedding light on the type of nautical chart from which Waldseemüller drew these images. First it was a luxury nautical chart, as the hinterlands contained illustrations. Also, as most nautical charts show the rectangular area defined by a diagonal from the Red Sea northwest to Ireland, plus a bit more of the Atlantic, we know that Waldseemüller's chart was larger than average, as it included lands north to Scandinavia and east to Armenia (Mount Ararat), and in fact to the Caspian. In particular, the surviving charts that have the illustration of a man riding a reindeer date from 1413 to about 1460. It would not surprise me if the chart fragment of c. 1375 which is in Istanbul, Topkapi Saray, H. 1828, once contained such an illustration,[161] but in any case there are no charts later than c. 1460 that have this illustration. So in addition to the quite recent and no doubt very expensive Caverio chart, Waldseemüller had an older large luxury nautical chart, on which the hinterlands were much more elaborately decorated than on the Caverio chart. This is one case in which Waldseemüller was content to make occasional use of a somewhat older source.

[158] Florence, BNCF Portolano 16 is reproduced in Pujades, *Les cartes portolanes* (see note 31), pp. 270–271, and on the accompanying CD, number C41.

[159] On the Catalan-Estense map see note 150.

[160] The manuscript is in Genoa, Biblioteca Universitaria, MS B. I. 36. The legends in the manuscript are transcribed and translated into French in Jacques Paviot, "Une mappemonde génoise disparue de la fin du XIVe siècle," in Gaston Duchet-Suchaux, ed., *L'Iconographie: études sur les rapports entre textes et images dans l'Occident médiéval* (Paris: Le Léopard d'Or, 2001), pp. 69–97. Paviot indicates that the lost map is from the late fourteenth century, but as one of its legends cites Antoniotto Usodimare, it must be from 1455 or later. The legend about the reindeer is legend 7, pp. 79 (Latin) and 89 (French).

[161] For discussion of this chart fragment see Marcel Destombes, "Fragments of Two Medieval World Maps at the Topkapu Saray Library," *Imago Mundi* 12 (1955), pp. 150–152; and Philipp Billion, *Graphische Zeichen auf mittelalterlichen Portolankarten* (Marburg: Tektum Verlag, 2011), pp. 184–188.

Waldseemüller's use of so many recent sources, both textual and iconographical, clearly reflects his ambition to create a thoroughly up-to-date image of the world. It also provides insight into the cartographer's schedule of work on the *Carta marina*: he was actively working on the map right until it was printed in 1516.

1.6 The Development of Waldseemüller's Cartographic Thought

Waldseemüller's use of Ptolemaic cartographic principles in his 1507 map and his repudiation of them in his 1516 map merit tracing in more detail. As we saw earlier, the title of the 1507 map describes an essentially Ptolemaic world map, with the addition of the New World. Waldseemüller's description of his project in the accompanying *Cosmographiae introductio* is similar, except that he says he was influenced not only by verbal accounts of the new discoveries but also by nautical charts[162]:

> Haec pro inductione ad Cosmographiam dicta sufficiant si te modo ammonuerimus prius, nos in depingendis tabulis typi generalis non omnimodo sequutos esse Ptholomeum, presertim circa nouas terras, ubi in cartis marinis aliter animaduertimus, equatorem constituti, quam Ptholomeus foecerit. Et proinde non debent nos statim culpare qui illud ipsum notauerint. Consulto enim foecimus quod hic Ptholomeum, alibi cartas marinas sequuti sumus.
>
> All that has been said by way of introduction to cosmography will be sufficient, if we merely advise you that in designing the sheets of our world-map we have not followed Ptolemy in every respect, particularly as regards the new lands, where on nautical charts we observe that the equator is placed otherwise than Ptolemy represented it. Therefore those who notice this ought not to find fault with us, for we have done so purposely, because here we have followed Ptolemy, and elsewhere nautical charts.

So there was some tension between the Ptolemaic and nautical chart models in the 1507 map. In fact, expressions of doubt about or criticism of Ptolemy go back to the first Latin translation of the work, which was made in 1409 by Jacopo Angeli da Scarperia,[163] who mentions other authors *qui et alia quedam habent quae ab auctore hoc Ptolomeo uidentur pretermissa*, "who have other things which seem to have been omitted by this author, Ptolemy."[164] A manuscript of Ptolemy's *Geography* from c. 1436–1455 shows marked influence of nautical charts, using nautical chart data for the coastlines, but Ptolemaic place names in the interior.[165] Fra Mauro on his world map of c. 1455 notes that Ptolemy's information about various regions is incomplete, and declines to employ his system of latitude and longitude.[166] The addition of *Tabulae modernae*, or modern maps, to both manuscripts and printed editions of Ptolemy represents a profound if unarticulated criticism of Ptolemy's data,[167] and such maps existed in the 1486 Ulm edition of Ptolemy, of which Waldseemüller owned a copy.

---

[162]See Joseph Fischer and Franz von Wieser in *The 'Cosmographiae introductio' of Martin Waldseemüller*, Chap. 9, p. xxxvii (Latin), and p. 78 (English); and Hessler, *The Naming of America* (see note 7), p. 106.

[163]On Angeli da Scarperia and his translation of Ptolemy's *Geography* see Roberto Weiss, "Jacopo Angeli da Scarperia (c. 1360–1410/11)," in *Medioevo e Rinascimento: Studi in onore di Bruno Nardi* (Florence: Sansoni, 1955), vol. 2, pp. 801–827; reprinted in Weiss's *Medieval and Humanist Greek: Collected Essays* (Padua: Antenore, 1977), pp. 255–277.

[164]Jacopo Angeli's introduction was published in the Vicenza, 1475, and Bologna, [1477] editions of the *Geography*. The version of the introduction in Harvard University, Houghton Library, MS Typ 5, ff. 1r–2v, has been published by James Hankins in "Ptolemy's *Geography* in the Renaissance," in Rodney G. Dennis with Elizabeth Falsey, eds., *The Marks in the Fields: Essays on the Uses of Manuscripts* (Cambridge, Mass.: The Houghton Library, 1992), pp. 118–127; and reprinted in Hankins' *Humanism and Platonism in the Italian Renaissance* (Rome: Edizioni di storia e letteratura, 2003-), vol. 1, pp. 457–468. This text is reprinted and translated into English by Charles Burnett in Zur Shalev and Charles Burnett, eds., *Ptolemy's 'Geography' in the Renaissance* (London: Warburg Institute, and Turin: Nino Aragno Editore, 2011), pp. 225–229. For discussion of some other passages in which Renaissance authors express their doubts about the correctness of the ancients see R. Hooykaas, *Humanism and the Voyages of Discovery in 16th Century Portuguese Science and Letters* (Amsterdam and New York: Noord-Hollandsche U.M., 1979).

[165]See Marica Milanesi, "A Forgotten Ptolemy: Harley Codex 3686 in the British Library," *Imago Mundi* 48 (1996), pp. 43–64.

[166]See Falchetta, *Fra Mauro's World Map* (see note 143), pp. 452–453, *1405; pp. 698–701, *2834; and pp. 710–711, *2892.

[167]For discussion of the *Tabulae modernae* or *Tabulae novae* added to manuscripts and editions of Ptolemy see James Richard Akerman, "On the Shoulders of a Titan: Viewing the World of the Past in Atlas Structure," Ph.D. Dissertation, The Pennsylvania State University, 1991, pp. 228–253. On maps of the New World in printed editions of Ptolemy see Oswald A. W. Dilke and Margaret S. Dilke, "The Adjustment of Ptolemaic Atlases to Feature the New World," in Wolfgang Haase and Meyer Reinhold, eds., *The Classical Tradition and the Americas*, vol. 1.1, *European Images of the Americas and the Classical Tradition* (Berlin and New York: Walter de Gruyter, 1994), pp. 117–134; and for lists of the *tabulae modernae* in several printed editions of Ptolemy see Uta Lindgren, "Die Geographie des Claudius Ptolemaeus in München: Beschreibung der gedruckten Exemplare in der Bayerischen Staatsbibliothek," *Archives Internationales d'Histoire des Sciences* 35.114–115 (1985), pp. 148–239.

In 1505, Waldseemüller together with Matthias Ringmann and other colleagues began work on a new edition of Ptolemy's *Geography*, which after several long delays was published in 1513. Two years earlier, Bernardus Sylvanus had published an edition of the *Geography* in Venice.[168] In his introduction, Sylvanus says that in studying Ptolemy,[169]

> Cum Ptolemaeum inter alios geographiae scriptores diligentissime et situs et distantias locorum scripsisse conspicerem admirabar profecto cur illius tabulae paucis admodum in rebus cum nostri temporis navigationibus consentirent: eoque magis admirabar quod Ptolemaeum quoque navigationibus comprimis innixum ea quae scripserit scripsisse arbitrabar.
>
> Although I used to view Ptolemy as having recorded the sites and distances of places more diligently than other writers of geography, I was none the less puzzled as to why his tables agreed in very few instances with the navigations of our times: and I was all the more puzzled because I used to think that Ptolemy too had recorded what he recorded relying on navigations more than anything else.

Sylvanus revised Ptolemy's maps and data in accordance with recent nautical charts, convincing himself that in doing so he was actually restoring them to what Ptolemy had originally intended. Specifically, he copied the coastlines from nautical charts (where available), but retained the place names and hinterland geographical details of Ptolemy, then extrapolated latitude and longitude values for the new coastlines, and adjusted the data in Ptolemy's text appropriately.[170] Thus Sylvanus's edition clearly shows that nautical charts were seen as superior to then-available Ptolemaic maps at least in some quarters in the early sixteenth century, and from a comment that Waldseemüller makes in the introduction to the index in his edition of Ptolemy, it seems very likely that he had seen Sylvanus's edition.[171]

On the title page of his 1513 edition of Ptolemy,[172] Waldseemüller says that it consists of two parts; first the text of the *Geography*, an index, a brief account of the Greek numbering system, and twenty-seven Ptolemaic maps, and then:

> Pars secunda moderniorum lustrationum viginti tabulis veluti supplementum quoddam antiquitatis obsoletae suo loco quae vel abstrusa vel erronea videbantur resolutissime pandit.
>
> The second part, through twenty maps of modern explorations, boldly offers a kind of supplement to obsolete antiquity [i.e. obsolete ancient authors] wherever it seems to be obscure or erroneous.

The separate one-page introduction to the second part of the work discusses how time changes many things, and how the names of many cities and regions are different than what they had been previously—a passage very similar to part of the introductory paragraph on the *Carta marina* (compare Legend 9.3)[173]:

> Ptolemaei Geographiam prima parte clausimus operis: ut incorruptior & selecta stet antiquitas sua. Verum quia temporis lapsus multa quidem labilitate quoque sua indies mutat: plaerisque visus est auctor notabilius a modernioribus deuiasse. Id quod cernere licet in utraque Pannonia, quae nunc Hungaria & Austria vocatur. Et quae regio dum floruit unica appellatione Sarmatia, siue Sauromatia dicebatur: nunc diuisim Poloniam, Russiam, Prussiam, Moscouiam & Lituaniam nominamus. Populorum denique usui placuit transmutatio vocabulorum. Quos enim vetustas Eluetios & Sequanos, nunc vulgo Burgundiones Suitensesque vocamus. Quaedam & ciuitates primitiuis nominibus orbati sunt. Quis enim iuxta Rhenum Flauium, Canodurum, Augustam rauricum, Elcebum & Berthomagum urbes a Ptolemaeo comemoratas digito mostrabit?

---

[168]Sylvanus's edition has been published in facsimile as Ptolemy, *Geographia: Venice, 1511* (Amsterdam: Theatrum Orbis Terrarum, 1969). For discussion of Sylvanus's work as a cartographer see Robert W. Karrow Jr., *Mapmakers of the Sixteenth Century and Their Maps* (Chicago: Speculum Orbis Press, 1993), pp. 520–524.

[169]This passage is from the first page of Sylvanus's introduction; R. A. Skelton translates the passage into English in his "Bibliographical Note," in Ptolemy, *Geographia: Venice, 1511* (Amsterdam: Theatrum Orbis Terrarum, 1969), pp. v–xi, at vii. But here we borrow the translation from Margaret Small, "Warring Traditions: Ptolemy and Strabo in the Geography of Sebastian Münster," in Zur Shalev and Charles Burnett, eds., *Ptolemy's Geography in the Renaissance*, pp. 167–185, at 177.

[170]For discussion of Sylvanus's methods for updating Ptolemy's maps and text see Giulia Guglielmi-Zazo, "Bernardo Silvano e la sua edizione della Geografia di Tolomeo," *Rivista Geografica Italiana* 32 (1925), pp. 37–56 and 207–16; and 33 (1926), pp. 25–52; and R. A. Skelton, "Bibliographical Note," in Ptolemy, *Geographia: Venice, 1511* (Amsterdam: Theatrum Orbis Terrarum, 1969), pp. v–xi.

[171]Waldseemüller writes *Quod tanta ordinis sui confusione scatet, ut in plerisque locis an modernioribus an Ptolomaeo ipsi conquadret, lector etiam studiosissimus nesciat*, that is, "There is so much confusion in the presentation, that in many places even very learned readers cannot tell which data comes from modern authors, and which from Ptolemy himself."

[172]For bibliography on the 1513 edition of Ptolemy's *Geography* see note 17.

[173]The translation is from Henry N. Stevens, *The First Delineation of the New World and the First Use of the Name America on a Printed Map* (London: H. Stevens, Son & Stiles, 1928), p. 40.

We have confined the Geography of Ptolemy to the first part of the work, in order that its antiquity may remain intact and separate. But since the course of time changes many things from day to day as it passes, it has become generally evident that the author deviates notably from those more modern, as may be seen in the two Pannonias, which are now called Hungary and Austria; and the region which was called while it flourished, by the sole appellation of Sarmatia or Sauromatia, we now name in its divisions, Poland, Russia, Prussia, Muscovy and Lithuania. Change in the names of nations has also come into use. For those whom the ancients called Helvetii and Sequani, we now commonly call Burgundians and Swiss. Certain cities, too, have lost their primitive names, for who with his finger will point out on the River Rhine the cities Canodorum, Augusta Rauricum, Elcebus and Berthomagus mentioned by Ptolemy?

He then continues[174]:

Haec vel his similia non est qui Auctoris imperitiae subscribat. Quin potius hoc Supplemento modernioris lustrationis discat seipsum certius informare. Qua tripartiti orbis explanationem planius ad tempora nostra videbit. Charta autem Marina, quam Hydrographiam vocant, per Admiralem quondam serenissi. Portugaliae regis Ferdinandi, caeteros denique lustratores verissimis peragrationibus lustrata....

These or similar [inaccuracies in place names] let no one attribute to the ignorance of the author [i.e. Ptolemy], but rather from this supplement let him learn to inform himself more accurately about modern explorations, in which he will see an image of the three parts of the world more clearly adapted to our times. Specifically the nautical chart which they call a hydrography, which was surveyed by the very authentic explorations of a former Admiral of Ferdinand, the Most Serene King of Portugal, and of other explorers....

There is a slip of the pen here, as Ferdinand was not king of Portugal, but rather of Aragon, and through his wife, Isabela, of Castile.[175] The Admiral in question must be Columbus,[176] but it is misleading to call the map (as a number of scholars have done) "The Admiral's Map," implying a particularly close connection with Columbus. Waldseemüller clearly states that the map is based on the discoveries not only of Columbus but also of other explorers. In any case, we see that by 1513 Waldseemüller had realized that Columbus, rather than Vespucci, was the first to reach the New World.

The new world map in the 1513 Ptolemy (Fig. 1.27) serves as an important indicator of the development of Waldseemüller's thinking about cartography at that time. The title of the map is *Orbis Typus Universalis Iuxta Hydrographorum Traditionem*, "General Map of the World According to the Tradition of the Hydrographers." By "hydrographers" Waldseemüller means the makers of nautical charts: the map has a system of rhumb lines like a nautical chart, and no Ptolemaic grid of latitude and longitude, though it does indicate the equator and tropics. Thus the cartographer is clearly proclaiming that a world map "more clearly adapted to our times" must be based on nautical charts, and the depiction of southern Asia shows the influence of the Caverio chart, while the shape of Africa is also clearly based on recent Portuguese cartographic data. Moreover, and this is very important, of the twenty modern charts in the 1513 Ptolemy, all but one are made using a nautical chart projection, rather than one of Ptolemy's projections—a strong confirmation of Waldseemüller's recognition of the value of nautical cartography.[177]

At the same time, even while Waldseemüller proclaims that his new world map in the 1513 Ptolemy is based on nautical cartography, certain elements of the map do not derive from that genre: his depiction of Scandinavia and the sweeping rounded coast of eastern Asia clearly derive from one of the world maps by Henricus Martellus (Fig. 1.28).[178] Waldseemüller had based his 1507 map on a large world map by Martellus similar to that at Yale, so the elements from

[174]This passage is translated into English by Stevens, *The First Delineation of the New World* (see note 173), p. 40, and part is quoted in Robert W. Karrow Jr., "Intellectual Foundations of the Cartographic Revolution," Ph.D. Dissertation, Loyola University of Chicago, 1999, pp. 184–185; I modify that translation slightly here.

[175]R. A. Skelton suggests the error was caused by words having dropped out of the text: see his "Bibliographical Note" in Ptolemy, *Geographia, Strassburg, 1513* (see note 172), p. xvi.

[176]See Samuel McCoskry Stanton, "The Admiral's Map: What Was It? And Who the Admiral?" *Isis* 22.2 (1935), pp. 511–515.

[177]Some of Waldseemüller's inclination to abandon Ptolemy is visible even in the Ptolemaic part of the 1513 edition of the *Geography*: in the Ptolemaic world map, Waldseemüller omits the traditional Ptolemaic land bridge along the map's southern edge that (so Ptolemy believed) joined southern Africa and southern Asia, and would render it impossible to sail around Africa to Asia. Even while depicting the world according to Ptolemy, Waldseemüller could not bring himself to depict a landmass he knew did not exist.

[178]This map by Martellus is in Florence, Biblioteca Nazionale Centrale, Landau Finaly, Carte Rosselli, planisfero. It is very similar to the world maps that illustrate manuscripts of Martellus's *Insularium illustratum*. For discussion of the map see Sebastiano Crinò, "I planisferi de Francesco Rosselli dell'epoca delle grandi scoperte geografiche," *La Bibliofilía* 41 (1939), pp. 381–405, esp. 393–401; Roberto Almagià, "On the Cartographic Work of Francesco Rosselli," *Imago Mundi* 8 (1951), pp. 27–34, esp. 3–2; and Tony Campbell, *The Earliest Printed Maps, 1472–1500* (London: British Library, 1987), pp. 70–78, esp. 72–74.

Fig. 1.27 "Modern" world map by Waldseemüller, the so-called "Admiral's Map", in the 1513 edition of Ptolemy's *Geography*. Library of Congress, Rare Book and Special Collection Division, Rosenwald Collection 624. Courtesy of the Library of Congress

Martellus in this new world map in the 1513 Ptolemy show that in some ways Waldseemüller was still holding onto this older style of cartography.

The 1513 Ptolemy shows Waldseemüller at a point of transition. Certainly he would not have devoted the time and energy to creating a new edition of Ptolemy if he did not believe in the value of Ptolemaic geography and cartography. The elements of Ptolemy and Martellus in his new world map also reflect that belief. But his twenty modern maps in the book,[179] almost as many as Ptolemy's twenty-seven, and all but one of which are made according to the principles of nautical charts, constitute a cartographic parallel universe to Ptolemy's. Further, his declarations about the value of nautical charts clearly indicate that if one has to choose a cartographic system for a modern world map, it will be that of nautical charts—manifestly anticipating the purer expression of that philosophy in the *Carta marina*.

[179]The large number of modern maps in the 1513 Ptolemy recall the collection of twelve such maps in a manuscript of Ptolemy's *Geography* made between 1480 and 1496 (Florence, Biblioteca Nazionale Centrale, Magliabechiano XII 16) by a cartographer whose works were clearly well known to Waldseemüller, namely Henricus Martellus. For discussion of this manuscript see Sebastiano Gentile, ed., *Firenze e la scoperta dell'America: umanesimo e geografia nel'400 fiorentino* (Florence: L. S. Olschki, 1992), pp. 240–243 with plates 47–48, including a good list of the *tabulae modernae* and bibliography; and Cavallo, *Cristoforo Colombo e l'apertura degli spazi* (see note 134), vol. 1, pp. 517–521, with a good color reproduction of the world map on pp. 518–519. The manuscript has been published in facsimile as *Ptolomei cosmographia* (Florence: Vallecchi, 2004), with studies by Sebastiano Gentile and Angelo Cattaneo. The "Introduzione" by Cattaneo, pp. 23–53, is a valuable description and analysis of the manuscript and its maps.

Fig. 1.28 World map by Henricus Martellus in a manuscript of his *Insularium illustratum*, c. 1490. British Library, Add. MS 15760, ff. 68v–69r. © The British Library Board

1.7 The Cutting of the Woodblocks for the *Carta Marina*

Careful examination of the sheets of the *Carta marina* shows that multiple artisans were involved in cutting the woodblocks, and also a lack of close coordination among those workers. This contrasts with the situation with Waldseemüller's 1507 map, where the signs of different hands and of lack of coordination between those cutting adjacent blocks are more subtle. Illustrations supporting the points made in the following paragraphs can be found in Fig. 1.4, of the whole *Carta marina*, and in the plates of each sheet of the map below at the beginning of the relevant sections of the transcription.

On the *Carta marina* there are significant differences of style in the rendering of the wind-heads and associated decorations in the map's margins. For example, the lines representing the wind blown by two of the wind-heads in the margin of sheet 9, in the lower left corner of the map, cross the border into the map proper, but this is not the case on the other sheets. The margins of some of the sheets have stars as part of their decoration (1, 2, 4, 8, and 12), while the others do not; the styles of rendering the clouds varies, with those in the border of sheet 9 being particularly puffy; and the styles of drawing the heads themselves are inconsistent, with those in the border of sheet 12 being more stern in appearance, for example.

The style of rendering the oceans also varies from sheet to sheet. On sheet 1 in the upper left corner of the map the texture of the surface of the water is intermittently depicted, and some clouds are shown above the water, but the ocean on the adjacent sheet 2 has neither of these features—the difference is quite dramatic. On sheet 10 the block cutter gives some texture to the surface of the water, and shows a few clouds (differently than in sheet 1), and both sheet 6 above it, and sheet 11 to its right, have notably more plain styles of rendering the ocean.

The vast majority of the cartouches on the map are simple frames; a few have small geometrical decorations at their tops or sides, namely those on sheet 2 by the southern tip of Greenland, on sheet 6 by the Canary Islands, on sheet 7 off the coast from Mogadishu, on sheet 10 off the coast of Brazil, and on sheet 12 the large cartouche east of Java. And two other cartouches have very elaborate artistic decoration: the large cartouche on sheet 9 is embellished with vegetal motifs, knots, scrollwork, and two dragons; and the small cartouche at the right edge of sheet 12 with vegetal motifs and knots.

The block cutters also rendered differently the very simple compass roses at the nodes of the rhumb line network. These compass roses consist of two concentric circles and pointers to the north and east. On sheet 10 they are small and there is very little gap between the two circles, while on sheet 6 directly above they are larger and have a larger gap between their circles. These many stylistic differences among the sheets of the *Carta marina* demonstrate clearly that multiple block cutters were working on the map.

There are additional differneces between the sheets that point not so much to differences of style between the block cutters as to a lack of coordination among them. The most egregious example is found in the scale of latitude at the left-hand edge of the map, which runs down sheet 1 and sheet 5, but is not continued on sheet 9. Also on sheet 9, it is surprising that the decorative cartouche is cut off by the right-hand edge of the sheet—this sheet shows the most differences from its neighbors of any on the map, and is quite problematic. In addition, there are a few cases of rivers and mountain ranges that are discontinuous from one sheet to another, for example the mountains in Africa just north of the equator at the left edge of sheet 7 that do not continue onto sheet 6. Mention should also be made of the mountain range that extends from sheet 4 in India south onto sheet 8: it consists of sharp peaks north of that point, and rounded peaks south of that point.

What emerges from this examination of the details of the map is the fact that the production of the *Carta marina* was chaotic, with inadequate coordination among the artisans cutting the blocks for the map. Was the problem that the cutting was done hastily? Or could it have been the opposite, that the production was drawn out due to lack of funds (for example), and that the different blocks were cut at different times, and that was what reduced the consistency among them? In Legend 9.1 Waldseemüller offers thanks to Hugues des Hazards, Bishop of Toul from 1506 to 1517, presumably for his financial contribution to the production of the *Carta marina*, but this could have been funds that allowed the project to be brought to completion after a period of difficulties. Thus it does not seem possible to know the nature of the difficulties in the production of the *Carta marina* without additional evidence, but the inconsistencies among the sheets of the map show that it was indeed a challenging process.

1.8 Evidence for the Diffusion of the *Carta Marina*

Hildegard Binder Johnson has argued that the *Carta marina* was never published or sold, and that the only surviving copy was not part of the map's print run, but rather a special proof printing[180]; to my knowledge, no evidence to the contrary has been presented by other scholars.

The fact that only one exemplar each of Waldseemüller's 1507 and 1516 maps survives has been used to raise questions about the degree to which both maps were diffused. But this fact does not tell at all against the maps' diffusion: wall maps are notorious for their low survival rates, and there are many sixteenth-century printed maps, both wall maps and in smaller formats, that do not survive at all, or survive in only one or two exemplars. Such maps include:

Giovanni Contarini, world map of 1506 (one exemplar)[181]
Francesco Rosselli, printed nautical chart, c. 1508 (two)[182]
Waldseemüller, *Carta itineraria Europae* of 1511 (zero)[183]
Louis Boulengier, globe gores of c. 1514 (one)[184]
Dürer and Stabius, globe map of 1515 (none)[185]

[180]For the assertion that the 1516 *Carta marina* never reached the market see Johnson, *Carta marina* (see note 21), pp. 57–59. R. A. Skelton in his review of Johnson's book in *Geographical Review* 55.2 (1965), pp. 307–308, accepts this conclusion.

[181]The unique surviving exemplar of Contarini's map is in London, British Library, Maps C.2.cc.4., and is reproduced in facsimile with commentary in *A Map of the World, Designed by Gio. Matteo Contarini, Engraved by Fran. Roselli 1506* (London: Printed by Order of the Trustees, Sold at the British Museum, 1926).

[182]Rosselli's map is in Florence, Biblioteca Nazionale Centrale, Landau Finaly carte Rosselli.

[183]For references on the 1511 printing of Waldseemüller's *Carta itineraria Europae* see note 14 above.

[184]The unique surviving exemplar of Boulengier's globe gores are in the New York Public Libray, Rare Book Division, *KB 1517 (Waldseemüller, M. Cosmographiae Introdvctio/Cvm Qvibvsdam/Geometriae).

[185]Although no contemprary printings of the Dürer-Stabius globe-map survive, the woodblocks are in the Albertina in Vienna, with reference numbers HO2006/676–678; for discussion of the map see Günther Hamann, "Die Stabius-Dürer Karte von 1515," *Kartographische Nachrichten* 21.6 (1971), pp. 212–223.

Waldseemüller, *Carta itineraria Europae* of 1520 (one)[186]
Giovanni Vespucci, world map of 1524 (two)[187]
Lorenz Fries, *Carta marina* of 1525 (zero)[188]
Lorenz Fries, *Carta marina* of 1530 (one)[189]
Lorenz Fries, *Carta marina* of 1531 (one)[190]
Sebastian Cabot, world map of 1544 (two)[191]
Giacomo Gastaldi, *Univesale* of 1546 (two)[192]
Caspar Vopel, world map of 1558 (one)[193]
Giacomo Gastaldi, world map of c. 1561 (one)[194]
Diego Gutiérrez, map of the New World, 1562 (two)[195]
Gerard Mercator, world map of 1569 (three)[196]
Caspar Vopel, world map of 1570 (one)[197]
Georg Braun, world map of 1574 (one)[198]
Petrus Plancius, world map, Amsterdam/Antwerp, 1592 (one)[199]
Jodocus Hondius, wall map of 1595–96 (one)[200]

This list might very easily be expanded, so it is not at all unusual that Waldseemüller's two wall maps survive in only one exemplar each, and the fact that more exemplars do not survive cannot be adduced as evidence of a small print run or low diffusion.

The main difficulty in finding good evidence for the dissemination of Waldseemüller's *Carta marina* is distinguishing between its influence, and the influence of its re-edition by Lorenz Fries.[201] In 1525, Fries and the publisher Johann Grüninger produced a new version of the map, on a somewhat reduced scale (1876 × 1031 mm, or 74 × 40.6 inches, versus 2330 × 1280 mm, or 91.7 × 50.4 inches for Waldseemüller's map), with most of the legends translated into German, and accompanied by a booklet with more detailed descriptions of various parts of the world than there was room for on the map itself.[202] The booklet that Fries wrote is titled *Uslegung der mercarthen oder Charta Marina* (Explanation of the

[186]For references on the 1520 printing of Waldseemüller's *Carta itineraria Europae* see note 14 above.

[187]The two exemplars of Vespucci's map are at Harvard, Houghton Library, shelfmark p 51-2573; and Wolfenbüttel, Herzog August Bibliothek, shelfmark 15 Astron. 2°. For discussion of the discovery of the Wolfenbüttel exemplar see Christian Heitzmann,"Wem gehören die Molukken? Eine unbekannte Weltkarte aus der Frühzeit der Entdeckungen," *Zeitschrift für Ideengeschichte* 1.2 (2007), 101–110.

[188]On Fries's work as a cartographer see Robert W. Karrow, Jr., *Mapmakers of the Sixteenth Century and Their Maps: Bio-Bibliographies of the Cartographers of Abraham Ortelius, 1570* (Chicago: Published for The Newberry Library by Speculum Orbis Press, 1993), pp. 191–204.

[189]Fries's 1530 *Carta marina* is in Munich, Bayerische Staatsbibliothek (Mapp. I, 9 m-1). Digital images of the map's sheets are available at http://daten.digitale-sammlungen.de/bsb00012490/image_1.

[190]Fries's 1531 *Carta marina* is in Schaffhausen, Switzerland, Museum zu Allerheiligen (Inv. 6102).

[191]The two exemplars of Sebastian Cabot's world map are in Paris, BnF, Rés. Ge AA 582; and Klassik Stiftung Weimar, Kt 020-31 S.

[192]Gastaldi's *Universale* is at Harvard, Houghton Library, pf 51-2492; and London, British Library, Maps K.Top.4.6.

[193]Vopel's 1558 map is at Harvard, Houghton Library, p 51-2577.

[194]The unique surviving exemplar of Gastaldi's world map of c. 1561 is in London, British Library, Maps C.18.n.1.

[195]Diego Gutiérrez's map is in Washington, Library of Congress, Geography and Map Division, G3290 1562 .G7 Vault Oversize; and London, British Library, Maps * 69810.(18.).

[196]The three surviving exemplars of Mercator's 1569 map are in Paris, BnF, Rés Ge. A 1064; Basel, Universitätsbibliothek, Kartenslg AA 3-5; and Rotterdam, Maritiem Museum "Prins Hendrik," Atlas51.

[197]Vopel's world map of 1570 is in Wolfenbüttel, Herzog-August-Bibliothek, K 3,5.

[198]Braun's world map of 1574 is in Wolfenbüttel, Herzog-August-Bibliothek, K 2,6.

[199]Plancius's world map of 1592 is in Valencia, Colegio del Corpus Christi.

[200]Hondius's world map of c. 1595–96 is in Dresden, Sächsisches Hauptstaatsarchiv, 12884.

[201]For biographical information on Fries see Charles Schmidt, "Laurent Fries de Colmar, médecin, astrologue, géographe à Strassbourg et à Metz," *Annales de l'Est* 4 (1890), pp. 523–575; and Peter Weidisch, "Lorenz Fries—eine biographische Skizze," in Ulrich Wagner, ed., *Lorenz Fries (1489–1550), fürstbischöflicher Rat und Sekretär: Studien zu einem fränkischen Geschichtsschreiber* (Würzburg: F. Schöningh, 1989), pp. 23–43. On his writings see Josef Benzing, "Bibliographie der Schriften des Colmarer Artztes Lorenz Fries," *Philobiblon* 6 (1962), pp. 121–140.

[202]For a good discussion of this publishing project see Johnson, *Carta marina* (see note 21); Johnson compares many details of Waldseemüller's and Fries's maps on pp. 51–71. Also see Meret Petrzilka, *Die Karten des Laurent Fries von 1530 und 1531 und ihre Vorlage* (Zurich: Neuen Zürcher Zeitung, 1970). On Fries's work as a cartographer see Robert W. Karrow, Jr., *Mapmakers of the Sixteenth Century and Their Maps: Bio-Bibliographies of the Cartographers of Abraham Ortelius, 1570* (Chicago: Published for The Newberry Library by Speculum Orbis Press, 1993), pp. 191–204.

Sea Chart or *Carta marina*), and was published by Grüninger in Strasbourg in 1525.[203] No copy of the 1525 edition of Fries's *Carta marina* is extant, but single copies of two later editions do survive: one of the 1530 edition in Munich in the Bayerische Staatsbibliothek (Mapp. I,9m-1),[204] and one of the 1531 edition in Schaffhausen, Switzerland, in the Museum zu Allerheiligen (Inv. 6102), on which the legends are in Latin.[205]

While it may well be true that the copy of Waldseemüller's *Carta marina* now in the Library of Congress was specially printed, and while the existence of Fries's re-edition makes it challenging to demonstrate the diffusion of Waldseemüller's map rather than of Fries's, there is in fact good evidence that the 1516 map was published and disseminated.[206] The Huntington Library in San Marino, California, has a manuscript atlas of nautical charts known as the Vallard Atlas, made in the Dieppe region of France in approximately 1547,[207] a product of the so-called Dieppe School of cartography.[208] A study of the images in this atlas makes it clear that the cartographer had both Waldseemüller's *Carta marina* and the later edition by Fries in his workshop. The image in the Vallard Atlas of the King of France riding a sea monster south of Africa (Fig. 1.29) is clearly copied from the image on Fries's map (Fig. 1.30), and not from Waldseemüller's (Fig. 1.7). But there are also images that prove that the creator of the atlas had a copy of Waldseemüller's *Carta marina*. The image of Mecca in the atlas (Fig. 1.31) is strikingly similar to that on Waldseemüller's map (Fig. 1.17), and it is essentially impossible that the cartographer would have painted an image of Mecca so much like Waldseemüller's if he had been working from the simpler image on Fries's map (Fig. 1.32). As shown above, Waldseemüller copied his image of Mecca from the image of Medina in the 1515 illustrated edition of Varthema. Although a detailed image of the city very similar to Waldseemüller's was thus in theory available to the maker of the Vallard Atlas independently of Waldseemüller's map, it is all but inconceivable that the cartographer of the atlas would have consulted Fries's *Carta marina* and then the 1515 edition of Varthema, and yet not have followed the 1515 book in using the image for Medina, rather than Mecca. So the image of Mecca in the Vallard Atlas is indeed good evidence of the diffusion and influence of Waldseemüller's *Carta marina*.

There is also strong evidence that the cartographer of the Vallard Atlas used Waldseemüller's *Carta marina* in his depiction of the Indian Ocean. Specifically, on the third map in the atlas there is a scene of cannibal butchery on *lille de geans* or the Island of Giants[209]: a man who looks European holds a large cleaver and is chopping up a human body that is laid out on a butcher's table, with a small conduit to drain the blood into a bucket below (Fig. 1.33). This scene is not similar to the cannibalistic scene on the island of Java on Fries's *Carta marina*: that scene is much more complex, as Fries has added a woman and a baby, the woman holding a plate for the meat; the stance of the man with the cleaver is different, the body's head is on the ground, the bucket for the blood is off to the side rather than under the table, and there is no sign of the conduit (Fig. 1.34). But the scene in the Vallard Atlas is very similar to Waldseemüller's on Java: the stance and arm positions of the man with the cleaver are the same, and both tables have the conduit and the bucket for blood

---

[203]Johnson gives a good account of the *Uslegung* in *Carta marina* (see note 21), pp. 85–116. The full text of the *Uslegung* is translated into modern German by Petrzilka, *Die Karten des Laurent Fries* (see note 202), pp. 115–162.

[204]A facsimile of the 1530 copy was published in Munich by the bookseller Ludwig Rosenthal c. 1926; digital images of the sheets of the map at the BSB are available at http://daten.digitale-sammlungen.de/bsb00012490/image_1.

[205]The Schaffhausen copy of Fries's *Carta marina* is hand-colored. For discussion of the map see Henry J. Bruman, "The Schaffhausen Carta Marina of 1531," *Imago Mundi* 41 (1989), pp. 124–132. Some fragments of what seem to be proof sheets of a Latin edition of Fries's map also survive: see Leo Bagrow, "Fragments of the 'Carta Marina' by Laurentius Fries, 1524," *Imago Mundi* 14 (1959), pp. 110–112.

[206]Fischer and von Wieser indicate that Mercator borrowed from Waldseemüller's *Carta marina* in creating his famous 1569 world map, particularly in his legends in India and the topography and hydrography of southern Africa: see Joseph Fischer and Franz Ritter von Wieser, *Die älteste Karte mit dem Namen Amerika* (see note 9), p. 40. Johnson is skeptical of Fischer and von Wieser's claim in her *Carta marina* (see note 21), p. 58, and the notes on p. 136. My own study of the matter has revealed only one legend on Mercator's 1569 map that is similar to any on Waldseemüller's *Carta marina*, and that is the description of the opossum—information that Mercator could have obtained from another source.

[207]The Vallard Atlas is in San Marino, California, Huntington Library HM 29. It has been reproduced in facsimile as *Atlas d'ancienne cartes: Atlas Vallard* (Barcelona: M. Moleiro Editor, 2008–2010), with a volume of commentary by Luís Filipe F. R. Thomaz, Dennis Reinhartz, and Carlos Miranda García-Tejedor. High-resolution images of the maps are available via the Digital Scriptorium at http://www.digital-scriptorium.org/.

[208]On the "Dieppe School" see Henry Harrisse, "La cartographie Américano-Dieppoise," in his *Découverte et évolution cartographique de Terre-Neuve et des pays circonvoisins, 1497–1501–1769* (London and Paris, 1900; Amsterdam: N. Israel, 1968), Part 2, Chaps. 2–11; and Gayle K. Brunelle, "Dieppe School," in David Buisseret, ed., *The Oxford Companion to World Exploration* (New York: Oxford University Press, 2007), pp. 237–238. There is a list of all of the maps produced by the Dieppe cartographers in the Appendix to Sarah Toulouse, "Marine Cartography and Navigation in Renaissance France," in David Woodward, ed., *The History of Cartography*, Volume 3, *Cartography in the European Renaissance*, Part 2 (Chicago: The University of Chicago Press, 2007), pp. 1550–1568.

[209]On the Island of Giants in the Vallard Atlas and in Jean Rotz's *Boke of Idrography* of 1542 see W. A. R. Richardson, "Enigmatic Indian Ocean Coastlines on Early Maps and Charts," *The Globe: Journal of the Australian Map Circle* 46 (1998), pp. 21–41, esp. 34.

Fig. 1.29 Detail of the King of France riding a sea monster south of Africa from the Vallard Atlas. San Marino, California, Huntington Library, MS HM 29, f. 5. Courtesy of the Huntington Library

Fig. 1.30 King Manuel of Portugal riding a sea monster south of Africa on Laurent Fries's *Carta marina* of 1530. Bayerische Staatsbibliothek, Mapp. I,9m-1. Courtesy of the Bayerische Staatsbibliothek

Fig. 1.31 Detail of Mecca from the Vallard Atlas. San Marino, California, Huntington Library, MS HM 29, f. 4. Courtesy of the Huntington Library

Fig. 1.32 Mecca on Laurent Fries's *Carta marina* of 1530. Bayerische Staatsbibliothek, Mapp. I,9m-1. Courtesy of the Bayerische Staatsbibliothek

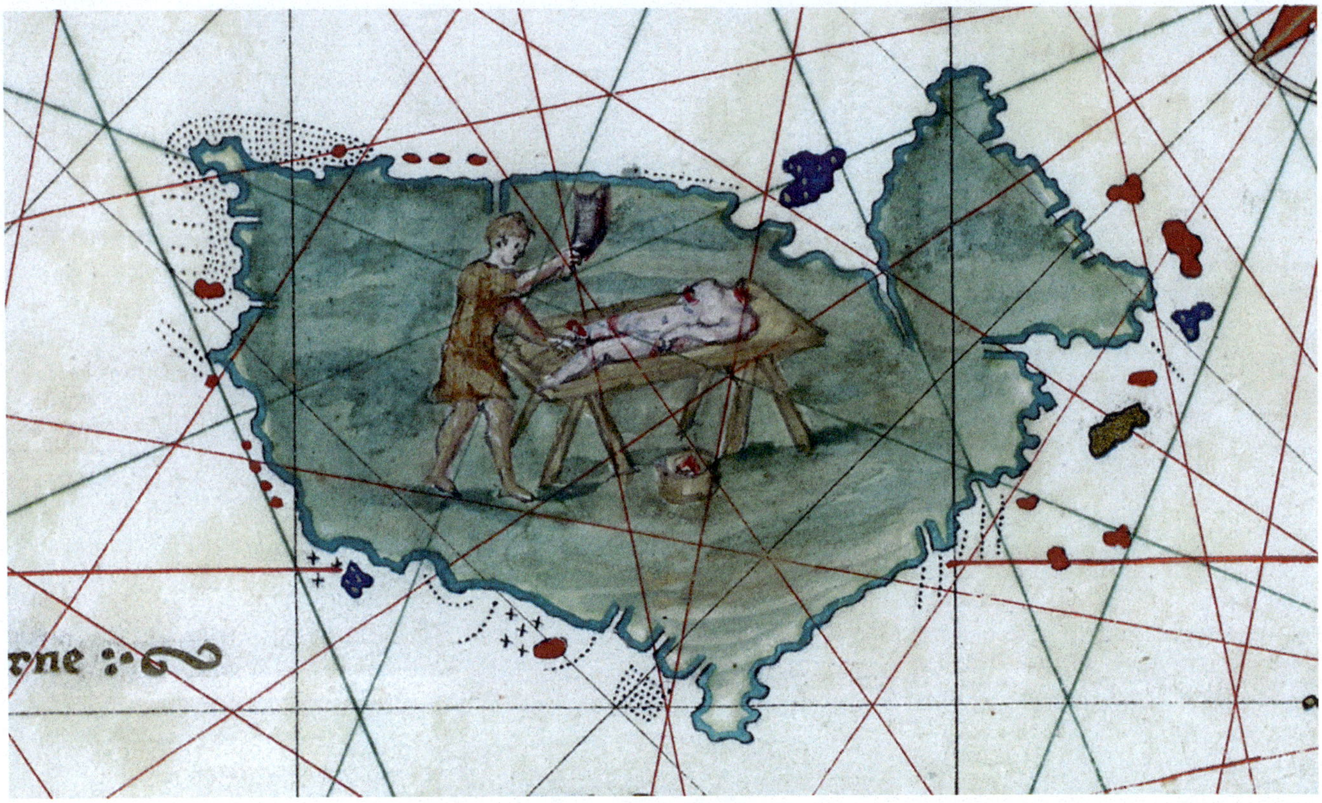

Fig. 1.33 Cannibal butchery on *lille de geans* or the Island of Giants from the Vallard Atlas. San Marino, California, Huntington Library, MS HM 29, f. 3. Courtesy of the Huntington Library

beneath them (Fig. 1.35).[210] Again, it is inconceivable that the cartographer of the Vallard Atlas could have arrived at an image so similar to Waldseemüller's if he were only working from Fries's map.

There is additional corroboratory evidence in Northern Europe that the maker of the Vallard Atlas used of Waldseemüller's *Carta marina* as a source. As we saw earlier, Waldseemüller portrays the walrus as a creature that looks very much like an elephant (see Fig. 1.10). Lorenz Fries copies this image in the 1522 edition of Ptolemy in the *Tabula moderna Gronlandie et Rusie* and also in his version of the *Carta marina* (1530, 1531), and there is a similar image in the Vallard Atlas in the map of Western Europe and the Mediterranean (ff. 7v–8r). But again, only Waldseemüller's map can have been the source. The walrus in the 1522 Ptolemy has a long elephantine trunk, which the image on the Vallard Atlas does not have; the image on Fries's *Carta marina* is much more similar to that on the Vallard Atlas, but it shows the elephant's far ear sticking up above the elephant's head to some extent, which is the case on Waldseemüller's map, but not on Fries's.

There is also good evidence that Waldseemüller's *Carta marina* was available to the sixteenth-century Norman cartographer Pierre Desceliers. Elsewhere I have argued that Desceliers copied his image of the former Hindu practice of *suttee* or *sati* (in which a widow threw herself on her husband's funeral pyre) on his 1546 world map from Waldseemüller's *Carta marina*, rather than from the *Tabula moderna Indiae* in Fries' 1522 Ptolemy or from Fries' edition of the *Carta marina*.[211] It also seems very likely that Desceliers copied his image of Jean-François de La Rocque de Roberval's 1542 settlement in

[210] There is no similar scene of cannibalism in the Indian Ocean in the 1513 edition of Ptolemy's *Geography* that might have influenced the cartographer of the Vallard Atlas, and the scene of cannibalistic butchery in the *Tabula moder[na] Indie Orientalis* in the 1522 (Strasbourg), 1525 (Strasbourg), 1535 (Lyon), and 1541 (Vienna) editions of Ptolemy's *Geography*, whose maps were made by Fries, is quite different, and certainly did not influence the cartographer of the Vallard Atlas. For descriptions of these editions of Ptolemy see Wilberforce Eames, *A List of Editions of Ptolemy's Geography 1475–1730* (New York, 1886) (reprinted from Joseph Sabin's *Bibliotheca Americana*), pp. 15–17, 17–18, 18–19, and 20–21; and Carlos Sanz, *La Geographia de Ptolomeo, ampliada con los primeros mapas impresos de América, desde 1507* (Madrid: Librería General V. Suárez, 1959), pp. 150–155, 156–164, 169–178, and 187–189.

[211] Desceliers' 1546 map is in Manchester, John Rylands Library, French MS 1*; for discussion of the image of *sati* see Chet Van Duzer, *The World for a King: Pierre Desceliers' Map of 1550* (London: British Library, 2015), pp. 43 and 46.

Fig. 1.34 The cannibalistic scene on the island of Java on Laurent Fries's *Carta marina* of 1530. Bayerische Staatsbibliothek, Mapp. I,9m-1. Courtesy of the Bayerische Staatsbibliothek

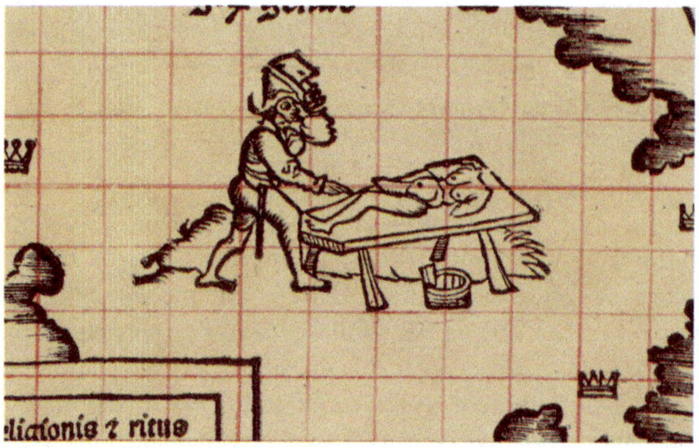

Fig. 1.35 The cannibalistic scene on the island of Java on Waldseemüller's *Carta marina* (sheet 12). Courtesy of the Library of Congress

Canada from Waldseemüller's image of Mecca,[212] discussed just above (see Fig. 1.17), but it is at least possible that Desceliers copied his image from that of Mecca in the Vallard Atlas (see Fig. 1.31), which was made three years before his 1550 map.

There is one other very clear piece of evidence confirming that Waldseemüller's *Carta marina* was disseminated: Abraham Ortelius cites the map as a source in his *Theatrum orbis terrarum* of 1570. In his *Catalogus auctorum tabularum geographicarum* (Catalog of mapmakers), on p. Cii, Ortelius writes:

> Martinus Ilacomilus Friburgensis, Europam; eam alicubi in Germania impressam habemus.
>
> Martinus Waldseemuller, Universalem navigatoriam (quam Marinam vulgo appellant) in Germania editam. Puto hunc eundem esse cum Ilacomilo praedicto.
>
> Martin Ilacomylus of Freigburg, a map of Europe; we have a copy printed somewhere in Germany.[213]

---

[212] Van Duzer, *The World for a King* (see note 211), pp. 10–11.

[213] The reference is to Waldseemüller's *Carta itineraria Europae*, first published in 1511; for bibliography on the map see note 14 above.

Martin Waldseemüller, Universal nautical chart (which is commonly called a marine chart) published in Germany. I think that this cartographer is the same as the Ilacomylus just mentioned.

This passage is the clearest possible corroboration that the *Carta marina* did in fact circulate and influence other cartographers, and not just in the Dieppe region.

However, some other claims that Waldseemüller's *Carta marina* influenced later globes and maps, and thus implicitly must have been well diffused, cannot be accepted as proven, largely because of the difficulty presented in distinguishing between the influence of Waldseemüller's map and Fries's. It has been asserted that Gerard Mercator used Waldseemüller's *Carta marina* as a source in his depiction of southern Africa on his terrestrial globe of 1541,[214] but without any attempt to determine whether the influence was from Waldseemüller's *Carta marina* or from Fries's.[215] In fact the place names in southern Africa are very similar on Waldseemüller's map and Fries's,[216] so the similarities between Mercator's globe and Waldseemüller's map cannot be taken as providing additional evidence that the 1516 map was disseminated.

Fischer and von Wieser, in their introduction to their facsimile edition of Waldseemüller's 1507 and 1516 maps, indicate that Gerard Mercator borrowed from Waldseemüller's *Carta marina* in creating his famous 1569 world map,[217] particularly in his legends in India and the topography and hydrography of southern Africa.[218] But they do not provide details, and in fact after examining the legends on the 1516 and 1569 maps,[219] I find a close correspondence in only one place, in the legends describing the opossum in South America—and Mercator could have borrowed that legend from the 1531 (Latin) edition of Fries's *Carta marina*, where it is the same as on Waldseemüller's *Carta marina*. With regard to southern Africa, as just mentioned, the place names in this region are very similar on Waldseemüller's map and Fries's, so the very similar place names on Mercator's 1569 map and Waldseemüller's *Carta marina* do not establish that Waldseemüller's map reached Mercator.[220]

Another aspect of the diffusion of Waldseemüller's *Carta marina* is its copying by Lorenz Fries in his editions of 1525, 1530, and 1531, which have been mentioned several times now. It is unlikely that Fries made use of Johann Schöner's exemplar of the map—the only one that now surives—as the model for his maps, so his editions suggest the existence of at least one other exemplar of Waldseemüller's *Carta marina*. As Fries's maps are smaller than Waldseemüller's, they contain fewer legends and illustrations; they are also of a substantially lower artistic quality than Waldseemüller's map,[221] and in fact introduce a number of errors. But as the historian Henry Bruman has noted, the Fries-Grüninger *Carta marina* had a rather different aim than Waldseemüller's map. Waldseemüller aimed to bring the latest geographical scholarship to a broad audience, while the Fries-Grüninger map was "an object of popularization and commerce, merchandised to a wide public," designed "to disseminate reasonably recent, reasonably accurate information about different parts of the world in a picturesque, decorative way."[222] On the 1530 edition of the map, and presumably in the lost 1525 edition as well, most of the legends are translated into German—an effort at the democratization of cartographic knowledge that is a logical step forward from Waldseemüller's own *Carta marina*, itself a democratization of an expensive manuscript nautical chart. But in the 1531

---

[214]On Mercator's terrestrial globe see Edward L. Stevenson, *Terrestrial and Celestial Globes* (New Haven, CT: Yale University Press, 1921), vol. 1, pp. 124–135; and Elly Dekker and Peter van der Krogt, "Les Globes," in Marcel Watelet, ed., *Gérard Mercator cosmographe: le temps et l'espace* (Anvers: Fonds Mercator Paribas, 1994), pp. 243–267, esp. 246–258.

[215]See Josef Fischer, "Die Hauptquelle für die Darstellung Afrikas auf dem Globus Mercators von 1541," *Mitteilungen der Geographischen Gesellschaft Wien* 87.4-6 (1944), pp. 65–69; and W. G. L. Randles, "South East Africa and the Empire of Monomotapa as Shown on Selected Printed Maps of the 16th Century," *Studia* 2 (1958), pp. 103–163, esp. 150.

[216]See Petrzilka, *Die Karten des Laurent Fries* (see note 202), pp. 54–55.

[217]For the three surviving exemplars of Mercator's 1569 map see note 196. The third of these is hand-colored and has been reproduced in facsimile as *Gerard Mercator's Map of the World (1569) in the Form of an Atlas in the Maritiem Museum 'Prins Hendrik' at Rotterdam* (Rotterdam, 1961); and more recently as *Atlas van der Wereld: De wereldkaart va Gerard Mercator uit 1569* (Zutphen: Walberg Pers, 2011). There is a good, easily accessible reproduction of the whole map in Kenneth Nebenzahl, *Atlas of Columbus and the Great Discoveries* (Chicago: Rand McNally, 1990), pp. 128–129.

[218]See Joseph Fischer and Franz Ritter von Wieser, *Die älteste Karte mit dem Namen Amerika* (see note 9), p. 40. Johnson is also skeptical of Fischer and von Wieser's claim: see her *Carta marina* (see note 21), p. 58, and the notes on p. 136.

[219]The legends on Mercator's map are transcribed and translated into English in Gerard Mercator, "Text and Translation of the Legends of the Original Chart of the World by Gerhard Mercator, Issued in 1569," *Hydrographic Review* 9.2 (1932), pp. 7–45.

[220]Incidentally there is good evidence that Waldseemüller's edition of Ptolemy, either the 1513 or the 1520 reprint, reached Mercator: in Mercator's edition of Ptolemy, *Tabulae Geographicae* (Cologne: Typis Godefridi Kempensis, 1578), there is a ship in the Mediterranean on the map of the Holy Land positioned very much as the ship on the corresponding map in Waldseemüller's edition, and there is no ship on this map in any other edition of Ptolemy before Mercator's.

[221]Johnson, *Carta marina* (see note 21), p. 66, claims that "Fries's world map is altogether artistically superior [to Waldseemüller's]," but most of Fries's images on his map are plainly inferior to Waldseemüller's.

[222]Henry J. Bruman, "The Schaffhausen Carta Marina of 1531," *Imago Mundi* 41 (1989), pp. 124–132, p. 130.

edition, the legends are in Latin again, so either the edition in German was not well received, or the publishers wanted to sell a new version of the map to scholars.

Waldseemüller's *Carta marina* also influenced another early sixteenth-century cartographer, a topic that has been little discussed by map historians, and this influence explains some features of the surviving copy of the *Carta marina*. As mentioned above, it was Johann Schöner who preserved the only surviving copies of Waldseemüller's 1507 and 1516 maps. Schöner's printed globe of 1515 was heavily influenced by Waldseemüller's 1507 map,[223] and his magnificent but largely unstudied manuscript globe of 1520[224] borrows a number of legends from the *Carta marina*.[225]

The copy of the *Carta marina* that Schöner preserved has a number of corrections made by hand, in accordance with the list of corrections that was printed on the lower of two escutcheons in the southwest corner of the map, but is now covered by a small piece of paper (Legend 9.2).[226] Thus Schöner seems to have taken care that the map was as correct as possible. The grid of red parallels and meridians drawn on much of the map bespeaks careful study and analysis of the map's geography, no doubt by Schöner himself.[227] Moreover, stored in the Schöner Sammelband together with the twelve printed sheets of the *Carta marina* was a careful manuscript copy that Schöner made of sheet 6 of the map (here labeled sheet 6A), which covers western Africa. The existence of this manuscript copy has not been previously explained, but we can be quite certain that it was made as part of Schöner's preparations for using data from the *Carta marina* on his 1520 globe.[228] This is confirmed by a difference between the printed sheet 6 and manuscript sheet 6A: on the printed sheet no legend appears in the Gulf of Guinea, but on the manuscript sheet there is a legend describing the islands in the São Tomé group (see Legend 6A.1)—and there is a similar legend in the same location on Schöner's 1520 globe.

We have few records that tell us much about Martin Waldseemüller, but his two multi-sheet world maps shed important light on his character. The maps are products of a cartographer with a great creative vision, and a great ambition to disseminate the latest cartographic knowledge to scholars throughout Europe. The fact that he and his colleagues were able to gather in the small town of Saint-Dié the diverse array of cartographic and geographic information necessary to produce these maps—rare manuscript maps, codices of Ptolemy's *Geography*, recent travel narratives, images of exotic animals—testifies to a remarkable drive and experience in research. Waldseemüller's willingness to cast aside all of the work that had gone into his 1507 map, and to create less than a decade later a new world map based on a new cartographic philosophy and almost entirely new sources, demonstrates a wonderful open-mindedness, energy, and thirst for knowledge. His *Carta marina* represents the culmination of more than a decade of thought about how the world should be mapped, and much painstaking research into the latest texts and images that could be used to create a rich and detailed image of the earth as it was known and traveled by human beings.

---

[223]See Chet Van Duzer, *Johann Schöner's Globe of 1515: Transcription and Study* (Philadelphia: American Philosophical Society, 2010).
[224]Schöner's globe of 1520 is in Nürnberg, Germanisches Nationalmuseum, WI 1, and is 87 cm in diameter. A tracing of the east Asian, American, and Atlantic portions of the globe may be found in Konrad Kretschmer, *Die Entdeckung Amerika's in ihrer Bedeutung für die Geschichte des Weltbildes* (Berlin: W. H. Kühl, 1892), plate 13; and in Franz Ritter von Wieser, *Magalhães-Strasse und Austral-Continent auf den Globen des Johannes Schöner* (Innsbruck: Verlag der Wagner'schen Universitaets-Buchhandlung, 1881; Amsterdam: Meridian Publishing, 1967), plate 1. A brief discussion of the globe, with transcription of some legends, is in Friedrich Wilhelm Ghillany, *Der Erdglobus der Martin Behaim vom Jahre 1492 und der des Johann Schöner vom Jahre 1520* (Nürnberg: Druck der Campeschen Officin, 1842), pp. 13–18. See also Konrad Kretschmer, "Der Globus Johannes Schöner's vom Jahre 1520," in *Beiträge zur alten Geschichte und Geographie: Festschrift für Heinrich Kiepert* (Berlin: D. Reimer [Ernst Vohsen], 1898), pp. 113–23; and Norbert Holst, *Mundus, Mirabilia, Mentalität: Weltbild und Quellen des Kartographen Johannes Schöner; eine Spurensuche* (Frankfurt [Oder]: Scrîpvaz-Verlag, 1999), pp. 57–63.
[225]Holst in his *Mundus, Mirabilia, Mentalität* (see note 224) alludes to Schöner's borrowings from the *Carta marina* on Schöner's 1520 globe. I hope to explore these borrowings in more detail in the future, but can confirm that several legends in Africa on the 1520 globe come from the *Carta marina*.
[226]These corrections are listed by Joseph Fischer and Franz Ritter von Wieser, *Die älteste Karte mit dem Namen Amerika* (see note 9), pp. 20–21; and John W. Hessler, "Correcting Waldseemüller: Analysis of Hyperspectral Images of the Pastedown Shield on the 1516 Carta Marina," written in 2008 and available at http://www.kislakfoundation.org/download/Hessler-Analysis_of_Hyperspectral_Images.pdf.
[227]R. A. Skelton, in his "Bibliographical Note" in Ptolemy, *Geographia, Strassburg, 1513*, p. xx, says that the red lines were drawn on the sheets of the map before the sheets were printed. But Elizabeth Harris, in "The Waldseemüller World Map: A Typographic Appraisal," *Imago Mundi* 37 (1985), pp. 30–53, at 31, as well as multispectral images made by the Library of Congress of both the 1507 and 1516 maps, confirm that, as one would expect, the grid of red lines was drawn after the sheets had been printed.
[228]In fact, Schöner copied by hand a brief study that discusses the process of transferring cartographic data from maps to globes. This text is in Vienna, Österreichische Nationalbibliothek, MS Vind. 3505, ff. 124v–125r, and is transcribed by Dana Bennett Durand, *The Vienna-Klosterneuburg Map Corpus of the Fifteenth Century: A Study in the Transition from Medieval to Modern Science* (Leiden: E. J. Brill, 1952), pp. 364–367, with discussion on pp. 163–164.

2 The Long Legends: Transcription, Translation, and Commentary

The long legends on the *Carta marina* will be addressed sheet by sheet, and within each sheet, from left to right and top to bottom. The legends are numbered with a two-number system, the first indicating the sheet, and the second the number of the legend on that sheet. The commentary aims to identify the sources of the legends whenever possible, and remarks will also be offered on images associated with the legends when necessary.

The Electronic Supplementary Material for this book, available via www.springer.com, include high-resolution images of each sheet of the *Carta marina* and a high-resolution image of the whole map. These will allow the reader to zoom in and better see the details being discussed. The ESM also includes an index PDF of the whole map that indicates with a number the location of each of the long texts on the map that are transcribed, translated, and studied in the sections that follow. This PDF is searchable, so that if the reader is having difficulty determining where exactly on the map Legend 8.7 is located, a search in the PDF for "8.7" will find it.

Electronic supplementary material
The online version of this chapter (https://doi.org/10.1007/978-3-030-22703-6_2) contains supplementary material, which is available to authorized users.

C. Van Duzer, *Martin Waldseemüller's 'Carta marina' of 1516*,
https://doi.org/10.1007/978-3-030-22703-6_2

2.1 Sheet 1. North America, Caribbean, North Atlantic (Plate 2.1)

The toponyms on this sheet are transcribed by Petrzilka, *Die Karten des Laurent Fries*, pp. 42–43; they are transcribed and compared with those on Caverio's chart in Stevenson, *Marine World Chart*, pp. 84–85. As discussed above in the introduction, in his depiction of the New World on the *Carta marina*, Waldseemüller changes from the Vespuccian conception that he depicted on his 1507 map to a Columbian conception, particularly in regard to the absence of the name "America" on the 1516 map and in the indication here on sheet 1 that the newly discovered lands are part of Asia. The island at the eastern edge of the sheet represents Newfoundland; on the 1507 map, and also on Caverio's chart, a Portuguese flag indicates that it is under the control of that country, but here the flag is Spanish. This is a curious mistake on Waldseemüller's part, as the legend that describes the island (see Legend 2.1), clearly states that it was discovered for Portugal. It is likely that Waldseemüller's style of depicting the waters of the oceans in the 1507 map, by a dense and uniform covering of closely-spaced lines running parallel to the lines of latitude, was abandoned in the 1513 Ptolemy and the *Carta marina* in favor of shading near the coastlines and in patches in the open ocean, no doubt in an effort to use less ink, and also perhaps out of a desire to make the map more amenable to hand-coloring.

1.1
TERRA DE CVBA · ASIE PARTIS

The land of Cuba, part of Asia.

Plate 2.1 Sheet 1 of the *Carta marina*: North America, Caribbean, North Atlantic. Courtesy of the Library of Congress

On his 1507 world map Waldseemüller famously depicted the New World as separated from Asia by water, in accordance with Vespucci's suggestion that it was an island, but on his *Carta marina* he reverts to a Columbian conception of the New World as part of Asia.[1] There is an ample bibliography tracing the history of the idea that the newly discovered lands in the West were connected with Asia, and the cartographic expression of that idea, so there is no need to review the matter in more detail here.[2]

[1]For discussion of the background of Waldseemüller's depiction of the "Terra de Cuba," see Donald L. McGuirk Jr., "The Depiction of Cuba on the Ruysch World Map," *Terrae Incognitae* 20 (1988), pp. 89–97, esp. 97.

[2]On Columbus's belief that his discoveries were part of Asia see George E. Nunn, *The Geographical Conceptions of Columbus: A Critical Consideration of Four Problems* (New York: American Geographical Society, 1924); expanded edition with an essay titled "The Test of Time" by Clinton R. Edwards (Milwaukee: American Geographical Society Collection of the Golda Meir Library, University of Wisconsin-Milwaukee; and New York: American Geographical Society, 1992); E. G. R. Taylor, "Idée Fixe: The Mind of Christopher Columbus," *The Hispanic American Historical Review* 11.3 (1931), pp. 289–301; and Folker Reichert, "Columbus und Marco Polo—Asien in Amerika. Zur Literaturgeschichte der Entdeckungen," *Zeitschrift für historische Forschung* 15 (1988), pp. 1–63. Also see Johann Georg Kohl, "Asia and America: An Historical Disquisition Concerning the Ideas which Former Geographers had about the Geographical Relation and Connection of the Old and New World," *Proceedings of the American Antiquarian Society*, n.s. 21 (1911), pp. 284–338; J. H. Parry, "Asia-in-the-West," *Terrae Incognitae* 8.1 (1976), pp. 59–72; Errol Wayne Stevens, "The Asian-American Connection: The Rise and Fall of a Cartographic Idea," *Terrae Incognitae* 21.1 (1989), pp. 27–39; and Marica Milanesi, "Arsarot oder Anian? Identitat und Unterscheidung zwischen Asien und der Neuen Welt in der Kartographie des 16. Jahrhunderts (1500–1570)," in Adriano Prosperi and Wolfgang Reinhard, eds., *Die Neue Welt im Bewusstsein der Italiener und Deutschen des 16. Jahrhunderts* (Berlin: Duncker & Humblot, 1993), pp. 15–68.

2.2 Sheet 2. Newfoundland and Europe (Plate 2.2)

On this sheet Waldseemüller retained and expanded his use of one of the decorative motifs on the relatively spare 1507 map, namely displaying the coats of arms of the various countries—and he had had practice with this motif on his map of Europe of 1511, which survives in one copy of a printing of 1520, whose border is filled with coats of arms[3]; and on his map of the Duchy of Lorraine in the 1513 edition of Ptolemy.[4] He had room to increase his use of this motif on the *Carta marina* because this map, though the same physical size as the 1507 map, shows less of the earth's surface, so that it presents a "zoomed in" view of the world, with more room for geographical detail, text, and images. In terms of geography, in the *Carta marina* Waldseemüller has moved well beyond his 1507 map. He retains some features of the modern world map in the 1513 Ptolemy (see Fig. 1.27), such as the large peninsula of Scandinavia curving down into the North Atlantic (which derives from Martellus, see Fig. 1.28), but in other respects he has innovated even with respect to the 1513 Ptolemy. For example, the shape of Spain is substantially different on the *Carta marina* than it is in the modern map of Spain in the 1513 Ptolemy, and various other examples might be adduced. On the 1507 map the Azores have a Portuguese flag; on the modern map of Spain in the 1513 Ptolemy, the islands have wandered north to what was the position of the Cassiterides on the 1507 map, while on the *Carta marina* they are back in their more or less correct position, and now have a Spanish flag. The islands have a Portuguese flag on both the Cantino and Caverio charts, and it is not clear on what authority Waldseemüller departed from this tradition. The place name *Islanda* on an island near the northern edge of the sheet was not printed by Waldseemüller, but was written there, probably by Schöner—in accordance with the list of corrections in Legend 9.2.

> 2.1
> Hec Terra Corterati inuenta est ex mandato regis Portugallie per Casparum Corterati capitaneum duorum nauium anno domini .1501. quam ob sui magnitudinem litoris plusquam 600. milliarium protendentis firmam esse opinabatur. habet hec pluralitatem magnorum fluminum et enim populosa gens ~~que~~ huius. habet domos ex maximis lignis constructis. quarum tectona ex coriamentis piscium compacta. vestes eorum sunt de pellibus ferarum quarum in estate pilos ab extra, in hieme vero ab intra vertentes portant. Sunt signati facie: tamquam indi. carent ferro et loco illius: lapideis instrumentis vutuntur. Magnam habent copiam lignorum de genere pini etiam multos pisces Salmones Aleca and Strumulos.
>
> This land of the Corte-Reals was discovered by order of the King of Portugal by Gaspar Corte-Real, the captain of two ships, in the year 1501. Because of the size of its shore, which extends more than six hundred miles, it was thought to be part of a continent. It has many large rivers and its people are numerous. They have houses made from very large logs, whose roofs are made from the skins of fish. Their clothes are the furs of wild beasts, which in the summer they wear with the fur on the outside, and in the winter with the fur turned inward. Their faces are painted, like those of people in India. They have no iron, and in its place they use stone tools. They have a great abundance of pine wood and many fish, including salmon, herring, and stockfish.

There is no corresponding legend on the Caverio chart or on Waldseemüller's 1507 map, but there is one on the earlier Cantino chart, albeit shorter than the one on the *Carta marina*.[5] The source of Waldseemüller's legend is a letter from Pietro Pasqualigo to his brothers, dated October 19, 1501, which was published *Paesi novamente retrovati*, Book 6, Chap. 126.[6]

---

[3]On the 1513 edition of Ptolemy's *Geography* see note 172 in Chap. 1 above.

[4]For discussion of Waldseemüller's map of the Duchy of Lorraine see Jean-Marie Gérardin, "1508–2008: A propos de la première carte imprimée du duché de Lorraine et du Vastum Regnum," *L'Annuaire de la Société du Val de Villé* 33 (2008), pp. 57–77.

[5]The legend on the Cantino chart is transcribed and translated by Armando Cortesão and Avelino Teixeira da Mota, *Portugaliae monumenta cartographica* (Lisbon: Comissão Executiva das Comemorações do Quinto Centenário da Morte do Infante D. Henrique, 1960–62), vol. 1, p. 11: *Esta terra he descoberta per mandado do muy alto exçelentissimo principe Rey dom manuell Rey de portuguall a qual descobrio gaspar de corte Real cavalleiro na cassa do dito Rey, o quall quãdo a descobrio mandou hũ naujo com çertos omes e molheres que achou na dita terra e elle ficou com outro náujo e nũca mais veo e crese que he perdido e aquj ha muitos mastos*; "This land is discovered by order of the very high, most excellent prince King Dom Manuel, King of Portugal, which was discovered by Gaspar de Corte Real, a knight in the house of the said King, and when he discovered it he send a ship with certain men and women whom he found in said land, and he remained with another ship and never more returned, and it is believed that he is lost, and there are here many masts."

[6]The letter from Pietro Pasqualigo is reprinted in Henry Harrisse, *Les Corte-Real et leurs voyages au Nouveau-monde d'après des documents nouveaux ou peu connus* (Paris: E. Leroux, 1883), pp. 211–212; and is translated into English in Arthur James Weise, *The Discoveries of America to the Year 1525* (New York: G. P. Putnam's Sons, 1884), pp. 209–211, and better by James A. Williamson, ed., *The Voyages of the Cabots and the English Discovery of North America under Henry VII and Henry VIII* (London: The Argonaut Press, 1929), pp. 40–41; Williamson's translation is reprinted in David B. Quinn, ed., *America from Concept to Discovery: Early Exploration of North America* (New York: Arno Press, 1979), vol. 1, pp. 149–151.

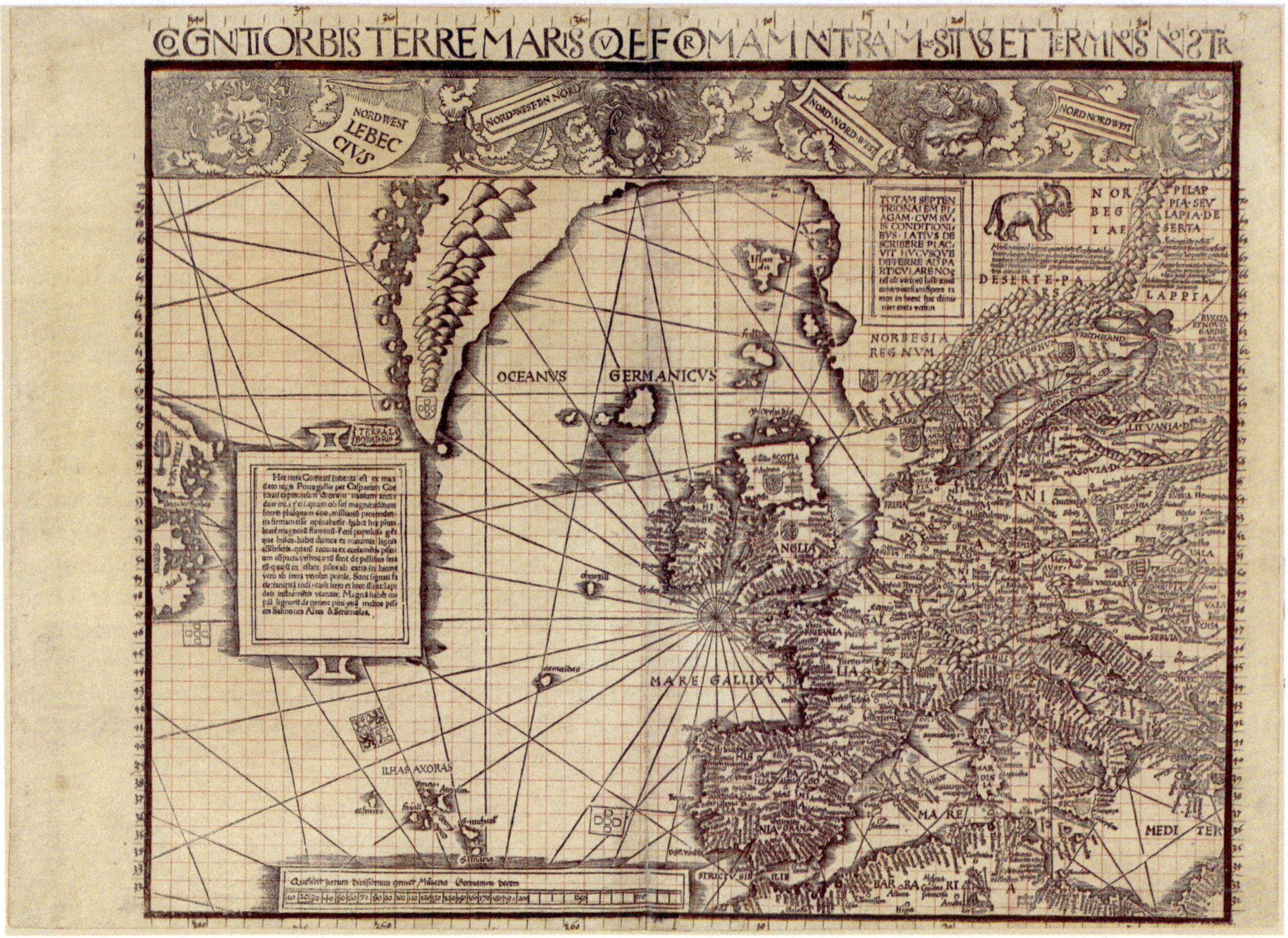

Plate 2.2 Sheet 2 of the *Carta marina*: Newfoundland and Europe. Courtesy of the Library of Congress

The island is thought to represent Labrador and Newfoundland.[7] The prominent trees on the island on both the Cantino map (where the legend does not mention trees) and the Caverio map (where there is no legend) indicate that the depiction on the Cantino map was made by a cartographer who had access to Pietro Pasqualigo's letter, which mentions the trees. Waldseemüller used the past tense, the land *was* thought to be [*terram*] *firmam*, that is, part of a continent,[8] to contrast with his depiction of it as an island. Again, Waldseemüller's placing of a Spanish flag on the island (visible on sheet 1) is a strange mistake, as his legend says that Gaspar Corte-Real was sailing for the king of Portugal; there is a Portuguese flag just south of the island, but it is not clear what land it is supposed to be marking.

2.2
TOTAM SEPTENTRIONALEM PLAGAM, CUM SUIS CONDITIONIBUS, LATIUS DESCRIBERE PLACUIT HUCUSQUE DIFFERRE AD PARTICULARE NOStrum ob variorum lustratorum controuersiam. Spero tamen in breui hec eliminare juxta verum

[7]For discussion of the cartographic history of this island see Heinrich Winter, "The Pseudo-Labrador and the Oblique Meridian," *Imago Mundi* 2 (1937), pp. 61–73.

[8]On the meaning of *terre ferme* in the sixteenth century see Wilcomb E. Washburn, "The Meaning of 'Discovery' in the Fifteenth and Sixteenth Centuries," *The American Historical Review* 68.1 (1962), pp. 1–21.

It has seemed good to postpone fully depicting the entire northern region and its characteristics till I can make a special map because of the controversies of various explorers. However, in a short time I hope to eliminate these [controversies] according to the truth.

This text was transcribed but not translated by Fischer and Björnbo[9]; Fischer rightly remarks that Waldseemüller apparently never did draw a new map of the northern regions, but Waldseemüller's remark here indicates both his interest in the area and continuing ambition to create more maps.

2.3
Morsus animal ingens quantitate Elephantis huius dentes longos duos et quadrangulares carensque iuncturis in pedibus. reperitur in promontoriis septentrionalibus Norbegie incedit gregatim agmine ducentorum animalium.

The walrus is a huge animal, the size of an elephant, and it has two long teeth which are quadrangular, and lacks joints in its legs. It is found in the northern promontories of Norway, and they travel together in groups of two hundred animals.

This legend was discussed above in the introduction; it certainly indicates Waldseemüller's access to a recent source from Scandinavia, but we do not know what that source was: it was not any of the books that he lists in the long text block on sheet 9 of the *Carta marina* (see Legend 9.3). In 1519, a few years after the *Carta marina* was printed, the Norwegian archbishop Erik Walkendorf sent Pope Leo X the salted head of a walrus, and it was displayed in the city hall in Strasbourg; a sketch of a whole walrus was made based on this head in 1519, and in 1521 Albrecht Dürer sketched the head of a walrus using an unknown model.[10] Despite the availability—at least in some circles—of these quite accurate renderings of walruses, Waldseemüller's elephant-like depiction was copied on a number of later maps, including the 1522 (Strasbourg), 1525 (Strasbourg), 1535 (Lyon), and 1541 (Vienna) editions of Ptolemy's *Geography*, Lorenz Fries's *Carta marina*, Olaus Magnus's *Carta marina* of 1539,[11] the Vallard Atlas of c. 1547,[12] and Pierre Desceliers's world map of 1550.[13]

2.4
Regiones iste pellium preciosorum feracissime sunt que portantur ad partes occidental[es].

These regions produce many valuable furs, which are transported to the west.

This legend represents one of the few pieces of information that Waldseemüller retained from his 1507 map, where there is a legend that reads:

hinc portantur pelles ferarum ad partes occidentales christianorum et qui habitant has regiones que longa est .10. dietarum habent regem de stirpe magni cham sunt idolatre et aliqui adorant natigas.

From here the pelts of wild beasts are carried to Christian regions in the West, and those who live in this region, which is a ten day's journey long, have a king who descends from the Great Khan. They are idolaters, and some worship Natigas [a Mongolian divinity].

---

[9]See Joseph Fischer, *The Discoveries of the Norsemen in America, with special Relation to their Early Cartographical Representation*, trans. Basil H. Soulsby (London: H. Stevens, Son & Stiles; St. Louis: B. Herder, 1903), p. 90; and Axel Anton Björnbo, *Cartographia Groenlandica* (Copenhagen: I kommission hos C. A. Reitzel, 1912) (=*Meddelelser om Grønland*, 48), p. 200.

[10]On the 1519 and 1521 sketches of the walrus see Valentin Kiparsky, "L'Histoire du morse," *Annales Academiae Scientiarum Fennicae*, Ser. B, 73.3 (1952), pp. 1–53, at 46–47; and Kirsten A. Seaver, "'A Very Common and Usuall Trade': The Relationship between Cartographic Perceptions and 'Fishing' in the Davis Strait circa 1500–1550," *British Library Journal* 22.1 (1996), pp. 1–26, esp. 6–15. The Dürer sketch is in the British Museum, Prints & Drawings, registration number SL,5261.167.

[11]Olaus Magnus's *Carta marina* survives in two copies: Munich, Bayerische Staatsbibliothek 12 Mapp VII, and Uppsala University Library (no shelfmark). The map has been reproduced in facsimile as *Olai Magni Gothi Carta marina et descriptio septemtrionalium terrarum ac mirabilium rerum in eis contentarum* (Malmö: In officina J. Kroon, 1949), and as Olaus Magnus, *Die Wunder des Nordens*, ed. Elena Balzamo and Reinhard Kaiser (Frankfurt am Main: Eichborn, 2006). For discussion of the map see Edward Lynam, *The Carta marina of Olaus Magnus, Venice 1539 & Rome 1572* (Jenkintown, PA: Tall Tree Library, 1949).

[12]On the Vallard Atlas see note 207 in Chap. 1 above.

[13]Pierre Desceliers's map of 1550 is in London, British Library, Add. MS 24065, and is conveniently reproduced in Nebenzahl, *Atlas of Columbus* (see note 217 in Chap. 1 above), pp. 114–115; and in twelve sheets in black-and-white as Map 3 in James Ludovic Lindsay Crawford and Charles Henry Coote, *Autotype Facsimiles of Three Mappemondes* (Aberdeen: Aberdeen University Press, 1898). The map is reproduced at full size and in color, in sections, with detailed discussion, in Chet Van Duzer, *The World for a King: Pierre Desceliers' Map of 1550* (London: British Library, 2015).

Waldseemüller took this information from one of the most important sources for his 1507 map, a large world map by Henricus Martellus very similar to that now at Yale, where there is a damaged legend in the same position that addresses the same subject.[14] However, I have not been able to find the source of this material on Martellus's map: it does not come from Marco Polo or John of Plano Carpini, for example.[15] The long-standing fur trade from Russia to Flanders had been under the control of the Hanseatic League, but around 1480 Russia began to win back trading rights in the West.[16] There is a legend about the flow of furs from Russia to Flanders on a nautical chart by Ottomano Freducci made in 1529 (British Library, Add. MS 11548),[17] but unfortunately it does not shed light on the possible sources of Waldseemüller's or Martellus's legend. Compare Legend 3.6 below.

> 2.5
> Regio hec mutua ditione Magni principis Russie et regis datie obtemperatur. Habitatores sunt homines inculti habentes facies adinstar Simearum
>
> This region is under the joint control of the great prince of Russia and the king of Denmark. The inhabitants are wild and have faces similar to those of apes.

This legend relates to Lappia, i.e. Lapland. I have not been able to find the source of either of the two parts of this legend. The 1326 Treaty of Novgorod specified which of the Sami in Finmark would pay tribute to Novgorod, and which to Norway. Following the union of Norway and Denmark in 1380, Norway's rights went to the King of Denmark, and with the Grand Duchy of Moscow's annexation of Novgorod in 1478, Novgorod's right to collect tribute from the Sami went to Moscow,[18] but I do not know a source earlier than 1516 that describes the situation in terms similar to Waldseemüller's.[19] With regard to the comparison with apes, Paolo Giovio, in his *Libellus de legatione Basilii Magni Principis Moschouiae ad*

---

[14]See Chet Van Duzer, *Henricus Martellus's World Map at Yale (c. 1491): Multispectral Imaging, Sources, Influence* (New York: Springer, 2018), p. 89.

[15]A lost mid fifteenth-century chart mentioned earlier had a legend about the fur trade from Russia to Flanders, so this information did exist in the nautical chart tradition, though there is no way to be certain that Martellus obtained the information from a nautical chart. For the legend on the fifteenth-century chart see Jacques Paviot, "Une mappemonde génoise disparue de la fin du XIVe siècle," in Gaston Duchet-Suchaux, ed., *L'Iconographie: études sur les rapports entre textes et images dans l'Occident médiéval* (Paris: Le Léopard d'Or, 2001), pp. 69–97, legend 52, pp. 83 (Latin) and 93 (French). The legend runs: *Item naues sunt Alamanorum, que huc veniunt, videlicet in Rosia. Et onerantur pelipariis, cera et aliis mercibus. Et ipsas in Fiandriam conducunt*, "These are German ships, which come here, in Russia. They are loaded with furs, wax and other products, which they bring to Flanders."

[16]The stalls of the Church of St. Nicolas in Stralsund, Germany, used to have fourteenth- or fifteenth-century wooden sculptures that depicted Russian hunters providing German Hanseatic traders with furs: see Janusz Sztetyłło, "Wokół snycerki ze Stralsundu (Strzałowa) zdobiącej stalle hanzeatów kupczących w Nowogrodzie" ["Regarding the Wooden Sculpture of Stralsund (Strzałów): The Case of the Stalls of the Hanseatic Merchants Dealing in Novgorod"], in Marta Młynarska-Kaletynowa and Jerzy Kruppé, eds., *O Rzeczach minionych: Scripta rerum historicarum Annae Rutkowska-Płachcińska oblata* (Warsaw: Instytut Arcehologii i Etnologii Polskiej Akademii Nauk, 2006), pp. 319–327. On Russia's efforts to win back trading rights from the Hanseatic League beginning in 1480 see Ju. A. Tihonov, "Feodaalisen Venäjän ulkomaankauppapolitiikka (ennen 1600-lukua)" ["The Foreign Trade Policy of Feudal Russia Before 1600"], *Turun Historiallinen Arkisto* 36 (1982), pp. 87–105. Unfortunately the standard work on the medieval Russian fur trade, Janet Martin, *Treasure of the Land of Darkness: The Fur Trade and its Significance for Medieval Russia* (Cambridge: Cambridge University Press, 1986), is not of help in connection with Waldseemüller's legend.

[17]See Van Duzer, "Nautical Charts, Texts, and Transmission" (see note 138 above in Chap. 1), p. 53.

[18]The Latin text of the 1326 treaty together with a Russian translation are provided by Sigizmund Natanovich Valk, *Gramoty Velikogo Novgoroda i Pskova* (Moscow and Leningrad: Izd-vo Akademii nauk SSSR, 1949), pp. 69–70. For discussion see I. P. Shaskol'skii, "Dogovory Novgoroda s Norvegiei," *Istoricheskie Zapiski* 14 (1945), pp. 38–61; and the same author's "Russko-norvezhskii dogovor 1326 goda," *Skandinavskii Sbornik* 15 (1970), pp. 63–71.

[19]For example, Hartmann Schedel in his *Liber chronicarum* of 1493, ff. 282v–283r, gives a brief account of the history of northern Europe, but does not mention the division of Lappia between Denmark and Russia.

Clemente. VII, Pont. Max., first published in Rome in 1525, says that beyond the Lapps in the darkness of the far north there are pygmies who are a fearful race of men, who speak by chattering and seem to be as close to apes as they are far from human beings of normal height in terms of their stature and intelligence.[20] Incidentally, this passage from Giovio is borrowed by Baron Sigismund von Herberstein in his *Notes upon Russia*, first published in 1549.[21] But it is not at all clear that Giovio and Waldseemüller were using the same source. Waldseemüller had access to an unknown source for this legend and a few others on sheet 3 (see Legends 3.1, 3.2, and 3.3) that relate to Russia and northwestern Asia; I have consulted all of the candidate sources that I can find, but I have not discovered the source that Waldseemüller was using.[22] Some fifty years ago Leo Bagrow briefly examined the geography of Waldseemüller's depiction of Russia on the *Carta marina*, focusing on the fact that the toponyms indicate that the data had been supplied in a series of itineraries to Moscow. Unfortunately he ignored the descriptive texts, thus omitting from his analysis an essential element of the map.[23]

[20]See Paolo Giovio, *Pavli Iovii Novocomensis Libellvs de legatione Basilij Magni principis Moschouiæ ad Clementem VII. pontificem max.* (Basel: [Johann Froben], 1527), p. 17: *...meticulosum genus hominum & garritu sermonem exprimens, adeo ut tam Simiae propinqui quam statura ac sensibus ab iustae proceritatis homine remoti uideantur.* This passage is translated into English in Baron Sigismund von Herberstein, *Notes upon Russia: Being a Translation of the Earliest Account of that Country, entitled 'Rerum Moscoviticarum commentarii'*, ed. and trans. R. H. Major (London: Printed for the Hakluyt Society, 1851–52), vol. 2, p. 239. For discussion of the tradition of ape-like men in the far north see Leonid S. Chekin, *Northern Eurasia in Medieval Cartography: Inventory, Texts, Translation, and Commentary* (Turnhout: Brepols, 2006), pp. 255–256.

[21]See Baron Sigismund von Herberstein, *Notes upon Russia: Being a Translation of the Earliest Account of that Country, entitled 'Rerum Moscoviticarum commentarii'*, ed. and trans. R. H. Major (London: Printed for the Hakluyt Society, 1851–52), vol. 2, p. 239: "Some men of great credite and aucthoritie, do testifie that in a region beyond the Lapones, betwene the West and the North, oppressed with perpetuall darknesse, is the nation of the people called Pigmei, who being growen to theyr ful grought, do scarcely excede the stature of our chyldren of ten yeeres of age. It is a fearefull kynde of men, and expresse theyr wordes in suche chatteryng sort, that they seeme to be so muche the more lyke vnto Apes, in howe muche they dyffer in sence and stature from men of iust heyght."

[22]There is no evidence that Waldseemüller had access to the work translated in this article: Robert M. Crosky and E. C. Ronquist, "George Trakhaniot's Description of Russia in 1486," *Russian History/ Histoire Russe* 17 (1990), pp. 55–64; nor that he made use of Albert Krantz's *Wandalia*, which, though not published until 1517, was finished by 1504; nor that he made use of Christian Bomhover's *Eynne schonne hystorie van vnderlyken gescheffthen der heren tho lyfflanth myth den Rüssen unde tataren*, published in Christian Bomhover, "Eynne Schonne hystorie...," ed. G. Schirren, *Archiv für die Geschichte Liv-, Est-, Curlands* 8 (1861), pp. 113–265. For bibliography on early accounts of Russia see Friedrich von Adelung, *Kritisch-literärische Übersicht der Reisenden in Russland bis 1700, deren Berichte bekannt sind* (St. Petersburg: Eggers, 1846; Amsterdam: N. Israel, 1960); Marshall Poe, *Foreign Descriptions of Muscovy: An Analytic Bibliography of Primary and Secondary Sources* (Columbus, OH: Slavica Publishers, 1995); and Marshall Poe, "Terra Incognita: The Earliest European Descriptions of Muscovy," in *'A People Born to Slavery': Russia in Early Modern European Ethnography, 1476–1748* (Ithaca and London: Cornell University Press, 2000), pp. 11–38.

[23]Leo Bagrow, "At the Sources of the Cartography of Russia," *Imago Mundi* 16 (1962), pp. 33–48, esp. 34–36; much of this account is repeated in Leo Bagrow, *A History of the Cartography of Russia up to 1600*, ed. Henry W. Castner (Wolfe Island, Ontario: Walker Press, 1975), pp. 44–48. Bagrow's failure to address the long legends led him to the incorrect conclusion that "Waldseemüller in fact provides us with new and interesting material only in respect of Muscovy proper, circumstantially describing the routes within it," in "At the Sources" p. 36, and *A History of the Cartography of Russia*, p. 48. I have published a discussion of the legends relating to Russia on the *Carta marina* in Chet Van Duzer, "Тексты о России на «Морской карте» Мартина Вальдзеемюллера 1516 года," *Istoricheskaia geografiia* 2 (2014), pp. 236–267.

2.3 Sheet 3. Northern Asia (Plate 2.3)

This sheet is dense with legends, reflecting a strong interest in Asia. Along the top edge of the sheet some monstrous races are enclosed by mountains; Waldseemüller retained the mountainous enclosures that on his 1507 map contain the *iudei clausi*, or Enclosed Jews, which on other maps with the same configuration of mountains are referred to as Gog and Magog,[24] and here he places entirely different races inside. There is a system of borderlines indicated on this sheet that does not immediately appear to the eye: it runs from the bit of the OCEANVS SEPTEN[TRIONALIS] that appears in the upper left-hand part of the sheet, down along the mountain range that Johann Schöner has labeled the Hyperborean and Riphaean mountains, perhaps by analogy with a range on the 1507 map, to the *Mare Maior* or Black Sea. From the eastern shore of the Black Sea it runs first to the south, and then weaves its way east across the sheet. The line is not labeled on this sheet, but on sheet 4 the line is explicitly said to separate India from Tartaria to the north, and on sheet 3 it also marks the limits of the Tartars' empire: most of the sovereigns to the east and north of the line are said to be under the control of the Great Khan. The nature of the northern part of the border is indicated by the series of crosses just to the west of it: here it separates Christians from heathens. But the area between the Black and Caspian Seas which is north of the borderline, and thus under Tartar control, is dense with crosses, the density no doubt an indication of concern about the Christians under Tartar control.

3.1
apud istum Laram [i.e. Lacum] reperiuntur Ursi albi

By this lake are found white bears.

I have not found the source of Waldseemüller's *Lacus Albus* or of the polar bears on its shores.[25] The *Lacus Albus* refers to *Beloe ozero* or White Lake: this identification is confirmed by the presence of the city *Kargopolis* or Kargopol north of the lake, and of *Beloser* or Beloozero on the east shore of the lake (the modern city is on the southern shore). Thus this legend is one of a few relating to this region (see also Legends 2.5, 3.2, and 3.3) that quite probably come from a single unknown source. There is a brief legend about polar bears on Waldseemüller's 1507 map further to the northeast, on the shore of the northern ocean, but it is not clear that there is any connection between the two legends.

3.2
Hic dominator Magnus princeps et Imperator Russie et Moscovie podolie ac plescovie rex

Here the ruler is the great prince and Emperor of Russia and Moscow, the king of Polodia and Plescovia.

This legend is right above an image of the Emperor, who is above an image of Moscow, which is where we are to understand that he resides. The use of the title "Emperor" is politically significant, as the use of this title by Princes of Moscow was controverted at this period.

The image of the Emperor, and also of the city of Moscow below (which is located on the *mosca fl.*) emphasize the Christianity of both: the Emperor wears a crown that looks rather like a bishop's miter, and is topped by a cross[26]; near his outstretched hand there is a staff topped by a cross, and the building that represents Moscow is topped by a cross.

The Moskva River is a tributary of the Oka which is a tributary of the Volga, and Waldseemüller shows the Moskva as a tributary of the Volga, so his representation is reasonably accurate. The coat of arms to the left of the Emperor, which has a lion rampant and a dragon rampant, combatant (see Fig. 2.1), is problematic. First, it is not clear whether it is intended to be the coat of arms of Novgorod, the Emperor, or Moscow; and second, although the sixteenth-century coat of arms of Moscow does involve a dragon or basilisk, it depicts a man on horseback slaying the creature below—nothing like a rampant dragon,

---

[24]For discussion see Andrew Runni Anderson, *Alexander's Gate, Gog and Magog, and the Inclosed Nations* (Cambridge, MA: The Medieval Academy of America, 1932); Andrew Gow, "Gog and Magog on Mappaemundi and Early Printed World Maps: Orientalizing Ethnography in the Apocalyptic Tradition," *Journal of Early Modern History* 2.1 (1998), pp. 61–88; and Scott D. Westrem, "Against Gog and Magog," in Sylvia Tomasch and Sealy Gilles, eds., *Text and Territory: Geographical Imagination in the European Middle Ages* (Philadelphia: University of Pennsylvania Press, 1998), pp. 54–75.

[25]I would like to offer my enthusiastic thanks to Leonid Chekin for remarks to me on the identification of this lake and on various points in my discussion of this and the following two legends.

[26]Joseph Fischer in "Der russische Zar als 'Kaiser' auf der Carta Marina Waldseemüllers vom Jahre 1516," *Stimmen der Zeit* 90 (1916), pp. 108–116, and "Die Entdeckung Russland durch Nikolaus Poppel in dem Jahren 1486–1489," *Stimmen der Zeit* 89 (1915), pp. 395–400, argued that the representation of the Emperor on the *Carta marina* indicated that Waldseemüller had made use of information from Niclas von Popplau (c. 1443–1490). But as Bagrow notes in "At the Sources" (see note 23), p. 34, and *A History of the Cartography of Russia* (see note 23), p. 48, there is no unique detail that connects Waldseemüller's image of Russia with von Popplau's account.

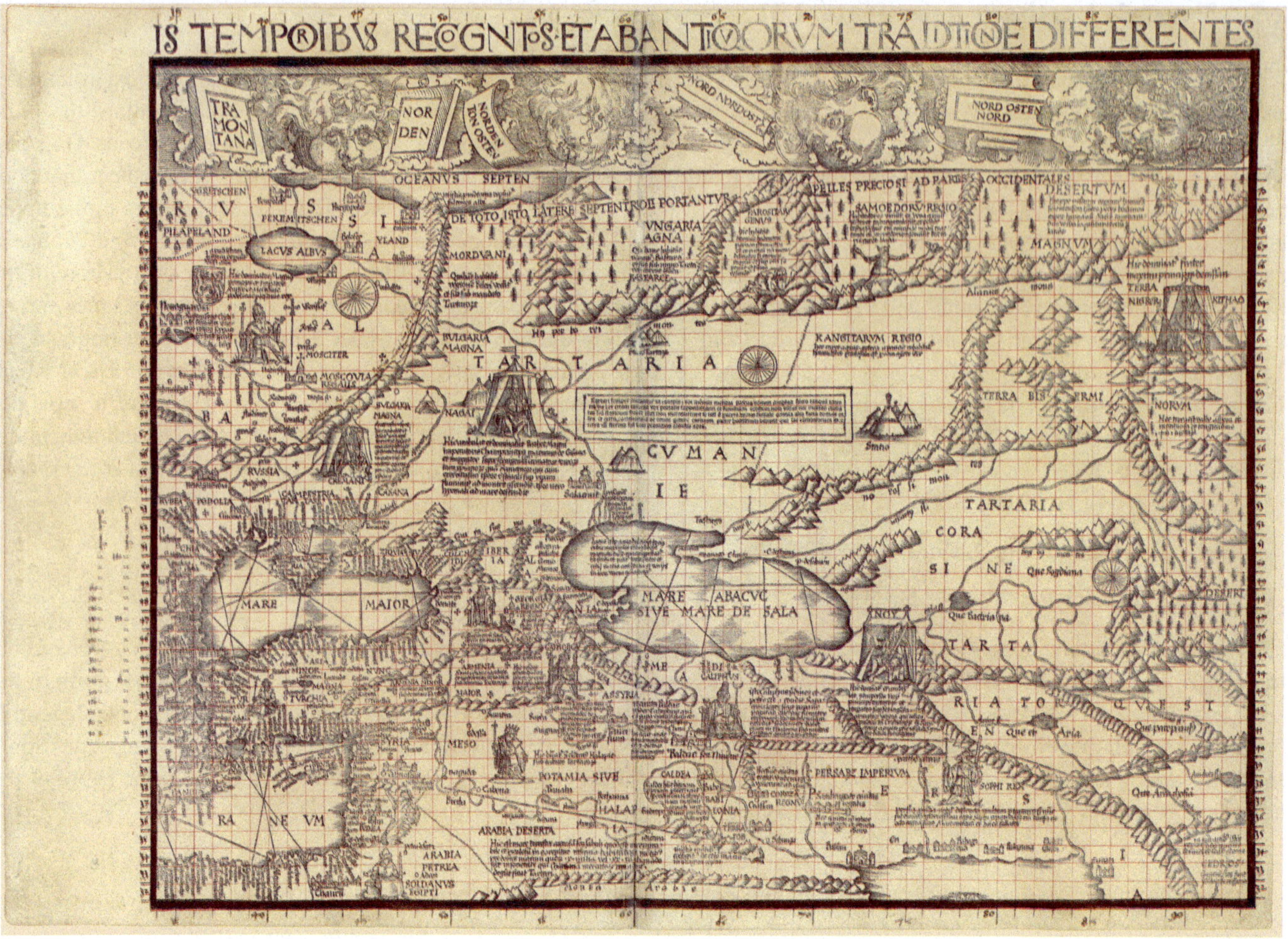

Plate 2.3 Sheet 3 of the *Carta marina*: Northern Asia. Courtesy of the Library of Congress

and the lion was not a part of the heraldic imagery of Novgorod, Moscow, or Russia at that time. Plescovia refers to the principality of Pskov (Pleskau), which fell to Vassily III and the Muscovite army in 1510, so Waldseemüller's source here is very recent, but I have not been able to determine what it is. Podolia is mentioned in Hartmann Schedel's *Liber chronicarum* of 1493 as the part of Poland closest to Russia,[27] but not as part of the Russian dominions—and indeed it was not part of Russia until 1793, so there is an error here by either Waldseemüller or the source he was using.[28] Podolia and Plescovia (Pleskovia) are mentioned in Maciej z Miechowa's *Tractatus de duabus Sarmatiis*, but this book was first published in Krakow in 1517, a year after the *Carta marina*,[29] so it does not seem that Waldseemüller could have used it as a source.

[27]See Schedel's *Liber chronicarum*, in the section "De Sarmacia regione Europe," in the chapter "De regno polonie et eius initio." For an English translation of this passage see Hartmann Schedel, *Sarmatia, the Early Polish Kingdom: From the Original Nuremberg Chronicle, Printed by Anton Koberger in 1493*, trans. Bogdan Deresiewicz (Los Angeles: Plantin Press, 1976), p. 17: "A few years earlier the Polish kingdom had been very large but wars with the perfidious nations Tartars and Turks brought great losses to all of Sarmatia. Thus a province bordering upon Russia, which the people called Podolia, was totally burned down and devastated so that it could not provide travelers with food, even though it was a most fertile land where the grass grew higher than a tall man."

[28]Christophe von Werdt suggests that Waldseemüller took this legend from a verbal (pre-publication) report of the voyage of Sigismund von Herberstein in his "Wo thront der Zar? Die Entdeckung Russlands im äussersten Norden auf den frühesten. Kartenbildem in der Sammlung *Rossica Europeana*," in Jörn Happel, ed., *Osteuropa kartiert - Mapping Eastern Europe* (Berlin: Lit, 2010), pp. 113–131, at 125.

[29]Podolia is mentioned in the *Tractatus* in Book 1, Tractate 1, Chap. 3, and Book 2, Tractate 1, Chap. 1; while Plescovia is mentioned in Book 2, Tractate 1, Chaps. 2 and 3. In the edition Maciej z Miechowa, *Traktat o dvuch Sarmatijach*, ed. S. A. Anninskij (Moskow: Izd. Akad. Nauk SSSR, 1936), in the Latin text Podolia is mentioned on pp. 131 and 174, and Pleskovia is mentioned on pp. 181–182, with a more detailed description on pp. 184–185. For discussion of Maciej z Miechowa's book see Konstanty Zantuan, "The Discovery of Modern Russia: *Tractatus de duabus Sarmatiis*," *Russian Review* 27.3 (1968), pp. 327–337; and W. Voisé, "The First Intellectual Exploration of Eastern Europe (Mathias de Miechovia: *Tractatus de duabus Sarmatiis*… Cracow 1517)," *Janus* 64 (1977), pp. 41–49.

Fig. 2.1 Detail of sheet 3 of Waldseemüller's *Carta marina* showing Russia and the regions to the north of it. Courtesy of the Library of Congress

At the western (left) edge of Sheet 3 of the *Carta marina* there is an image of the city *Nouoguardia* (Novgorod) on a lake (Ilmen?) through which flows the *Volga fl.*, with the legend:

3.3

Ruteni sunt Cristiani cismatici Grecorum ritum seruantes habent specialem literam.

The Rutheni [i.e. Russians] are schismatic Christians who follow the Greek Rite, and they have their own alphabet.

Waldseemüller knew something about Novgorod when he made his map of Europe in 1511, for although the city does not appear on that map, it is mentioned in the pamphlet that was written by Ringmann to accompany it, his *Instructio manuductionem prestans in cartam itinerariam Martini Hilacomili* (Strasbourg: Grüninger, 1511), f. 12r: *Russia ciuitatem permaximam Nogardiam appellatam: ad quam mercatores Theutonici magno labore perueniunt. huius regionis populi appellantur Rutheni: quos Strabo Crossones videtur nominare*, "The largest city of Russia is called Novgorod, to which German merchants come with great difficulty. The people of this region are called Rutheni, whom Strabo seems to call Crossones."[30] John of Plano Carpini does mention that the Rutheni have their own alphabet,[31] but does not explicitly mention that the Rutheni follow the Greek Rite. One might infer from Plano Carpini's text that the Rutheni follow the Greek Rite, but the fact that he does not state this explicitly makes it seem likely that Waldseemüller was using another source.

On the *Carta marina* there are two Novgorods, one at 63° N on a lake through which the Volga flows (the one just mentioned), and the other at 54° N on the Dnieper.[32] The northern of these two cities is Velikii Novgorod (58°33′ N, 31°17′ E), and the more southern city is Novgorod-Siversky (51°59′ N, 33°16′ E): the name of the city should be understood as being divided by the name of the Dnieper (*neper fl.*), and thus reads *nouoguardia seuerski*.[33] Waldseemüller shows the city on an unnamed tributary of the Dnieper, and Novgorod-Siversky is on the Desna, which is a tributary of the Dnieper, but Waldseemüller's tributary flows into the Dnieper from the northwest, while the Desna flows in from the northeast.[34] A good understanding of how Waldseemüller arrived at his perception of the region's geography is elusive.

The sources whose phrasing is closest to that in Waldseemüller's legend is in Maciej z Miechowa's *Tractatus de duabus Sarmatiis*, Book 2, Tractate 1, Chap. 1: *Ruteni habitu et ecclesiasticis officiis Graecos insequuntur habentque proprias litteras et abecedarium instar et proximum Graecis*, "The Rutheni follow the Greeks in both their clothing and their ecclesiastical offices, and they have their own language and an alphabet similar to that of the Greeks." The *Tractatus* was published a year after the *Carta marina*, so it is not clear what to make of this similarity: perhaps Waldseemüller and Miechowa made use of the same source.

It should be mentioned that while it seems that Waldseemüller's source contained some of the same information as Miechowa's *Tractatus de duabus Sarmatiis*, it does not seem that Waldseemüller had access to a manuscript of Miechowa's work or anything of that nature, as Miechowa locates Novgorod at 66° N (*Habet elevationem poli Nowygrod sexaginta sex graduum*),[35] whereas Waldseemüller has it somewhat further to the south.

3.4

Hyperborei and Riphes montes non sunt in rerum natura

The Hyperborean and Riphaean mountains do not exist.

This is a manuscript addition to the *Carta marina*, no doubt made by Johann Schöner, but it is of sufficient interest to list among the legends on the map. This denial of the mountains' existence, written right along a chain that it is reasonable to identify with the Riphaean mountains based on Waldseemüller's 1507 map and the *Tabula moderna Sarmatie* in the 1513 Ptolemy, comes almost verbatim from Maciej Miechowita's *Tractatus de duabus Sarmatiis*, Book 1, Tractate 2, Chap. 5: *Accipe quarto, quod montes Riphei et Hyperborei non sunt in rerum natura*, "And take this fourth [conclusion], that the

[30]Ringmann's source here would seem to be Schedel's *Liber chronicarum* of 1493, f. 280v, "De Russia": *Rutheni quos appellant crossanos vt Strabo videtur lituanis contermini sunt… In hac gente civitatem permaximam [esse] tradunt Nogardiam appellatam. ad quam theutonici mercatores magno labore perveniunt*, or else Schedel's source, which was Aeneas Silvius Piccolomini's *In Europam*. For discussion of this passage in Schedel see Leonid Chekin, "Открытие арктического острова русскими мореплавателями эпохи Колумба: Сводный анализ источников," *Вопросы истории естествознания и техники* 3 (2004), pp. 3–42, esp. 11–13. Yet Waldseemüller was using a different source on his *Carta marina*, as Schedel does not mention the use of the Greek Rite or a special alphabet.

[31]See Giovanni da Pian di Carpine, *Storia dei Mongoli*, ed. Enrico Menestò and trans. Maria Cristiana Lungarotti (Spoleto: Centro italiano di studi sull'alto Medioevo, 1989), pp. 325 (Latin) and 395 (Italian); and Christopher Dawson, ed., *The Mongol Mission: Narratives and Letters of the Franciscan Missionaries in Mongolia and China in the Thirteenth and Fourteenth Centuries* (New York: Sheed and Ward, 1955), p. 56.

[32]On Waldseemüller's map titled *Tabula moderna Sarmatie* in the 1513 edition of Ptolemy's *Geography* the city also appears twice, near the upper edge of the map at about 56° N (*nouogrado*), and about 53° N (*nouogradectus*). But neither of these cities is on a river, so it is clear that Waldseemüller was using a new source of information in making this part of his *Carta marina*.

[33]Petrzilka reads these two toponyms separately in *Die Karten des Laurent Fries* (see note 202 in Chap. 1 above), pp. 99 (*Seuerf.*) and 87 (*nouoguardia*).

[34]I thank Mitia Frumin for his advice regarding the geography of this region.

[35]Maciej z Miechowa, *Tractatus de duabus Sarmatiis Asiana et Europiana de contintis in eis* (Cracow: Ioannis Haller, 1517), Book 2, Tractate 1, Chap. 3. In the edition Maciej z Miechowa, *Traktat o dvuch Sarmatijach*, ed. S. A. Anninskij (Moskow: Izd. Akad. Nauk SSSR, 1936), the Latin text of this passage is on p. 184.

Riphaean and Hyperborean mountains do not exist."[36] This is a very interesting and previously unknown datapoint in the reception of Maciej Miechowita's work,[37] and also indicates that Schöner had access to Miechowita's book, though in his *Opusculum geographicum* (Nürnberg: [Joannes Petreius], 1533), Chap. 5, he still refers to the *Riphei montes*. This may indicate that Schöner only saw Miechowita's work—and learned that the Riphaean Mountains did not exist—later in his life, and made this annotation to the *Carta marina* sometime between 1533 and his death in 1547.

3.5
in istis promontoriis reperiuntur falcones albi

In these promontories white falcons are found.

This legend is at the northern tip of what is apparently the Riphaean mountain range, where it reaches the northern ocean. A number of nautical charts have legends about falcons that are found in the north, but legends about white falcons are less common. There is a legend to this effect on the Borgia metal *mappamundi* from the first half of the fifteenth century: in the north the text reads *Hic sunt ursi et falcones albi et consimilia*, "Here there are white falcons and bears and similar creatures,"[38] and there is a related legend on the lost mid fifteenth-century nautical chart mentioned earlier: the legend speaks of "aues albi" in Norway.[39] Albertus Magnus speaks of white falcons coming from the far north,[40] and Pero López de Ayala in his late fourteenth-century manual of falconry says that white falcons usually come from Norway.[41] But given Waldseemüller's reference to promontories, the likely source would seem to be an interpolated version of Claudius Clavus's description of the north, which is preserved in Vienna, ÖNB, Cod. Vindob. lat. 5277, ff. 271r-276r, which contains the statement *Liste promontorium, ubi capiuntur falcones albi*, "The Liste promontory [unidentified], where white falcons are captured."[42] This text was circulating in Germany and was used by Johann Schöner in his *Luculentissima quaedam terrae totius descriptio* (Nürnberg: Johannes Stuchs, 1515), and Franciscus Irenicus (Franz Friedlieb) in his *Germaniae exegeseos volumina duodecim* (Hagenoae: typis Thomae Anshelmi... sumptibus Ioannis Kobergii, 1518), and in fact Irenicus repeats the passage about the white falcons on Liste promontorium.[43]

3.6
DE TOTO ISTO LATERE SEPTENTRIO[NA]LI PORTANTVR PELLES PRECIOSI AD PARTES OCCIDENTALES

From this whole northern coast valuable furs are exported to the West.

See above on Legend 2.4: this material derives from Waldseemüller's 1507 map, and Waldseemüller took it from a large world map by Henricus Martellus, but I have not been able to find Martellus's source.

---

[36]On Miechowita's denial of the existence of the Riphaean mountains see Leo Bagrow, *A History of the Cartography of Russia up to 1600*, ed. Henry W. Castner (Wolfe Island, Ontario: Walker Press, 1975), p. 48; and Marshall Poe, "Muscovy in European Cosmographies, 1504–1544," *Russian History* 25.1–2 (1998), pp. 89–106, at p. 91. Incidentally Paulo Giovio also emphatically denies the existence of these mountains, not only in the text of his *Libellus de legatione Basilii Magni Principis Moschouiae ad Clementem VII*, but even in the latter part of the title, which runs: ... *Caeterum ostenditur error Strabonis, Ptolemaei, aliorumque Geographiae scriptorum, ubi de Rypheis motibus meminêre, quos hac aetate nusquam esse, plane compertum est*. The passage in the book proper, which is on p. 22 in the 1527 edition, runs: …*nulli omnino montes ea in regione multa etiam hominum peregrinatione reperiantur, ita ut Rypheos montes & Hyperboreos toties ab antiquis celebratos plerique Cosmographiae ueteris studiosi penitus fabulosos esse arbitrentur*.

[37]On the reception of Miechowita's book see Henryk Barycz, "Kilka glos do znajomości Macieja z Miechowa w XVI i XVII wieku" ["On the Knowledge of Maciej of Miechów in the Sixteenth and Seventeenth Centuries"], *Kwartalnik Historii Nauki i Techniki* 8 (1963), pp. 237–243.

[38]A. E. Nordenskiöld, "Om ett aftryck från XV: de seklet af den i metall graverade världskarta, som förvarats i kardinal Stephan Borgias museum i Velletri, Med 1 facsimile," *Ymer* 11 (1891), pp. 83–92, at p. 88.

[39]The manuscript is in Genoa, Biblioteca Universitaria, MS B. I. 36. The legends in the manuscript are transcribed and translated into French by Jacques Paviot, "Une mappemonde génoise disparue de la fin du XIVe siècle," in Gaston Duchet-Suchaux, ed., *L'Iconographie: études sur les rapports entre textes et images dans l'Occident médiéval* (Paris: Le Léopard d'Or, 2001), pp. 69–97. Paviot indicates that the lost map is from the late fourteenth century, but as one of its legends cites Antoniotto Usodimare, it must be from 1455 or later. The legend cited here is Paviot's no. 5, on his pp. 79 (Latin) and 89 (French).

[40]See Albertus Magnus, *De animalibus libri XXVI, nach der Cölner Urschrift*, ed. Hermann Stadler (Münster: Aschendorff, 1916), Book 23, Chap. 11, vol. 2, p. 1465; and Albertus Magnus, *On Animals: A Medieval Summa Zoologica*, trans. Kenneth F. Kitchell, Jr. and Irven Michael Resnick (Baltimore: Johns Hopkins University Press, 1999), vol. 2, pp. 1586–1587.

[41]See Pero López de Ayala, *Libro de la caza de las aves*, ed. José Fradejas Lebrero (Madrid: Castalia, 1969), Chap. 4, pp. 71–72.

[42]See Axel Anthon Björnbo and Carl S. Petersen, *Der Däne Claudius Claussøn Swart (Claudius Clavus): der älteste Kartograph des Nordens, der erste Ptolemäus-Epigon der Renaissance: eine monographie* (Innsbruck: Wagner, 1909), pp. 68, 81, and 176.

[43]On Schöner and Irenicus's use of Clavus see Anthon Björnbo and Carl S. Petersen, *Der Däne Claudius Claussøn Swart*, Chap. 4, "Die Ulmer Ausgaben, Schöner und Friedlieb," pp. 48–63; and for Irenicus's quotation of the material about the white falcons see his *Germaniae exegeseos*, Book 10, Chap. 18, f. 200r (a mistake for f. 198r): *Haud procul Lisce promontorium Noruegiae, ubi candidi falcones capiuntur, cuius dimensio 35, 62, 10 aestimatur*. I thank Kirsten Seaver for her advice on these subjects.

3.7
Qui hanc habitant regionem Bileri vocantur et sunt sub mandato Tartarorum

Those who inhabit this region are called Bileri [Bulgari] and are under the control of the Tartars.

This legend derives from John of Plano Carpini, who says that the Bileri were conquered by the Tartars.[44] Beazley explains the identity of the Bileri: "The Byleri or Bulgaria Magna are the Old or Black Bulgarians of Bolghar and the Volga below Kazan, at its junction with the Kama; they are called Bilar[s] in several Moslem geographers and historians, e.g. Abulfeda and Rashid-ed-din."[45] The place name *Bulgaria Magna* appears just to the south of this legend, and further to the southwest that place name is repeated with the following legend about this region.

3.8
BVLGARIA MAGNA A nepro flu[mine] usque huc ambulat et dominatur princeps tartarorum de Crema campestrie et Gazarie dominus estque imperator super armatorum 600000 pro custodia contra cristianos.

Great Bulgaria. From the Dneper to this point the prince of the Crema Tartars of the fields and the lord of Gazaria lives as a nomad and rules; he is the commander of 600,000 armed men who defend against the Christians.

The Cremani are the inhabitants of the Crimean Khanate. Much of this legend,[46] which is about the Mongol chief Corenza, comes from John of Plano Carpini, who explains in a bit more detail than Waldseemüller does that this king is charged with keeping watch on the Christians to the west, to prevent them from making an unexpected attack on the Tartars.[47] Plano Carpini, via Vincent of Beauvais, *Speculum historiale*, Book 32, Chap. 20, at least in the Strasbourg, 1473 edition of the text, lists the number of soldiers under this king as 60,000 rather than 600,000. But Waldseemüller was apparently convinced that 600,000 was the correct number, as he repeated it in the list of corrections in Legend 9.2. Also, Waldseemüller adds new information from an unknown source to what he found in Plano Carpini, for he extends Corenza's realm south to Gazaria, and gives him titles similar to those of the emperor of Moscow.[48]

3.9
Hic ambulat et dominatur Bathot Magni imperatoris Cham princeps maximus de Casana et imperator super sexingenta [*sic*] .m. armatorum virorum tam christianorum quam Saracenorum qui cum exercitu suo tempore Estiuali super ripam fluminis ad montes ascendit, tempero uero hyemali ad mare descendit.

Here Batu lives as a nomad and rules, the greatest prince of Casana [and descendant] of the great emperor Chan, and commander of sixty thousand soldiers, both Christians and Saracens, who with his army ascends over the riverbank to the mountains in the summer, but in the winter descends to the sea.

Batu or Baatu, better known today as Batu Khan of the Golden Horde, was son of Jochi, and thus a grandson of Genghis Khan. He conquered Russia and was Khan of that territory from 1227 until his death in 1255, and led the main contingent of the Mongol attack on Eastern Europe in 1241.[49] Most of Waldseemüller's legend comes from John of Plano Carpini,[50] but Waldseemüller goes somewhat further than Plano Carpini in indicating that Batu is ruler of Kazan, and again, I do not know the source of this additional information. Just above the legend is a large but generic image of Batu seated in his tent.

[44]See Giovanni da Pian di Carpine, *Storia dei Mongoli* (see note 31), pp. 272 and 298 (Latin) and 361 and 371 (Italian); Dawson, *The Mongol Mission* (see note 31), pp. 30 and 41; and Beazley, *The Texts and Versions of John de Plano Carpini* (see note 43), p. 122.
[45]Beazley, *The Texts and Versions of John de Plano Carpini* (see note 43 in Chap. 1 above), p. 285.
[46]This legend is transcribed and translated by András Róna-Tas, *Chuvash Studies* (Budapest: Akadémiai Kiadó; Wiesbaden: Harrassowitz, 1982), p. 194, but the translation here is my own.
[47]See Giovanni da Pian di Carpine, *Storia dei Mongoli* (see note 31), pp. 307 and 309 (Latin) and 383 and 384 (Italian); Dawson, *The Mongol Mission* (see note 31), pp. 54 and 55; and Beazley, *The Texts and Versions of John de Plano Carpini* (see note 43 in Chap. 1 above), pp. 129 and 130.
[48]I thank Leonid Chekin for shedding light on this legend.
[49]See for example Harold T. Cheshire, "The Great Tartar Invasion of Europe," *The Slavonic Review* 5.13 (1926), pp. 89–105; James Chambers, *The Devil's Horsemen: The Mongol Invasion of Europe* (New York: Atheneum, 1979), pp. 70–82 on Batu's invasion of Russia, and esp. 101–113 on his invasion of Hungary; Robert Marshall, *Storm from the East: From Ghenghis Khan to Khubilai Khan* (Berkeley: University of California Press, 1993), pp. 104–108 and 129–141; and Colmon Sodnomyn and Enchcimag Cenoma, "Podbój Polski przez wojska Batu" ["The Conquest of Poland by the Army of Batu"], in Wacław Korta, ed., *Bitwa legnicka: Historia i tradycja* (Wrocław: Volumen, 1994), pp. 84–89, with a German abstract pp. 88–89.
[50]See Giovanni da Pian di Carpine, *Storia dei Mongoli* (see note 31), pp. 309 (Latin) and 384 (Italian); and Dawson, *The Mongol Mission* (see note 31), p. 55.

3.10
GEORGIA REGNUM in eo sunt .18. episcopatus et sunt cristiani cismatici

The Kingdom of Georgia. In it there are eighteen bishoprics and they are schismatic Christians.

I do not see that Plano Carpini says that the people of Georgia are schismatics: Benedict the Pole, who traveled to the Mongol Empire 1245–1247, mentions that they follow the Greek rite,[51] but there is no other evidence that Waldseemüller made use of Benedict's work, and it may have simply been common knowledge that the Georgians were not adherents of the same branch of Christianity as the countries of Western Europe. The information about the eighteen bishoprics comes from Simon of Saint-Quentin.[52] As indicated above in the introduction, Waldseemüller in his long text block in the lower left corner of the map (see Legend 9.3) cites the traveler, Friar Ascelinus, rather than the author, Simon of Saint-Quentin. Georgia is on the Tartar side of the borderline on sheet 3 of the *Carta marina*, but is marked with several crosses, indicating a Christian population that is in danger.

3.11
Omnes sunt tonsi sicut clerici sunt boni bellatores

They all have their hair cut like monks and they are good fighters.

This legend comes from Plano Carpini's general description of the Tartars.[53] Noah's Ark is depicted just to the right of this legend.

3.12
Arach mons super quam requieuit Archa noe

Mount Ararat, upon which Noah's Ark rested.

This legend was discussed in the introduction (see p. 38), where I suggest that Waldseemüller drew the text from a large, elaborately decorated nautical chart that probably contained an illustration of the Ark as well, but that he gave the Ark a more boat-like shape than is typical of the illustrations of the Ark on nautical charts.

3.13
ARMENIA MINOR Hic sunt christiani cisma[tici] sub dominio Tarcorum [corrected to Tartarorum]

Lesser Armenia. Here there are schismatic Christians under the dominion of the Tartars.

Lesser Armenia was the part of Armenia that was west of the Euphrates, and that is how Waldseemüller depicts it, though one has to follow the river quite a ways to the south and then east to see its name. Marco Polo has a chapter on Lesser Armenia that was perhaps Waldseemüller's source here,[54] though his legend is so general that it is difficult to be certain. The Euphrates runs north and south here right where the borderline of the Tartar empire might be, which makes it difficult to notice that there is a break in the borderline so that Armenia Minor is included in the region under the control of Tartaria, as the legend indicates.

3.14
ARMENIA MAIOR Sunt christiani iacobite et nestoriani cismatici sub dominio Tartarorum

Greater Armenia. They are Jacobite Christians and schismatic Nestorians under the dominion of the Tartars.

Marco Polo has a chapter on Greater Armenia,[55] and he specifies that the people are under the dominion of the Tartars, and does mention Nestorians and Jacobites in the kingdom of Mosul two chapters later, but given that Mosul (*Mosalia*) is indicated some distance to the east of this legend, it seems likely that Waldseemüller was using a different source. That source was the account of Priest Joseph, whom Waldseemüller cites in the long text block in the lower left corner of the

[51]For the passage in Benedict the Pole see Dawson, *The Mongol Mission* (see note 31), p. 82.

[52]For the passage about the eighteen bishoprics in Simon of Saint-Quentin see his *Histoire des Tartares*, ed. Jean Richard (Paris: P. Geuthner, 1965), p. 58.

[53]See Giovanni da Pian di Carpine, *Storia dei Mongoli* (see note 31), pp. 232 (Latin) and 340 (Italian); and Dawson, *The Mongol Mission* (see note31), p. 6.

[54]See Marco Polo, *Marka Pavlova z Benátek, Milion: Dle jediného rukopisu spolu sprilusnym zakladem latinskym*, ed. Justin Václav Prásek (Prague: Nákl. Ceské akademie císare Frantiska Iozefa, 1902), pp. 16–17; and Marco Polo, *The Book of Ser Marco Polo*, ed. and trans. Henry Yule (New York: C. Scribner's Sons, 1903), Book 1, Chap. 1, vol. 1, pp. 41–42.

[55]See Marco Polo, *Marka Pavlova z Benátek, Milion* (see note 54), pp. 18–19; and *The Book of Ser Marco Polo*, Book 1 (see note 54), Chap. 3, vol. 1, pp. 45–46.

Carta marina (see Legend 9.3); Priest Joseph's account was published in the *Paesi novamente retrovati*, and the material cited here is from Chap. 133.[56] Waldseemüller mentions Jacobite Christians frequently in the texts on the *Carta marina*, so a few words about the history of this sect will not be out of place. The Jacobite Church of Syria, Iraq, and India was founded in Syria by Jacob Baradaeus in the sixth century with assistance from Empress Theodora. It is a Monophysite church and recognizes the Syrian Orthodox Patriarch of Antioch as its spiritual leader.[57] Waldseemüller probably read about the sect in Bernhard von Breydenbach's *Peregrinatio in terram sanctam* (Mainz: Peter Schöffer the Elder, 1486), ff. [80v]–[81r], in a section titled "De Jacobitis et eorum erroribus."[58]

3.15
Hic residet patriarcha omnium christianorum orientalium cismaticorum qui dicitur Catholica

Here lives the patriarch of all the eastern schismatic Christians who is called Catholica.

Marco Polo in his chapter on Mosul mentions this patriarch, but calls him "Iaholith" rather than "Catholica,"[59] so it seems that Waldseemüller's source here was not Marco Polo but rather Priest Joseph, whose account was published in the *Paesi novamente retrovati*, and who does use the name "Catholica."[60] Just to the right of the legend there is a labeled image of the Catholica, and he is dressed as a European bishop might be.

3.16
in Taurisio ciui[tate] que est opulentis sunt maxi[me] mercaciones de quibus recipit imperator plus quam rex francie de toco [*sic*] suo regno

In the city of Tabriz, which is wealthy, there are great markets from which the emperor receives more than the King of France receives from his whole realm.

Tabriz was once a major Silk Road market city. This legend comes from Odoric of Pordenone[61]; the image of the city to the right of the legend is a variant of a design that Waldseemüller uses quite often for important cities.

3.17
Hic dominatur Soldanus Halapie sub tributo tartarorum

Here rules the Sultan of Aleppo, who pays tribute to the Tartars.

This legend comes from Plano Carpini.[62] This legend tends to confirm that Waldseemüller was using the 1473 printed edition of Vincent of Beauvais's *Speculum historiale* as his source for material from Plano Carpini, rather than a manuscript of Plano Carpini's work, as most manuscripts of Plano Carpini mention the Sultan of Damascus in this passage,[63] but the printed edition of Vincent of Beauvais's *Speculum historiale* (32.16) mentions the Sultan of Aleppo, as Waldseemüller does. There are no crosses in this area on the *Carta marina*, since the people are under the control of a Sultan.

---

[56]Greenlee, "The Account of Priest Joseph" (see note 70 in Chap. 1 above), pp. 95–113, at 103.

[57]On the Jacobite Church see Peter Kawerau, *Die jakobitische Kirche im Zeitalter der syrischen Renaissance: Idee und Wirklichkeit* (Berlin, Akademie-Verlag, 1955); O. F. A. Meinardus, "The Syrian Jacobites in the Holy City," *Orientalia Suecana* 12 (1963), pp. 60–82; Andrew Palmer, "The History of the Syrian Orthodox in Jerusalem," *Oriens Christianus: Hefte für die Kunde des christlichen Orients* 75 (1991), pp. 16–43; Palmer's "The History of the Syrian Orthodox in Jerusalem, Part 2: Queen Melisende and the Jacobite Estates," *Oriens Christianus: Hefte für die Kunde des christlichen Orients* 76 (1992), pp. 74–94; and Herman G. B. Teule, "*It Is Not Right to Call Ourselves Orthodox and the Others Heretics*: Ecumenical Attitudes in the Jacobite Church in the Time of the Crusades," in Krijnie Ciggaar and Herman Teule, eds., *East and West in the Crusader States: Context, Contacts, Confrontations* (Louvain: Peeters, 1999), pp. 13–27.

[58]See Jeffrey Jaynes, *Christianity beyond Christendom: The Global Christian Experience on Medieval Mappaemundi and Early Modern World Maps* (Wiesbaden: Harrassowitz Verlag in Kommission, 2018), p. 287.

[59]See Marco Polo, *Marka Pavlova z Benátek, Milion* (see note 54), p. 19; and *The Book of Ser Marco Polo* (see note 54), Book 1, Chap. 5, vol. 1, pp. 60–61.

[60]The passage that Waldseemüller is using here is in Chap. 133 of the *Paesi*; for an English translation of the passage see Greenlee, "The Account of Priest Joseph" (see note 70 in Chap. 1 above), pp. 95–113, at 102–103.

[61]See Henry Yule, ed. and trans., *Cathay and the Way Thither: Being a Collection of Medieval Notices of China*, revised by Henri Cordier (London: The Hakluyt society, 1913–16), vol. 2, pp. 103–105 (English), 280 (Latin), and 338–339 (Italian).

[62]See Giovanni da Pian di Carpine, *Storia dei Mongoli* (see note 31), pp. 274–275 (Latin) and 362 (Italian); and Dawson, *The Mongol Mission* (see note 31), p. 32.

[63]See the critical apparatus in Giovanni da Pian di Carpine, *Storia dei Mongoli* (see note 31), p. 274.

3.18

ARABIA DESERTA Hic est mare terrestre arenosum seu sabuli quod est mare mirabile et periculosum in eo reperitur mumia. habitanti sunt[?] [i]n montibus predones maximi quorum. x. milia vel. xx. m [a]liquando sunt in societate qui Charvanam mercatorum inuadunt degunt sicut Tartari.

Desert Arabia. Here there is a terrestrial sea, which is of sand or gravel, and it is a marvelous and dangerous sea. Mummy is found in it. The inhabitants live in the mountains and are egregious thieves, of whom ten or twenty thousand sometimes unite and attack the caravans of merchants. They live like the Tartars.[64]

This is a very interesting composite legend: what Waldseemüller has done is to locate the sea of sand in accordance with Odoric, but as Odoric does not describe the sea in any detail, the cartographer took elements from the descriptions of other seas of sand from Marco Polo and Varthema. Odoric mentions the sea of sand in his chapter is about "De civitate magorum," "The city of the Three Wise Men," i.e. Cassan,[65] and Waldseemüller has Crassan, which he identifies as the city from which the Wise Men came, just to the northeast here. Odoric also mentions the land of Job that Waldseemüller has to the east, and the mountains with manna that Waldseemüller has just to the east. The material about the thieves comes from Marco Polo,[66] although Polo is describing a sea of sand near Kathmandu, which of course is nowhere near Arabia. The detail about the mummy being found in the sand comes from Varthema,[67] who is describing a sea of sand between Mecca and Medina, not far to the south of the area described by the legend on the *Carta marina*. Thus it is curious that Waldseemüller did not simply adopt Varthema's description, but it is true that Marco Polo's is more colorful—so our cartographer did have some taste for the dramatic.

The mummy mentioned in the legend means bodies of people who died crossing the desert and were then desiccated by the sun, which was used medicinally in Europe into the seventeenth century.[68]

The words *Charvanam* and *inuadunt* near the end of the legend are corrected by hand in accordance with the list of corrections in Legend 9.2

3.19

[H]VNGARIA [M]AGNA Qui hanc habitant vocantur Bastarci et sunt sub imperio Tartarorum carent blada BASTARCI

Great Hungary. Those who inhabit this region are called Bastarci and they are under the power of the Tartars. They have no wheat. Bastarci.

Great Hungary, the ancestral home of the Hungarians, was mentioned by Plano Carpini, who says that it lay north of *Bulgaria magna*. The Bascarts, the ancestors of the modern Bashkirs, are mentioned by Plano Carpini in conjunction with the races described in the following legends,[69] but Plano Carpini says nothing about their being under the control of the Tartars and lacking wheat. Given Waldseemüller's use of Plano Carpini for nearby legends, it seems most likely that the cartographer added these details from those other legends as pertaining to the region as a whole.

3.20

PAROSITAR[UM] GENUS. hic habitant homines habentes parvos stomachos et os parvum non manducantes sed carnes coquunt et super ollam se ponunt et fumo reficiuntur.

---

[64]Unfortunately Johnson, *Carta marina* (see note 21 in Chap. 1 above), p. 141 note 52, misinterprets Waldseemüller's legend, and claims that he says that "the mummies were thought to date back ten or twenty thousand years." She translates the corresponding shorter legend on Fries's *Carta marina* on p. 79.

[65]The passage from Odoric is in Yule, *Cathay and the Way Thither* (see note 61), vol. 2, pp. 107–108 (English), 281 (Latin), and 339–340 (Italian).

[66]See Marco Polo, *Marka Pavlova z Benátek, Milion* (see note 54), p. 25; for an English translation see *The Book of Ser Marco Polo* (see note 54), Book 1, Chap. 18, vol. 1, pp. 97–99, on the city of Camati or Camadi. The phrase *in societate* is also used of the thieves in a fifteenth-century manuscript, Vatican City, BAV MS Barb. lat. 2687: see *Der mitteldeutsche Marco Polo, nach der Admonter handschrift herausgegeben*, ed. Horst von Tscharner (Berlin: Weidmann, 1935), p. 8. This manuscript is briefly described by Tscharner, pp. x–xi, and see Consuelo Wager Dutschke, "Francesco Pipino and the Manuscripts of Marco Polo's Travels," Ph.D. Dissertation, University of California at Los Angeles, 1993, pp. 456–458 and 1135–1138.

[67]See Lodovico de Varthema, *The Travels of Ludovico di Varthema in Egypt, Syria, Arabia Deserta and Arabia Felix, in Persia, India, and Ethiopia, A.D. 1503 to 1508, translated from the original Italian edition of 1510... by John Winter Jones and edited, with notes and an introduction, by George Percy Badger* (London: The Hakluyt Society, 1863), p. 33.

[68]For discussion of the use of pieces of mummy in medicine see Warren R. Dawson, "Mummy as a Drug," *Proceedings of the Royal Society of Medicine* 21.1 (1927), pp. 34–39; and Karl H. Dannenfeldt, "Egyptian Mumia: The Sixteenth Century Experience and Debate," *The Sixteenth Century Journal* 16.2 (1985), pp. 163–180, esp. 167–168.

[69]See Giovanni da Pian di Carpine, *Storia dei Mongoli* (see note 31), pp. 272 (Latin) and 361 (Italian); and Dawson, *The Mongol Mission* (see note 31), p. 30. For discussion of Magna Hungaria see Czeglédy Károly, "Magna Hungaria," *Századok* 77 (1943), pp. 277–306, reprinted in his *Magyar őstörténeti tanulmányok* (Budapest: MTAK, 1985).

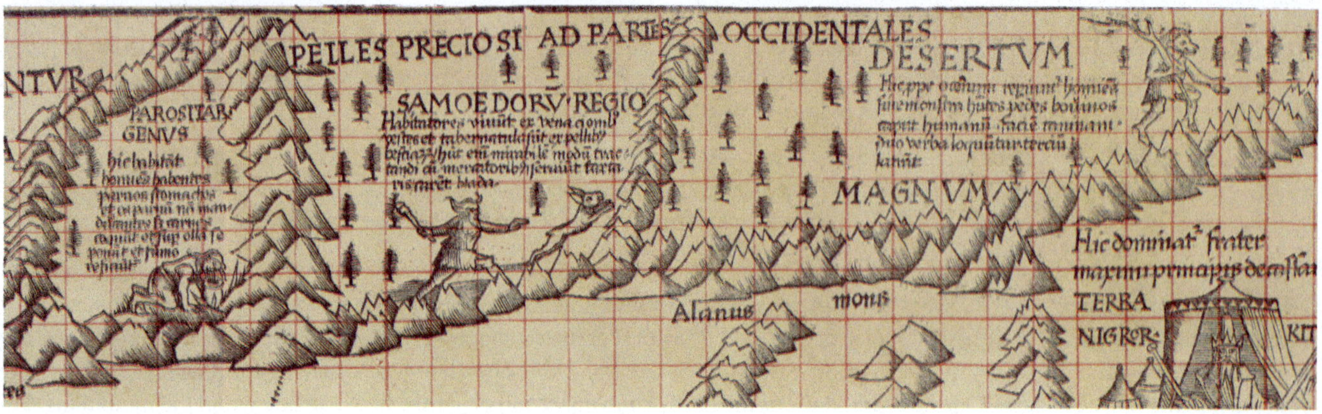

Fig. 2.2 Detail of sheet 3 of Waldseemüller's *Carta marina* showing the monstrous races of men enclosed by mountains along the northern edge of the map. Courtesy of the Library of Congress

> The Parossites. Here live men who have small stomachs and tiny mouths, and they do not eat; they cook meat and lean over the pot and refresh themselves with the steam.

I have discussed the three monstrous races along the top edge of this sheet (see Fig. 2.2) on a previous occasion,[70] when I showed that the races come from John of Plano Carpini, no doubt by way of the excerpts of Plano Carpini in Vincent of Beauvais's *Speculum historiale*.[71] I do not know the source for the illustration of this race.

> 3.21
> SAMOEDORUM REGIO. Habitatores vivunt ex venacionibus vestes et tabernatula sunt ex pellibus bestiarum, habent enim mirabilem modum tractandi cum mercatoribus, servunt tartaris, carent blada.

> The Region of the Samoyeds. The inhabitants live from hunting, and their clothes and tents are made of animal skins; they have a remarkable way of dealing with traders; they serve the Tartars, and have no wheat.

The term Samoyed was applied to some of the indigenous peoples of Siberia. This information about this people, like that described in Legend 3.20, comes from Plano Carpini.[72] There are illustrations similar to Waldseemüller's here in two fifteenth-century manuscripts of *Le livre des merveilles du monde*,[73] namely New York, Pierpont Morgan Library, MS M. 461, f. 41v,[74] and Paris, BnF, MS fr. 1378, f. 11v.[75] In the past I thought it unlikely that an image from a manuscript of this work might have influenced Waldseemüller, but the Morgan manuscript was commissioned by either René d'Anjou, Duke of Lorraine, or someone close to him,[76] and René's grandson René II gave at least some support to the publishing projects of

---

[70]Van Duzer, "A Northern Refuge" (see note 90 in Chap. 1 above), pp. 223–225 and 227.

[71]The Parossites are described in Vincent of Beauvais, *Speculum historiale*, Book 32, Chap. 15; see Beazley, *The Texts and Versions* (see note 43 in Chap. 1 above), p. 88, and for Plano Carpini's text, pp. 60–61. Plano Carpini's text is also supplied in Giovanni da Pian di Carpine, *Storia dei Mongoli* (see note 31), pp. 272 (Latin) and 361 (Italian). An English translation of the passage in Plano Carpini is given in Dawson, *The Mongol Mission* (see note 31), p. 30. This monstrous race seems to have originated with Strabo, *Geography* 15.1.57.

[72]The Samoyedes are described in Vincent of Beauvais, *Speculum historiale*, Book 32, Chap. 15; see Beazley, *The Texts and Versions* (see note 43 in Chap. 1 above), p. 88, and for Plano Carpini's text, p. 61. Plano Carpini's text is also supplied in Giovanni da Pian di Carpine, *Storia dei Mongoli* (see note 31), pp. 272–273 (Latin) and 361 (Italian). For an English translation of the passage in Plano Carpini see Dawson, *The Mongol Mission* (see note 31), p. 30.

[73]On this work see the important short article by John Block Friedman, "*Secretz de la Nature* [*Merveilles du Monde*]," in John Block Friedman and Kristen Mossler Figg, eds., *Trade, Travel and Exploration in the Middle Ages: An Encyclopedia* (New York and London: Garland, 2000), pp. 545–546. A substantial portion of the text of Paris, BnF, MS fr. 1377–1379 was edited by Anne-Caroline Beaugendre in *Les merveilles du Monde, ou Les secrets de l'histoire naturelle* (Paris: Bibliothèque nationale de France, 1996), and illustrated with images from Paris, BnF MS fr. 22971.

[74]This folio is illustrated in John Block Friedman, *The Monstrous Races in Medieval Art and Thought* (Cambridge, MA, and London: Harvard University Press, 1981; reprinted with expanded bibliography Syracuse: Syracuse University Press, 2000), p. 159.

[75]A low-resolution image of BnF, MS fr. 1378, f. 11v, is viewable through http://mandragore.bnf.fr.

[76]Marc-Édouard Gautier, "La bibliothèque du roi René," in Marc-Edouard Gautier, ed., *Splendeur de l'enluminure: le roi René et les livres* (Angers: Ville d'Angers, and Arles: Actes Sud, 2009), pp. 21–35, esp. 32 on René's interest in geography, noting his connection with the manuscript of *Le livre des merveilles du monde*, and see catalogue entry 35, pp. 324–325, for a more detailed account of the manuscript and its connection with René.

Waldseemüller and his colleagues,[77] so it seems quite possible that Waldseemüller had access to the manuscript and drew inspiration from the illustration on f. 41v.

3.22
Hic prope oceanum reperiuntur homines sive monstra habentes pedes bovinos caput humanum faciem caninam duo verba loquuntur tercium latrant.

Here near the ocean are found men or monsters who have the feet of cattle, a human head, but the face of dogs, and who speak two words, but bark the third.

This race, like those described in the preceding two legends, comes from Plano Carpini.[78] I do not know of any source for this image—I do not know of an illustrated manuscript of Plano Carpini or Vincent of Beauvais that contains such an illustration—so it may well have originated in Waldseemüller's workshop.

3.23
KANGITARUM REGIO her [*sic*] caret aquis proptera a paucis inhabitatur hominibus et periculosum est per eam agere iter

The Land of the Kangits. This [area] lacks water, and for that reason is thinly inhabited, and it is dangerous to cross the region.

The Kangits are the Kangli Turks. The legend comes from Plano Carpini[79]; it is copied by Schöner on his manuscript globe of 1520 in the Germanisches Nationalmuseum in Nuremberg.

3.24
Tartari semper morantur in campis: vbi inuenitur meliora pascua domus coopertas filero tanquam tentoria. habent etiam carucas vbi portant suppellectilem et familiam debilem. non villas nec multas ciuitates sed stationes habent. Uiri non intromittunt se nisi de guerra, vxores faciunt omnia alia facta domus intra et extra. commedunt de omni genere carnium preter hominum bibuntque lac iumentorum hec terra est sterelis sed solis pecoribus alendis apta.

The Tartars always remain in the fields, where there is better pasture. Their houses are covered with felt like tents. For they have carts in which they carry their furnishings and family members who are weak. They have no villages nor many cities, but rather stopping places. The men do not busy themselves, except about war, and the women do all of the other chores both inside the home and out. They eat all types of meat except human flesh, and they drink the milk of beasts of burden. The land is sterile, and only good for pasturing cattle.

[77]The account of Vespucci's voyages in Waldseemüller's *Cosmographiae introductio* is dedicated to René II, and in the letter to the reader in the 1513 edition of Ptolemy, Waldseemüller says that the edition was published *ministerio Renati dum vixit, nunc pie mortui Ducis illustriss. Lotharingiae*, "through the assistance, while he lived, of Rene most illustrious Duke of Lorraine, now piously deceased." This letter is reproduced and translated into English in Henry N. Stevens, *The First Delineation of the New World and the First Use of the Name America on a Printed Map* (London: H. Stevens, Son & Stiles, 1928), p. 40. In addition, Waldseemüller records René II's enthusiastic reception of his 1507 map and other works in the dedicatory letter in Ringmann's *Instructio manuductionem prestans in Cartam itinerariam* (Strasbourg: Grüninger, 1511), which implies an ongoing relationship between the scholars and the sovereign: for details see note 80 in Chap. 1 above.
[78]The race of men with cow's feet is described in Vincent of Beauvais, *Speculum historiale*, Book 32, Chap. 15; see Beazley, *The Texts and Versions* (see note 43 in Chap. 1 above), p. 88, and for Plano Carpini's text, p. 61. Plano Carpini's text is also supplied in Giovanni da Pian di Carpine, *Storia dei Mongoli* (see note 31), pp. 272 (Latin) and 361 (Italian). For an English translation of the passage in Plano Carpini see Dawson, *The Mongol Mission* (see note 31), pp. 30–31.
[79]See Giovanni da Pian di Carpine, *Storia dei Mongoli* (see note 31), pp. 313–314 (Latin) and 387 (Italian); and Dawson, *The Mongol Mission* (see note 31), pp. 58–59.

The legend is assembled from passages in Plano Carpini.[80] In the plain surrounding this legend three groups of tents are depicted, and one is labeled *statio tartarorum*, and another simply *statio*, to indicate the Tartars' nomadic lifestyle.[81] The brief description of Tartaria in Fries's *Uslegung der mercarthen oder Charta marina*, Chap. 110,[82] is quite similar to Waldseemüller's legend here. In addition, an expanded version of the text accompanies the *Tabula Superioris Indiae et Tartariae Maioris* in the 1522 and 1525 editions of Ptolemy's *Geography* that were published by Fries.

3.25
MARE ABACVC SIVE MARE DE SALA Lacus iste aque dulcis et tocius orbis maximus est et ob sui magnitudinem mare appellatur. habet portus et patitur tempestates et reliqua maris accidentia et varijs in locis varia nomina sortitur

The Abacuc Sea or Sala Sea [the Caspian]. This lake is of fresh water and is the largest in the world, and because of its size is called a sea. It has ports and it suffers storms and the other accidents that befall seas. The sea is called by different names in different places.

The names that Waldseemüller uses for the Caspian sea are different than those he used on his 1507 map (*Mare hircanum sive Caspium*) or the modern world map in the 1513 Ptolemy (*mare hircanum*). Paul Pelliot, in discussing the names that Waldseemüller uses on the *Carta marina*, suggests that *Abacuc* is a copyist's error for the name *Bachuc*, which is used for the Caspian by Odoric,[83] while the "Sea of Sarai" is a name used by Marco Polo.[84] However, this name for the sea is not used in the manuscript of Polo most similar to that Waldseemüller used in making his 1507 map,[85] which was probably still available when he made the *Carta marina*, and there can be little doubt that Waldseemüller took the two names for the sea on the latter map from a nautical chart. On the Catalan Atlas of 1375 the sea is labeled "Aquesta mar és appellada mar del Sarra e de Bacú," "This sea is called the Sea of Sarra and of Bacu,"[86] and on the Catalan-Estense *mappamundi* of c. 1460 it is labeled MAR DE SALA E DE BACU.[87] There are other large nautical charts that include the Caspian and indicate its names, such as the Pizzigani chart of 1367 and Mecia de Viladestes's chart of 1413, but the spellings on the Catalan Atlas and the Catalan-Estense *mappamundi* are the most similar to what we find on Waldseemüller's *Carta marina*. This confirms the evidence discussed in the introduction that Waldseemüller had a large, old, and elaborately decorated nautical chart in his workshop, in addition to the Caverio chart—which latter chart, we should mention, omits the Caspian Sea altogether. This is another example of Caverio's disregard for geographical features in the hinterlands that emphasizes Waldseemüller's detailed depictions of the hinterlands by contrast.

The rest of Waldseemüller's legend about the Caspian, about its size and similarity to a sea, comes from Pierre d'Ailly, who writes[88]:

[80]See Giovanni da Pian di Carpine, *Storia dei Mongoli* (see note 31), pp. 234–235 (Latin) and 341–342 (Italian) on the tents, pp. 230 and 339 on the lack of cities, pp. 248 and 349 on their eating all types of meat, pp. 249 and 350 on drinking milk of beasts of burden; and pp. 230 and 339 on the land being generally poor and only good for cattle. These passages are supplied in English by Dawson, *The Mongol Mission* (see note 31), p. 8 on the tents, p. 5 on the lack of cities, p. 16 on their eating all types of meat, p. 17 on drinking milk of beasts of burden; and p. 5 on the land being generally poor and only good for cattle. A significant difference between the text of Waldseemüller here and Plano Carpini (in Vincent of Beauvais) is that while Waldseemüller says that the Tartars eat all types except human, Plano Carpini says that they eat all types of flesh, including human (*& in necessitate carnes humanas* see Beazley, *Texts and Versions* [see note 43 in Chap. 1 above], p. 52). Since it is unlikely that Waldseemüller would want to change the text to make the image of the Tartars more positive, this difference in his text could be a good clue to which version of the text he was using.

[81]For discussion of the Tartars' tents and carts see Michael Gervers and Wayne A. Schlepp, "Felt and 'Tent Carts' in *The Secret History of the Mongols*," *Journal of the Royal Asiatic Society of Great Britain & Ireland*, Third Series 7.1 (1997), pp. 93–116, esp. 98–103.

[82]The text of this chapter of the *Uslegung* is translated into modern German by Petrzilka, *Die Karten des Laurent Fries* (see note 202 in Chap. 1 above), pp. 157–158.

[83]The passage is in Chap. 2 of Odoric's narrative; see Yule, *Cathay and the Way Thither* (see note 61), vol. 2, p. 280. According to the note there, the manuscript Venice, Biblioteca Marciana, MS 4326 (Lat. XIV.43), which is from the late fourteenth or early fifteenth century, reads *Abachuc*. There is a brief description of this manuscript in Yule's *Cathay* (see note 61), vol. 2, pp. 56–57.

[84]Paul Pelliot, *Notes on Marco Polo* (Paris: Impr. nationale, 1959–), vol. 1, p. 62; for Polo's use of the name Sea of Sarai see Marco Polo, *The Book of Ser Marco Polo* (see note 54), Book 4, Chap. 25, vol. 2, p. 495.

[85]The manuscript most similar that I know to that Waldseemüller used in making his 1507 map is Naples, Biblioteca Nazionale Vittorio Emanuele III, Vind. lat. 50, and the text of this manuscript has been published in Marco Polo, *Marka Pavlova z Benátek, Milion* (see note 54).

[86]See *Mapamundi del año 1375* (Barcelona: S.A. Ebrisa, 1983), p. 70.

[87]See Ernesto Milano and Annalisa Batini, *Mapamundi Catalán Estense, escuela cartográfica mallorquina* (Barcelona: M. Moleiro, 1996), p. 156.

[88]Pierre d'Ailly, *Ymago mundi*, ed. Edmond Buron (Paris: Maisonneuve frères, 1930), vol. 1, pp. 452–454.

Unde lacus et stagna describit Ysidorus nullam facit de mari Caspio mentionem licet tamen a quibusdam asseratur esse lacus aque dulcis, totius orbis maximus habens in circuitu portus et littora et nauigia uehit ingentia patiens insuper tempestates et maris accidentia reliqua. Propter quod et ob eius magnitudinem mare dicitur.

Where Isidore describes lakes and pools he makes no mention of the Caspian Sea, however, it is asserted by some to be a fresh water lake, the largest in the world, having ports and shores around it, and carrying huge ships, and, moreover, enduring storms and the other accidents of a sea. Because of this and because of its size it is called a sea.

This is one of the relatively few legends for which Waldseemüller uses d'Ailly, aside from for those about the monstrous races in India.

3.26
Baldac seu Niniue. In ciuita[te] Baldac residet Calyphus qui est papa omnium Saracenorum soluit tributem imperatori Cham. hec ciuit[as] in circuitu longitudinis est .3. dietarum. sunt ibi magne mercationes.

Baldock [Baghdad] or Nineveh. In the city of Baldock resides the Caliph who is the pope of all of the Saracens; he pays tribute to the Great Khan. This city is a three day journey in circumference, and there are great markets there.

Most of this legend comes from Marco Polo,[89] except for the part about the size of the city. It is possible that this detail was inspired by Jonah 3:3, which says that the city was three days across (rather than in circumference),[90] but more likely this information comes from a nautical chart, rather than directly from the Bible, as the Borgia metal *mappamundi* has a legend that reads *Ninive iij dierum longitudine*, "Nineveh is three days wide."[91] The great power of the Caliph of Baghdad earned him a mention in Wolfram von Eschenbach's *Parzival.*[92] The words *tributem imperatori* in this legend have been corrected by hand in accordance with the list of corrections in Legend 9.2.

3.27
iste Caliphus perdiues et potens est. condit Saracenis leges sicut papa noster precipit ab omnibus eas firmiter obseruari. Hic in die numquam egreditur sed de nocte; habet enim in thalamo suo multas virgines quibus commiscetur

This Caliph is very wealthy and powerful; he establishes laws for the Saracens just as our Pope [and] orders that they be strictly observed by everyone. He never goes out during the day, only at night; he has in his chamber many virgins with whom he consorts.

This legend gives additional description of the Caliph of Baghdad who was mentioned in the previous legend; in between these two legends and just above the image of the city (through which the Tigris River runs) there is an image of the Caliph. The source of this legend is somewhat surprising, as it is not a work that Waldseemüller lists among his sources in the long text block on sheet 9: it comes from a short, thirteenth-century account of a journey to the Holy Land by one Thetmar or Theitmar.[93] Some sixteen manuscripts of the work are known,[94] and the account was published a few times in the nineteenth

[89]See Marco Polo, *Marka Pavlova z Benátek, Milion*, pp. 19–20; and *The Book of Ser Marco Polo*, Book 1, Chap. 6, vol. 1, p. 63.

[90]The interpretation of Jonah 3:3 is controversial: see for example Charles Halton, "How Big Was Nineveh? Literal versus Figurative Interpretation of City Size," *Bulletin for Biblical Research* 18.2 (2008), pp. 193–207.

[91]See A. E. Nordenskiöld, "Om ett aftryck från XV:de seklet af den i metall graverade världskarta, som förvarats i kardinal Stephan Borgias museum i Velletri, Med 1 facsimile," *Ymer* 11 (1891), pp. 83–92, at 91.

[92]Wolfram, von Eschenbach, *Parzival*, ed. and trans. André Lefevere (New York: Continuum, 1991), Book 1, p. 7.

[93]On Thetmar see Karl Ernst Hermann Krause, "Zu Magister Thetmarus (Thietmarus)," *Forschungen zur deutschen Geschichte* 15 (1875), pp. 153–156; and Franz Joseph Worstbrock, "Magister Thietmar(us)," in Wolfgang Stammler, et al., eds., *Die deutsche Literatur des Mittelalters: Verfasserlexikon*, vol. 9 (Berlin: De Gruyter, 1995), cols. 793–795. There is a good English summary of Thetmar's narrative in Jaroslav Folda, *Crusader Art in the Holy Land: From the Third Crusade to the Fall of Acre, 1187–1291* (Cambridge, UK, and New York, NY: Cambridge University Press, 2005), pp. 118–123.

[94]The manuscripts are listed by Reinhold Röhricht, *Bibliotheca geographica Palaestinae* (Berlin: H. Reuther, 1890), p. 47, with editions and other bibliography listed on pp. 47–48. Röhricht includes indications of which manuscripts were used in the preparation of the various editions, with more accurate citations than the editors themselves sometimes supply.

century and also translated into French,[95] but since then it has attracted very little scholarly attention until a very good edition was published in 2011.[96] The relevant passage from that work runs[97]:

> Est eciam ab illo loco versus orientem in confinio Caldee, Ydumee et Persye ciuitas magna et munita, nomine Baydach, metropolis. Ubi est papa Sarracenorum, nomen habens Galiphel, predives et prepotens, et condit Sarracenis leges, et sub pena, sicut papa noster, precipit ab omnibus firmiter obseruari. Hic numquam egreditur de die, sed de nocte cuando placet.... Habet autem papa iste in thalamis et in domibus suis plurimas virgines, quibus commiscetur ubi vult.
>
> To the east of that place, in the borderlands of Chaldea, Idumea and Persia, is a great and strong city, Baghdad, a metropolis. That is where the wealthy and powerful Pope of the Saracens, who is named Galiphel, both establishes the laws for Saracens, and (like our Pope) orders them under penalty to be strictly observed by all. He never goes out during the day, but only at night, when he pleases... But that pope has in his chambers and houses many virgins, with whom he consorts whenever he wants.

Waldseemüller's use of this little-known work is another testament to the richness of his library.

3.28
CALDEA Caldei habent linguam propriam peruertunt quidem ordinem nature nam mascali ornati incedunt mulieres vero turpes

Chaldea. The Chaldeans have their own language, and in fact they pervert the natural order, for the men go finely decked out, but the women are unsightly.

This legend comes from Odoric.[98] Chaldea is in southern Babylonia; the kingdom had long since ceased to exist, but the name continued to be use becase it was used in the Septuagint and other translations of the Bible. According to the system of borderlines on sheets 3 and 4 of the *Carta marina*, Chaldea is the southernmost region that is under the control of the Tartars.

3.29
in istis montibus colligitur de celo manna in magna copia

In these mountains manna from the heavens is collected in great quantity.

This legend comes from Odoric.[99] The manna mentioned here and in various other texts, including the Bible, has been variously identified as the gum of a desert tree such as tamarisk, a lichen, or an excretion from an insect.[100]

3.30
Crassan Hec fuit ciuitas regalis unde magi venerunt ad Christum.

Kashan. This was the royal city from which the Three Wise Men came to Christ.

---

[95]For a full listing of editions see Röhricht (previous note); the main ones are: *Magister Thetmari iter ad Terram Sanctam anno 1217*, ed. Titus Tobler (St. Gall and Bern: Huber, 1851), with a very brief introduction, reprinted with an Italian translation in Sabino de Sandoli, *Itinera Hierosolymitana Crucesignatorum, saec. XII-XIII: textus Latini cum versione Italica* (Jerusalem: Franciscan Print. Press, 1978), vol. 3, pp. 254–295; *Voyages faits en terre-sainte par Thetmar, en 1217 et par Burchard de Strasbourg, en 1175, 1189 ou 1225*, ed. Jules de Saint-Genois (Brussels: Académie royale de Belgique, 1851) (an offprint from the *Mémoires de l'Académie royale de Belgique*, vol. 25), with a summary of Thetmar's voyage on pp. 8–12; and *M. Thietmari historia de dispositione terre sancte*, ed. Johann Christian Moritz Laurent (Hamburg: Joannis Augusti Meissneri, 1852), with a very brief introduction. A summary of Thetmar's narrative with French translation of some excerpts was published in L. de Saint-Aignan, "Les pèlerins célèbres: les pérégrinations de maître Thietmar," *La Terre Sainte* 8 (1882), pp. 538–534. A French translation of most of the work has been published in Christiane Deluz, "Le pèlerinage de Maître Thietmar," in Danielle Régnier-Bohler, ed., *Croisades et pélerinages: récits, chroniques et voyages en Terre Sainte, XIIe-XVIe siècle* (Paris: Laffont, 1997), pp. 928–958.

[96]Ulf Koppitz, "Magistri Thietmari Peregrenatio: Pilgerreise nach Palästina und auf den Sina in den Jahren 1217/1218," *Concilium medii aevi* 14 (2011), pp. 121–221.

[97]From Koppitz's edition (see previous note), Sect. 7, p. 142.

[98]See Yule, *Cathay and the Way Thither* (see note 61), vol. 2, pp. 110 (English), 282 (Latin), and 339 (Italian).

[99]See Yule, *Cathay and the Way Thither* (see note 61), vol. 2, pp. 109 (English) and 282 (Latin); I do not see the passage about manna in the Italian text that Yule edits.

[100]See F. S. Bodenheimer, "The Manna of Sinai," *The Biblical Archaeologist* 10.1 (1947), pp. 1–6; Harold N. Moldenke and Alma L. Moldenke, *Plants of the Bible* (Waltham, MA: Chronica Botanica Company, 1952), pp. 125–128; and R. A. Donkin, *Manna: An Historical Geography* (The Hague and Boston: Junk, 1980).

Kashan is in what is now central Iran; the legend comes from Odoric.[101] This is a case where Waldseemüller might have used Marco Polo,[102] but he seems to have regarded Odoric as being more authoritative.

3.31
Sambragate ciuitas regalis Hec ciuitas non minor in quantitate et contractu est cheiro

The royal city of Samarkand. This city is no less in size and commerce than Cairo.

Samarkand, in modern Uzbekistan, was at various times in its history the greatest city of Central Asia. This legend comes from Varthema.[103] Waldseemüller's decision to show Samarkand runs counter to earlier humanist geography, in which the city was typically removed from the world picture.[104]

3.32
TERRA NIGROR[UM] KITHAO Hic dominatur frater maximi principis de cassan

The land of the black Kythayans. Here rules the brother of the great prince of Casana.

This legend comes from Plano Carpini, who says that Siban, the brother of Bati (or Batu or Baatu, see Legend 3.9) is stationed among the Black Kitayans.[105] The image of Siban, who is not named here, is large and imposing.

3.33
TERRA BISERMINORUM Hec terra est valde aspera et montuosa et longitudine .40. dietarum

The land of the Bisermini. This land is very rough and mountainous, and takes forty days to cross.

This legend comes from Plano Carpini, but Waldseemüller gives the region much less extension to the southwest than Carpini indicates. The figure of forty days is Waldseemüller's calculation rather than something that Plano Carpini gives directly, and his calculation, at least according to the way the text runs in most versions, is not correct.[106] Plano Carpini writes that they traveled through the land of the Bisermini from about the feast of the Ascension, i.e. May 17, until eight days before the feast of St. John the Baptist, i.e. eight days before June 24, or June 16—for a total of just under a month. It seems likely that Waldseemüller miscalculated by forgetting to subtract the eight days, for all of the versions of Plano Carpini that I know, including the excerpts in the printed edition of Vincent of Beauvais that was Waldseemüller's most likely source for the text of Plano Carpini, are consistent in indicating that this land took a month to cross, rather than forty days.

3.34
Hic dominatur et ambulat contra persos Noy princeps tartarorum et imperator super sexingenta [milia added by hand] armatorum virorum qui omnes prouincias tam Cristianorum quam Saracenorum a capite persie usque ad Syriam sue ditioni subiungavat

Here rules and marches against the Persians Baiotnoy [Baiju], prince of the Tartars and commander of 60,000 soldiers who has brought under his control all of the countries, both Christian and Saracen, from the top of Persia to Syria.

This legend comes from Simon de Saint-Quentin, by way of Vincent of Beauvais.[107] It is appropriate, of course, that Waldseemüller should site Simon here, as Simon's mission was precisely to Baiju. Simon explains that "Noy" is a title,[108] and that the king's name is Baioth, but Waldseemüller chose not to explain this. It seems that Waldseemüller was working

---

[101]See Yule, *Cathay and the Way Thither* (see note 61), vol. 2, pp. 106 (English), 281 (Latin), and 339 (Italian).

[102]See V. Williams Jackson, "The Magi in Marco Polo and the Cities in Persia from Which They Came to Worship the Infant Christ," *Journal of the American Oriental Society* 26 (1905), pp. 79–83.

[103]See Varthema, *The Travels of Ludovico di Varthema*, trans. John Winter Jones (London: The Hakluyt Society, 1863), pp. 103–104.

[104]See Margaret Meserve, "From Samarkand to Scythia: Reinventions of Asia in Renaissance Geography and Political Thought," in Zweder von Martels and Arjo Vanderjagt, eds., *Pius II 'el più expeditivo pontifice': Selected Studies on Aeneas Silvius Piccolomini (1405–1464)* (Leiden and Boston: Brill, 2003), pp. 13–39, esp. 16–17.

[105]See Giovanni da Pian di Carpine, *Storia dei Mongoli* (see note 31), pp. 314 (Latin) and 388 (Italian); and Dawson, *The Mongol Mission* (see note 31), p. 59.

[106]See Giovanni da Pian di Carpine, *Storia dei Mongoli* (see note 31), pp. 314 (Latin) and 388 (Italian); and Dawson, *The Mongol Mission* (see note 31), p. 59.

[107]See Simon de Saint-Quentin, *Histoire des Tartares*, ed. Jean Richard (Paris: P. Geuthner, 1965), p. 93, which Waldseemüller no doubt saw in Vincent of Beauvais, *Speculum historiale*, Book 32, Chap. 34.

[108]On the title "Noy" see Francis W. Cleaves, "The Mongolian Names and Terms in the *History of the Nation of the Archers* by Grigor of Akanc," *Harvard Journal of Asiatic Studies* 12 (1949), pp. 400–443, esp. 411–413.

quickly when he read this passage: Vincent of Beauvais says that Baiju had 60,000 soldiers, but then goes on to say that he had 160,000 Tartar soldiers and 450,000 soldiers who were part Christians and part infidels, so the total should be 610,000. On the *Carta marina*, Baiju sits in a distinctive tent with his title, "Noy," indicated above the entrance. Schöner copies part of this legend on his manuscript globe of 1520 in the Germanisches Nationalmuseum in Nuremberg.

3.35
persia prouincia nobilis destructa multum per tartaros sed nunc sub ditione victoriosssimi [*sic*] regis Sophi reparata est enim diuisa in octo regna sunt Macometani et homines fallaces

The noble country of Persia was largely destroyed by the Tartars, but now, under the control of the unstoppable king Sophi [i.e. the Shah], it has been restored and divided into eight realms. The people are followers of Mohammed and are deceitful.

This legend was briefly discussed earlier in the introduction: "Sophi" is Shah Isma'il es-Sufi (1487–1524), the founder of the Safavid dynasty, who gained control over Persia and Khorasan (now Iran and adjoining territories to the east) around the year 1500.[109] The source of this legend is two-fold: the first part of the legend, including the part about the division of Persia into eight realms, comes from Marco Polo,[110] while the details about Sophi and his activities come from Varthema.[111] Waldseemüller's image of Sophi is distinctive and energetic, but I have not found a source for it—it does not derive from the 1515 edition of Varthema from which Waldseemüller borrowed other images, for example. Johann Schöner has a much abbreviated version of this legend on his 1520 manuscript globe.

3.36
in montibus istis reperiu[n]tur adamantes corniol et Calcedonie

In these mountains are found diamonds, cornelian, and chalcedony.

This legend is from Varthema.[112] Waldseemüller also mentions diamonds found in mountains in Legends 4.26 and 8.19.

3.37
GEDROSIA Guzerantes sunt ydolatre caffrani
Gedrosia. The Guzerantes are Caffrani idolaters.

The "IA" of GEDROSIA are on sheet 4. Varthema gives a brief account of the Guzerati,[113] but he says that they are neither moors nor heathens, so Waldseemüller's attitude is somewhat less tolerant. The designation Caffrani idolaters is one that Waldseemüller uses frequently on the *Carta marina*, but which I do not find in his sources, so it is reasonable to think that he added the designation here. In fact I do not know where Waldseemüller found the designation, but as the word will be appearing with some frequency, I cite a post-Waldseemüller sixteenth-century definition of it here, from Thomas Cooper's *Thesaurus linguae Romanae et Britannicae* (London, 1565), in the in the supplementary *Dictionarium historicum* and *Poeticum propria locorum* and *Personarum vocabula breviter complectens*, s.v.:

Caffrani, Idolatours dwelling in Indie the more, which worship Devils in most terrible figure, beleeving, that they are permitted of God, to punishe or spare men at their pleasure, wherefore unto them they sacrifice their children, and sometimes themselves. They have many wives, but they companye not with them, untill they be defloured by other hyred for that purpose. Also they suffer their priestes to have carnal companie with their wives in their absence. They have Bulles and Kine in greate reverence, but they never eate fleshe, their sustenaunce is rice, suger, & diverse sweet roots, they drynke the lycour that commeth of ripe dates.

---

[109]On Sophi see Ghulam Sarwar, *History of Shah Isma'il Safawi* (Aligarh: Muslim University, 1939; New York: AMS Press, 1975); Roger M. Savory and Ahmet T. Karamustafa, "Esmāʿīl i Ṣafawī, Shah Abu'l-Moẓaffar," in Ehsan Yarshater ed., *Encyclopaedia Iranica* (Costa Mesa, CA: Mazda Publishers, 1992–2010), vol. 8, pp. 628–636; and Palmira Brummett, "The Myth of Shah Ismail Safavi: Political Rhetoric and 'Divine' Kingship," in John Victor Tolan, ed., *Medieval Christian Perceptions of Islam: A Book of Essays* (New York: Garland, 1996), pp. 331–359.
[110]See Marco Polo, *Marka Pavlova z Benátek, Milion* (see note 54), p. 22; for an English translation see *The Book of Ser Marco Polo* (see note 54), Book 1, Chap. 13, vol. 1, pp. 78–79.
[111]See Varthema, *The Travels of Ludovico di Varthema* (see note 103), p. 103.
[112]See Varthema, *The Travels of Ludovico di Varthema* (see note 103), p. 107.
[113]See Varthema, *The Travels of Ludovico di Varthema* (see note 103), pp. 108–109.

2.4 Sheet 4. Northeastern Asia (Plate 2.4)

The borderline that demarcates the limits of the Tartar empire continues on this sheet, and here its main function is to separate India, which is depicted as a realm of monstrous races, from Tartaria, and on this sheet there are legends that make this function clear. There are few place names on this sheet, which is quite a change from Waldseemüller's coverage of this same area on his 1507 map. It is tempting to take this difference as reflecting Waldseemüller's realization that little is known about the geography of the region, and that one can really only guess about the relative positions of places mentioned by Marco Polo, Plano Carpini, etc. One symptom of Waldseemüller's disinclination to offer many geographical details on this sheet is his gratuitous repetition of TERRA MONGAL ET QUE VERA TARTARIA DICITUR, which appears at the top of the sheet, in TERRA MONGAL NVNC TARTAIA VERA to the southwest.

4.1
Quod extra ambitur hac linea et mari clauditur hoc maximi imperateris Gog Chaam ditioni subiatur

The land between this line and the sea is under the power of the great emperor Gog Khan.

The title "Gog Khan" simply meant emperor, but to European ears the word "Gog" would inevitably recall the mythical Gog, an evil people who was supposed to break out of their confinement in the northeastern reaches of the world at the end of time and join the armies of Antichrist (see Legend 4.18). The second word of the legend originally read *intra*, but was corrected by hand to *extra* in accordance with the list of corrections in Legend 9.2

4.2
MAGNVM INDIE C DIETARVM IN QVO DIVERSARUM SPECIERUM HOMINVM MONSTRA

Greater India, 100 days [wide], in which there are various monstrous species of men.

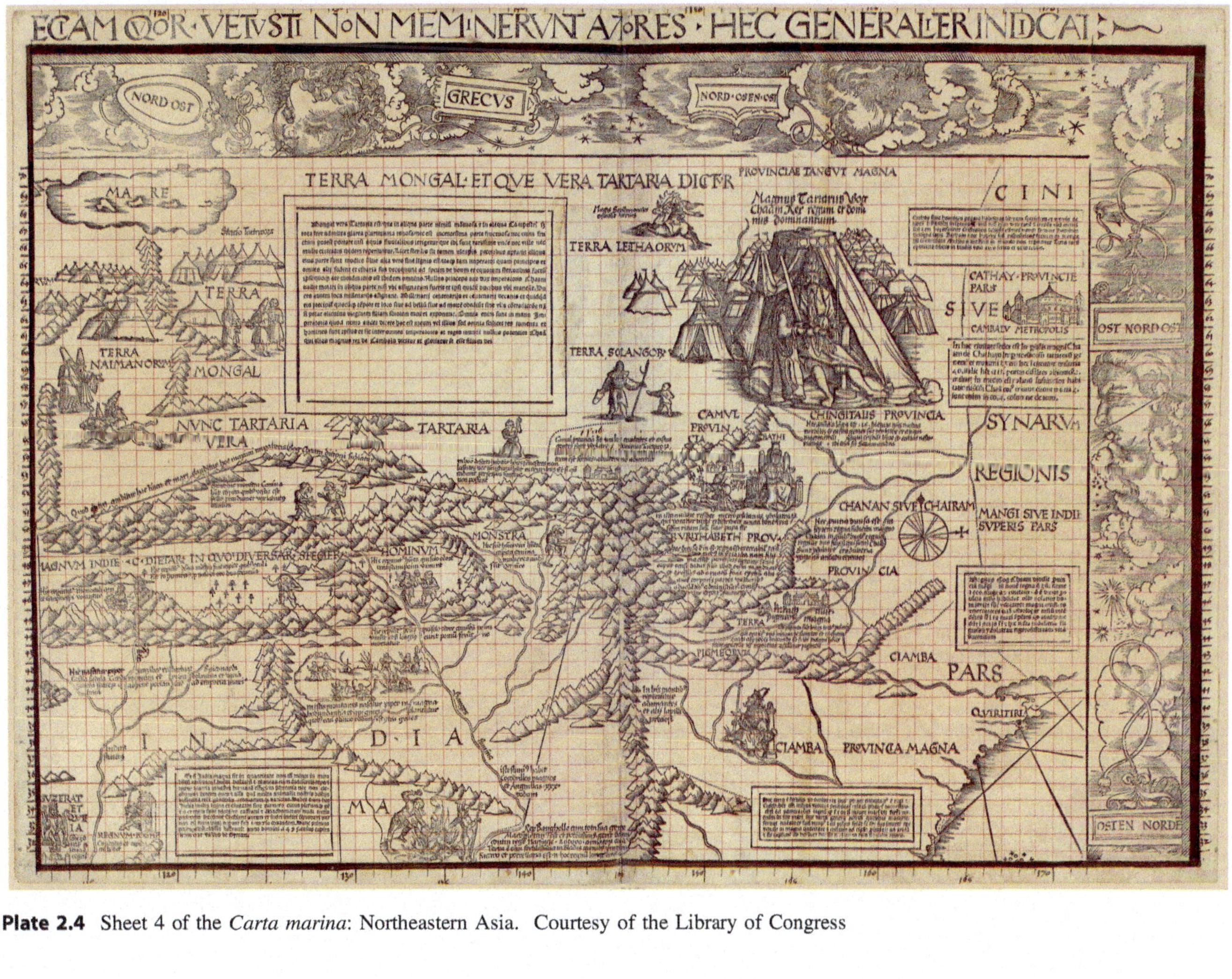

Plate 2.4 Sheet 4 of the *Carta marina*: Northeastern Asia. Courtesy of the Library of Congress

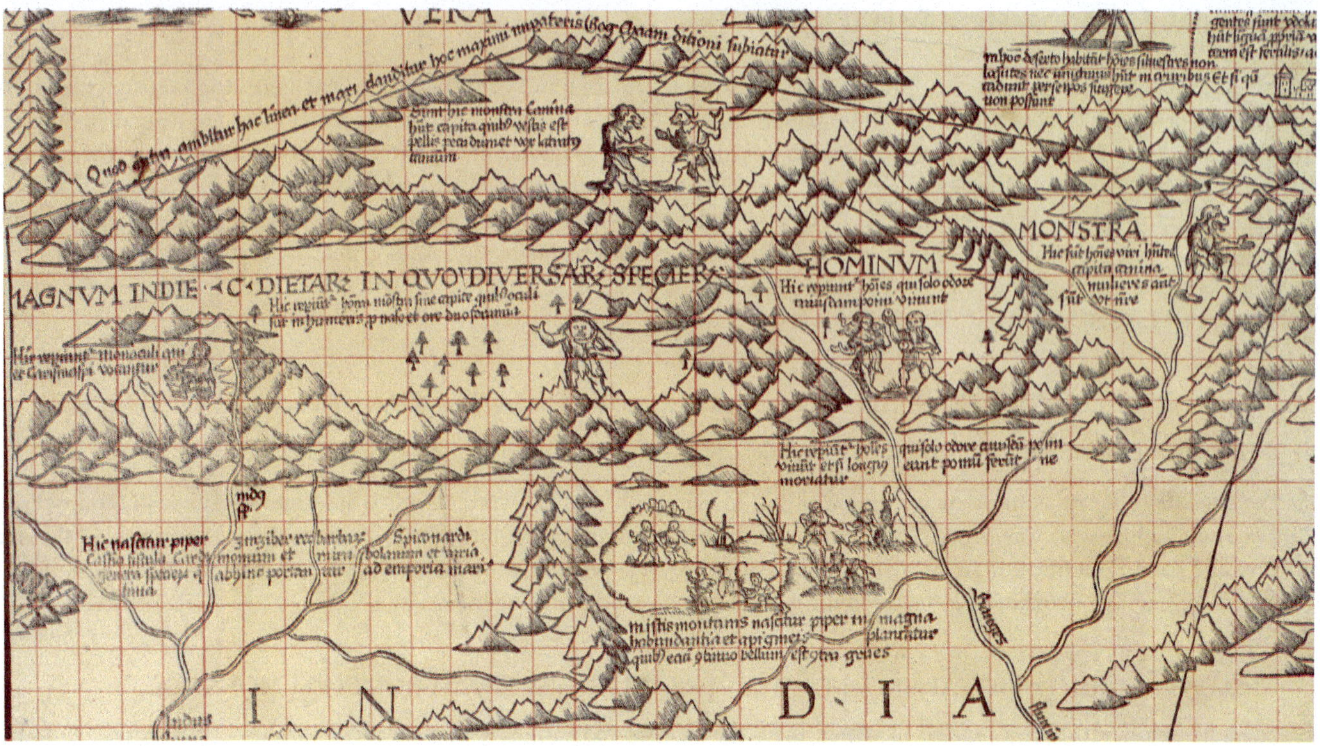

Fig. 2.3 Detail of India and its monstrous races of men from sheet 4 of Waldseemüller's *Carta marina*. Courtesy of the Library of Congress

I do not know the source of the statement that India is a hundred days wide. India had been the locus of monsters and wonders since classical times,[114] but remarkably Waldseemüller's map is the first to locate a full complement of the monstrous races in that region (see Fig. 2.3). As mentioned previously, I studied the monstrous races described in the following legends in an earlier article.[115]

4.3
Sunt hic monstra Canina habent capita quibus vestis est pellis pecudum et vox latratus caninum.

There are monsters here who have dogs' heads; their clothes are made of the skins of sheep, and their voice is a dog's bark.

The text comes from Pierre d'Ailly's *Ymago mundi*.[116] The illustration of two cynocephali together is fairly frequent in other sources and probably alludes to the idea that they are communicating with each other by barking.[117]

4.4
Hic reperiuntur monocoli qui et Carismaspi vocantur.

---

[114]See Rudolf Wittkower, "Marvels of the East: A Study in the History of Monsters," *Journal of the Warburg and Courtauld Institutes* 5 (1942), pp. 159–197, reprinted in his *Allegory and Migration of Symbols* (Boulder, CO: Westview Press, 1977), pp. 45–74; James Romm, "Belief and Other Worlds: Ktesias and the Founding of the 'Indian Wonders'," in George E. Slusser and Eric S. Rabkin, eds., *Mindscapes: The Geography of Imagined Worlds* (Carbondale: Southern Illinois University Press, 1989), pp. 121–135; and Chet Van Duzer, "*Hic sunt dracones*: The Geography and Cartography of Monsters," in Asa Mittman and Peter Dendle, eds., *The Ashgate Research Companion to Monsters and the Monstrous* (Farnham, England, and Burlington, VT: Ashgate Variorum, 2012), pp. 387–435, esp. 402–409.

[115]Van Duzer, "A Northern Refuge" (see note 90 in Chap. 1 above).

[116]See Pierre d'Ailly, *Ymago mundi*, ed. Edmond Buron (Paris: Maisonneuve frères, 1930), vol. 1, Chap. 16, "De mirabilibus Indie," p. 266.

[117]See Wittkower, "Marvels of the East" (note 114), p. 175 and plate 43d. Other illustrations of two or more cynocephali together include that on the Hereford *mappamundi*, where the creatures are curiously labeled giants (see Westrem, *The Hereford Map* [see note 143 in Chap. 1 above], pp. 40–41, no. 80); a manuscript of the *Libro del conosçimiento* (Munich, Bayerische Staatsbibliothek, Cod. hisp. 150, f. 18v); an early fifteenth-century manuscript of the *Secrets de l'histoire naturelle* (Paris, Bibliothèque nationale de France, MS fr. 1378, f. 11v); a manuscript of Mandeville's *Travels* (New York Public Library, Spencer MS 37, fol. 106v); two manuscripts of a Dutch translation of Thomas of Cantimpré (The Hague, KB 76 E 4, f. 4r, and KB KA 16, f. 41r); and the Piri Re'is map of 1513, illustrated in Nebenzahl, *Atlas of Columbus* (see note 217 in Chap. 1 above), p. 63.

> Here are found men with one eye who are also called Carimaspians.

The text comes from d'Ailly, *Ymago mundi*, Chap. 16.[118] The Carimaspians (usually known as the Arimaspians), a race of one-eyed men like the Cyclopes of Homer's *Odyssey* and many other early myths.[119] There are many medieval and early Renaissance images of Arimaspians, but I have not found an image closely similar to Waldseemüller's in any manuscript or early printed text.

> 4.5
> Hic reperiuntur hominum monstra sine capite quibus oculi sunt in humeris pro naso et ore duo foramina.
>
> Here are found monstrous men who have no heads; their eyes are in their shoulders, and they have two holes which serve as nose and mouth.

Again the text comes from d'Ailly, *Ymago mundi*, Chap. 16. There are two variants of the headless race whose faces are on their chests: the *blemmyae*[120] and the *epiphagi*.[121]

Since Waldseemüller specified that their eyes were in their shoulders, he must have had in mind the *epiphagi*, although in his illustration the creature's eyes are not located in its shoulders. The well-defined face high on the chest of Waldseemüller's creature is unusual but not unprecedented.

> 4.6
> Hic reperiuntur homines qui solo odore cuiusdam pomi vivunt.
>
> Here are found men who live just on the odor of a certain apple.

Again the text comes from d'Ailly, *Ymago mundi*, Chap. 16. This race is often called the *astomoi* or "mouthless men," and while Strabo 15.1.57 and Pliny 7.2.25 say that they do not have mouths, many other authors do not say that they lack mouths, only that they live just on the odor of apples.[122] Usually they are represented with the trees that supply their apples,[123] but one can imagine that Waldseemüller decided that he simply did not have room for trees with apples large enough to be visible. Pierre d'Ailly and other writers locate this race near the source of the Ganges, and although Waldseemüller does not mention the Ganges in his legend here, the river that he shows originating near these creatures is identified further to the south as the Ganges, indicating that he was following his source closely. This legend is repeated in a slightly longer version in Legend 4.10 below.

> 4.7
> Hic sunt homines viri habentes capita canina mulieres autem sunt ut nostre.
>
> Here are people where the men have dogs' heads, but the women are like ours.

This legend comes not from d'Ailly, but rather from Plano Carpini.[124] There is a perfectly logical explanation for the presence of this race in India, rather than in Tartaria (where the other races that derive from Plano Carpini appear)—and thus for the representation of cynocephali in India twice. Plano Carpini recounted a fierce battle between the Tartars and these cynocephali in which the Tartars were defeated. As a result, the Tartars did not incorporate the cynocephali into their empire,

---

[118]See Pierre d'Ailly, *Ymago mundi*, ed. Edmond Buron, vol. 1, Chap. 16, "De mirabilibus Indie," p. 267.

[119]For discussion of the Carimaspians/Arimaspians see David Malcolm Blamires, *Herzog Ernst and the Otherworld Voyage: A Comparative Study* (Manchester, Manchester University Press, 1979), pp. 99–112.

[120]On the *blemmyae* see Ivar Hallberg, *L'Extrême Orient dans la littérature et la cartographie de l'Occident des XIIIe, XIVe, et XVe siècles; étude sur l'histoire de la géographie* (Göteborg: W. Zachrissons boktryckeri a.-b., 1907), pp. 78–79; and Asa Mittman, *Maps and Monsters in Medieval England* (New York: Routledge, 2006), pp. 85–106.

[121]On the distinction between *blemmyae* and *epiphagi* see Friedman, *The Monstrous Races* (see note 74), p. 15 and the image on p. 10. The difference between the *blemmyae* and the *epiphagi* is illustrated among the monstrous races in southern Africa on the Hereford *mappamundi*, for example: see Westrem, *The Hereford Map* (see note 143 in Chap. 1 above), pp. 382–383, nos. 971 and 973.

[122]There is some discussion of the *astomoi* in H. Hosten, "The Mouthless Indians of Megasthenes," *Journal and Proceedings of the Asiatic Society of Bengal*, New Series 8 (1912), pp. 291–301.

[123]The apple-smellers are represented with their trees, for example, in Paris, Bibliothèque nationale de France, MS fr. 2810, f. 219v, an illustrated manuscript of Mandeville; and in New York, Pierpont Morgan Library, MS 461, f. 41v, fifteenth century, reproduced in Friedman, *The Monstrous Races* (see note 302 in Chap. 1 above), p. 159.

[124]These cynocephali are described in Vincent of Beauvais, *Speculum historiale*, Book 32, chap. 11; see Beazley, *The Texts and Versions* (see note 43 in Chap. 1 above), pp. 83–84, and for Plano Carpini's text, p. 56. For an English translation of the passage in Plano Carpini see Dawson, *The Mongol Mission* (see note 31), p. 23.

Fig. 2.4 The cynocephalus in Hartmann Schedel's *Buch der Croniken* (Nuremberg, 1493), f. 12r. Library of Congress, Rare Book and Special Collection Division, Rosenwald Collection 166. Courtesy of the Library of Congress

and so Waldseemüller located them not in Tartaria, but just across the border in India. In this case we can identify the source of Waldseemüller's illustration: his cynocephalus here is very similar to that in Hartmann Schedel's *Liber chronicarum*, f. 12r (Fig. 2.4). We thus know another book that Waldseemüller had on his shelf, although he makes no mention of it in his list of sources.

4.8
Hic nascitur piper zinziber reobarbarum Spiconardi cassia fistula Cardimonuium et mira vol[umen] et varia genera specierum que abhinc portantur ad emporia maritima.

Here grows pepper, ginger, rhubarb, spikenard, cassia pods, cardamom, and a wonderful quantity and a great variety of spices which from here are carried to markets by the sea.

I have not succeeded in finding the source of the list of spices that grow by the Indus River. Pepper is shown growing just to the east of this legend, and that pepper is mentioned in the following legend.

4.9
In istis montanis nascitur piper in magna habundantia et a pigmeis plantatur quibus eciam continuo bellum est contra grues.

In these mountains pepper grows in great abundance, planted by the pygmies, who are in continual battle against cranes.

The final monstrous race in India is to the south of all the others, almost in the middle of the territory demarcated as India. The legend comes from d'Ailly, *Ymago mundi*, Chap. 16. Pierre d'Ailly places more emphasis than Waldseemüller does on the pygmies' small stature. D'Ailly explains that pepper is naturally white, but the pygmies use fire to drive away the serpents that live among the plants, and it is the fire that turns the pepper black. Waldseemüller's depiction of the pepper harvest rather than the battle with the cranes is unusual, and could reflect contemporary interest in the spice trade. It was more common to portray the pygmies' battle with the cranes, as on the Catalan Atlas of 1375, the so-called 'Genoese' map of 1457, and Pierre Desceliers's world map of 1550.[125] I have not found a source for Waldseemüller's image of the pygmies, so it may be another creation of Waldseemüller's workshop.

4.10
Hic reperiuntur homines qui solo odore cuiusdam pomi vivunt et si longius eunt pomum ferunt ne moriatur

Here are found men who live just on the odor of a certain apple, and if they go far, they bring the apple lest they die.

This legend is a slightly longer version of Legend 4.6 above; see the discussion of the source there.

4.11
GVZERAT ET QVE GEDROSIA Guzerat ciuitas regal

Gujarat or Gedrosia. Gujarat is a royal city.

Part of this legend is in the lower right corner of sheet 3 (see Legend 3.37). The information about Gujarat comes from Varthema.[126] Although it is not clear on Waldseemüller's map, Gujarat is on the coast and was an important center of trade, which is why it was of interest in Europe.[127] There is a brief discussion of Gujarat in Fries's *Uslegung*, Chap. 54.[128]

4.12
REGNUM IOGHE Ciuitates et oppida non habent

The kingdom of the Ioghe. The people have no cities or towns.

The Joghees are a sect rather than a people, and are described by Varthema.[129] The king of the Joghees is illustrated in the 1515 edition of Varthema on f. 29r, and Waldseemüller's image is similar with regard to his crown and bare feet but not similar enough to be able to assert with confidence that Waldseemüller used the Varthema illustration as a model.[130]

4.13
Et si India magna sit in quantitate non tamen minor in mirabilium varietate scilicet hominum bestiarum et plantarum nam indeserris [i.e. *in desertis*] reperiuntur diuersa monstra humanam effigiem preferentia nec non elephantes leones aues et allia quam multa animalia nostris bestiis dissimilia eciam gemmarum aromatumque varietas. Habet enim hec

[125]For references on the Catalan Atlas see note 38 in Chap. 1 above. The Genoese map is in Florence, Biblioteca Nazionale Centrale, Portolano 1; it is discussed by Edward Luther Stevenson, *Genoese World Map, 1457* (New York: American Geographical Society and Hispanic Society of America, 1912), and is conveniently reproduced in color in is reproduced in Cavallo, *Cristoforo Colombo e l'apertura degli spazi* (see note 134 in Chap. 1 above), vol. 1, pp. 492–493, and the map has been reproduced in facsimile with a new transcription and translation of the legends by Angelo Cattaneo, in *Mappa mundi 1457* (Roma: Treccani, 2008). For references on Desceliers's map of 1550 see note 13.
[126]See Varthema, *The Travels of Ludovico di Varthema* (see note 103), pp. 105–107.
[127]See Geneviève Bouchon, "Pour une histoire du Gujarat au XVe au XVIIe siècle," *Mare Luso-Indicum* 4 (1980), pp. 145–158; reprinted in her *Inde découverte, Inde retrouvée, 1498–1630: études d'histoire indo-portugaise* (Lisbon: Fundação Calouste Gulbenkian; and Paris: Centre Culturel Calouste Gulbenkian and Commission Nationale pour les Commémorations des Découvertes Portugaises, 1999); and K. S. Mathew, "Maritime Trade of Gujarat and the Portuguese in the Sixteenth Century," *Mare Liberum* 9 (1995), pp. 187–194.
[128]Fries's text about Gujarat is translated into modern German by Petrzilka, *Die Karten des Laurent Fries* (see note 202 in Chap. 1 above), p. 139.
[129]See Varthema, *The Travels of Ludovico di Varthema* (see note 103), pp. 111–113.
[130]For discussion of the image of the king in the 1515 edition of Varthema see Stephanie Leitch, "Recuperating the Eyewitness: Jörg Breu's Images of Islamic and Hindu Culture in Ludovico de Varthema's *Travels* (Augsburg: 1515)," in *Mapping Ethnography in Early Modern Germany: New Worlds in Print Culture* (Basingstoke and New York: Palgrave Macmillan, 2010), pp. 101–145, esp. 142.

India multa regna et ciuitates sine numero cuius plerumque gentes et reges sunt ydolatre caffrani et Machometani nudi tectis pudendis incedunt Cristiani autem et iudei indici (quorum paruus est numerus) degunt sub imperio eorundem. Hanc primus portugallensis classis lustrauit anno domini .1495. cuius capitaneus erat vascus de Gyman.

And if India is great in size, it is nonetheless no smaller in terms of variety of wonders, namely of men, beasts, and plants, for in its deserts are found various monsters of human form, and also elephants, lions, birds, and numerous other animals unlike those of Europe, and further, a variety of gems and spices. For this India has many kingdoms and cities without number, and most of its people and kings are Caffrani idolaters and followers of Mohammed. They go nude but do cover their private parts. However, Christians and some Indian Jews (who are very few) do live under their rule. A Portuguese fleet first sailed along the coast of India in 1495, and the captain of the fleet was Vasco da Gama.

The beginning of this legend draws from the beginning of Pierre d'Ailly's chapter on the marvels of India (*Ymago mundi*, Chap. 16), which was the source of Waldseemüller's legends about the monstrous races: *Ex premissis manifestum est quam magna est India in quantitate. Se ex sequentibus patet que ipsa non est minor in mirabilium varietate*, "From what preceded it is clear how large India is, and from what follows it will be clear that it is not less in variety of wonders." The other parts of the legend are general enough that it is difficult to assign a source; for example, Marco Polo, Varthema, and Springer all mention that there are animals in India very different from those in Europe.[131] The legend thus seems to be Waldseemüller's gathering of his impressions from various sources about India. The reference to there being few Indian Jews comes either from Caspar the Jew of India, also called Gaspar de Gama, one of Waldseemüller's sources, or Joseph the Indian.[132] Contrary to the date of 1495 mentioned in the legend, Vasco da Gama did not reach India until 1498. Just to the right of this legend is the image of *suttee* or *sati* discussed above in the introduction (see pp. 27–29). On the Caffrani idolaters see Legend 3.37 e.

4.14
iste fluvius habet cocodrillos magnos et Anguillas. xxx. podum

This river has big crocodiles and eels 30 feet long.

The river is the Ganges; the tradition of long eels in the Ganges goes back to Pliny 9.2.4, and was repeated by many authors, including Honorius Augustodunensis, *Imago mundi* 1.12, and Gervase of Tilbury, *Otia imperialia* 2.3,[133] but in all of these cases the eels are 300 feet long, rather than thirty. It seems most likely that Waldseemüller either mistakenly wrote "xxx" in place of "ccc," or else made the change intentionally in order to make the creature's length more believable.

4.15
Rex Banghelle cum tota sua gente Macometanus est et potentissimus gerit bellum contra regem Narsinge .200000. armatorum copia Terra est omnium fertilissima in Bladis carnibus zinziro sacaro et porcellana Est .11. hoc regnum longe extensum[?]

The King of Bengala, together with all his people, is a follower of Mohammed and is very powerful. He wages war against the king of Narsinga with an army of 200,000. The land is extremely fertile in grain, meat, ginger, sugar, and porcelain. This kingdom is eleven [days] wide.

This legend comes from Varthema;[134] however, Varthema says that it was an eleven-day sea journey to reach this kingdom, not that it was eleven days wide, and he does not mention porcelain. So it seems that Waldseemüller took some liberties here.

[131]See Marco Polo, *Marka Pavlova z Benátek, Milion* (see note 54), Book 3, Chap. 32, p. 179, and *The Book of Ser Marco Polo* (see note 54), Book 3, Chap. 17, vol. 2, pp. 344–345; Varthema, *The Travels of Ludovico di Varthema* (see note 103), p. 122; and Balthasar Springer, *The Voyage from Lisbon to India, 1505–6: Being an Account and Journal*, ed. C. H. Coote (London: B. F. Stevens, 1894), p. 14 in the chapter on Cochin. Note that Springer's work is misattributed to Vespucci in this edition.

[132]For the remark by Caspar see the *Paesi* or *Itinerarium*, Chap. 60; translated into English in Ravenstein, *A Journal of the First Voyage* (see note 57 in Chap. 1 above), pp. 137–141, at 137. For the remark by Joseph see the *Paesi* or *Itinerarium*, Chap. 130, and for an English translation, Greenlee, "The Account of Priest Joseph" (see note 70 in Chap. 1 above), pp. 95–113, at 99.

[133]See Honorius Augustodunensis, *Imago mundi*, ed. Valerie Flint in *Archives d'histoire doctrinale du Moyen Age* 57 (1982), pp. 48–93, at p. 55; and Gervase of Tilbury, *Otia imperialia: Recreation for an Emperor*, ed. and trans. S. E. Banks and J. W. Binns (Oxford: Clarendon Press, 2002), pp. 190–191.

[134]Varthema, *The Travels of Ludovico di Varthema* (see note 103), pp. 210–212.

4.16
Mongal vera Tartaria est que in aliqua parte nimium montuosa et in aliqua campestris sed tota fere admixta glaria plurimum arenosa nec est in centesima parte fructuosa nec enim fructum potest portare nisi aquis fluuialibus irrigetur que ibi sunt rarissime vnde nec ville nec multe ciuitates ibidem reperiuntur. Licet sterilis sit tamen alenpis [emended to *alendis*] pecoribus apta, in aliqua eius parte sunt modice silue alia vero sine lignis est. itaque tam imperator quam principes et omnes alij sedent et cibaria sua decoquunt ad focum de boum et equorum stercoribus factum ipse quoque aer inordinatus est ibidem omnino. Nullus princeps aut dux imperatoris Chaam audet morari in aliqua parte nisi vbi assignatum fuerit et ipsi quidem ducibus vbi maneant. Duces autem loca millenariis asignant. Millenarii centenariis et centenarii decanis et quidquid eis precipitur quocumque tempore et loco siue ad bellum siue ad mortem obediunt sine vlla contradictione nam si petat alicuius virginem filiam sororem mox ei exponunt. Omnia enim sunt in manu Imperatoris quod nemo audet dicere hoc est meum vel illius sed omnia scilicet res iumenta et homines sunt ipsius et sic inter omnes imperatores et reges mundi nullus potentior Chaam qui alias magnus rex de Cambalu dicitur et gloriatur se esse filium dei.

Mongolia is the true Tartaria, which in some parts is extremely mountainous, while in others it is flat, but almost the whole of it is composed of very sandy gravel. Not one hundredth part of the land is fertile, nor can it bear fruit unless it is irrigated by running water, and streams are very rare there, so there are no towns and few cities found there. Although it is sterile it is nonetheless apt for raising cattle; in some districts there are small woods, but otherwise it is bare of trees. And so the Emperor as well as the nobles and everyone else sit and cook their food at a fire made of the dung of oxen and horses. The weather there is very irregular. No prince or general of the Great Khan dares to stay anywhere except in the place assigned to them. And they tell the generals where to stay, and the generals fix the positions of the captains of a thousand, the captains of a thousand those of the captains of a hundred, and the captains of a hundred those of the captains of ten. Moreover, whatever command is given to them, whatever the time, whatever the place, be it to battle or to death, they obey without a word of objection. Even if he asks for someone's unmarried daughter or sister, they give her to him quickly. For all things belong to the Emperor, such that no one dares to say "This is mine" or "That is his," but everything—property, beasts of burden and people—belong to him, and so among all of the emperors and kings of the world, none is more powerful than the Khan, who is also called the Great King of Cambalu [Khanbaliq], and boasts that he is the son of god.

Most of this legend comes from Plano Carpini,[135] except for the last part about the Khan's great power and his claim to be the son of god, which comes from Simon of Saint-Quentin,[136] so Waldseemüller has combined passages from Vincent of Beauvais's extracts from those two authors.[137]

4.17
magis Septentrionales equitant ceruos

In the far north they ride deer.

This legend and illustration were discussed in the introduction (see pp. 37–38): they come from the nautical chart tradition, and demonstrate that Waldseemüller had a large, elaborately decorated older nautical chart in his workshop.

4.18
Magnus Tartarum Gog Chaam Rex regum et dominus dominantium

The great Tartar Gog Khan, king of kings and lord of lords.

[135]See Giovanni da Pian di Carpine, *Storia dei Mongoli* (see note 31), pp. 230 and 267–277 (Latin) and 339 and 358 (Italian); and Dawson, *The Mongol Mission* (see note 31), pp. 5 and 27.
[136]See Simon de Saint-Quentin [fl. 1245–1248], *Histoire des Tartares*, ed. Jean Richard (Paris: P. Geuthner, 1965), p. 92, which is from Vincent of Beauvais, *Speculum historiale*, Book 32, Chap. 34: *Et hoc quidem noman chan sive chaam est appellativum idemque sonat quod rex vel imperator, sive magnificus vel magnificatus, sed hoc Tartari singulariter attribuunt domino suo, nomen ejus proprium reticendo. Ipse quoque gloriatur se esse filium Dei, seque sic ab hominibus appellari.*
[137]Incidentally Fries discusses the Great Khan in Chap. 112 of the *Uslegung*; this passage is translated into modern German by Petrzilka, *Die Karten des Laurent Fries* (see note 202 in Chap. 1 above), p. 158.

This legend is just above a large and imposing image of the Great Khan in his tent, an image discussed in the introduction (see pp. 17–18), where I suggest that it is to be compared with the large image of King Manuel of Portugal south of the southern tip of Africa, and also note that some details of the portrait, particularly his beard, are not consistent with Waldseemüller's textual sources. The phrase "Rex regum et dominus dominantium" comes from 1 Timothy 6:15 and Revelation 19:16,[138] and the use of this phrase to describe the Great Khan is certainly intended to indicate that he is Antichrist—and though the title "Gog Khan" meant "emperor," the word Gog would inevitably suggest the apocalyptic peoples Gog and Magog to European ears.[139] It is tempting to imagine that the application of this regal phrase to the Khan drew some inspiration from the Khan's boasting description of his own power as recorded by Simon of Saint-Quentin, and retold by Vincent of Beauvais, *Speculum historiale*, 32.46.[140]

4.19
Chingitalis Provincia Hec provincia longa est .16. dierarum habens multas ciuitates et castra gentes sunt ydolatre et aliqui macomethani aliqui cristiani habente .3. ecclesias nestorianas. ibi eciam fit Salamandra

The province of Chingintalas. This province is a sixteen day journey in extent, having many cities and encampments. The people are idolaters and some are followers of Mahommed, while others are Christians. There are three Nestorian churches. And the salamander grows there.

The three Nestorian churches are depicted on Waldseemüller's 1507 map, so this is one of the few elements from that map he retained on his *Carta marina*, and also one of the relatively few legends on the latter map that is based on Marco Polo.[141] Although Polo is often associated with marvels, there are fewer outright marvels in his narrative than one might think—but the salamander is one of them.[142]

4.20
Camul provincia habet multas ciuitates et castra. gentes sunt ydolatre [sub] dominio Tartarorum habent linguam propriam vacant omni voluptati et lascivie. terra est fertilis, adulterum non aduertunt

The province of Kamul has many cities and encampments. The people are idolaters under the power of the Tartars; they have their own language, and are unconcerned about all pleasure and lasciviousness. The soil is fertile. They do not pay attention to adultery.

Camul refers to a region in what is now eastern China, around the city now called Hami, whose Mongolian name was Qamil. The legend comes from Marco Polo.[143] The necessity of adding the word *sub* is indicated by hand, in accordance with the instructions in Legend 9.2

4.21
In hoc deserto habitant homines silvestres non loquentes nec iuncturas habent in cruribus et si quis radunt per se ipsos surgere non possunt.

---

[138]The same phrase is used to describe the Great Khan in the map titled *Tabula superioris Indiae et Tartariae maioris* in the four editions of Ptolemy's *Geography* that have maps by Lorenz Fries: the 1522 (Strasbourg), 1525 (Strasbourg), 1535 (Lyon), and 1541 (Vienna).

[139]On other attempts to associate the Great Khan with Antichrist see Charles Burnett, "An Apocryphal Letter from the Arabic Philosopher al-Kindi to Theodore, Frederick II's Astrologer, Concerning Gog and Magog, the Enclosed Nations, and the Scourge of the Mongols," *Viator* 15 (1984), pp. 151–168; and Charles Burnett and Patrick Gautier Dalché, "Attitudes Towards the Mongols in Medieval Literature," *Viator* 22 (1991), pp. 153–167.

[140]A phrase that also conveys the Khan's great power, but lacks the implication that he is Antichrist, opens a text about the Khan that was written in about 1330 for Pope John XXII: *Rex iste magnificus et potens est inter omnes reges mundi.* See Christine Gadrat, "*De statu, conditione ac regimine magni Canis*: l'original latin du *Livre de l'estat du grant Caan* et la question de l'auteur," *Bibliothèque de l'École des Chartes* 165.2 (2007), pp. 355–371, at 366.

[141]The province of Chingintalas is described in Marco Polo, *Marka Pavlova z Benátek, Milion* (see note 54), Book 1, Chap. 47, p. 48, with some mention of the Nestorians there, but the three churches are mentioned in another passage, namely Book 1, Chap. 64, p. 65. For the corresponding passages in Yule's English translation see *The Book of Ser Marco Polo* (see note 54), Book 1, Chap. 42, vol. 1, pp. 212–213, and Book 1, Chap. 58, vol. 1, p. 281—though in the latter passage Yule chose to follow a manuscript that does not indicate that there are specifically three churches.

[142]For discussion of the medieval myth of the salamander see Christoph Gerhardt, "*Daz werc von salamander* bei Wolfram von Eschenbach und im *Brief des Priesters Johannes*," in Hans-Walter Stork, Christoph Gerhardt, and Alois Thomas, eds., *Ars et Ecclesia: Festschrift für Franz J. Ronig zum 60. Geburtstag* (Trier: Paulinus-Verlag, 1989), pp. 135–160.

[143]See Marco Polo, *Marka Pavlova z Benátek, Milion* (see note 54), p. 46; and *The Book of Ser Marco Polo* (see note 54), Book 1, Chap. 41, vol. 1, pp. 209–211.

In this desert there live wild men who do not speak, and they have no joints in their knees, and if they fall they cannot get back up by themselves.

This kneeless race is another of the monstrous races described by Plano Carpini.[144] The race may have derived from the classical and medieval belief that the elephant had no knees and that if it fell, other elephants had to help it to its feet again.[145] I know of no earlier illustration of this race, and it is almost certain that Waldseemüller's image was the creation of his workshop. The staff with which the cartographer has equipped the man to raise himself should he fall is not mentioned by Vincent of Beauvais. The race is also represented in the 1522 (Strasbourg), 1525 (Strasbourg), 1535 (Lyon), and 1541 (Vienna) editions of Ptolemy in the northwest corner of the *Tabula Moderna of Indiae Superioris*; in Lorenz Fries's *Carta marina* of 1530 and 1531 (but here mistakenly depicted with knees); and on Pierre Desceliers's map of 1550 (London, British Library Add. MS 24065), in northeastern Asia, just east of the text describing the Great Khan.[146]

4.22
Antropophagi sunt. in ista ciuitate residet metropolitanus ydolatrarum qui vocatur bathi et distribuit cuncta beneficia secundum ritum suum sicut papa noster.

They are cannibals. In this city lives the archbishop of the idolaters who is called Bathi and he distributes all benefits according to his rite just like our pope.

This legend comes from Odoric.[147] The scene of the man being chopped up and fed to the birds illustrates the manner of disposing of the dead among this people, which Odoric describes in the same chapter:

Suppose such a one's father to die, then the son will say, "I desire to pay respect to my father's memory," and so he calls together all the priests and monks and players in the country round, and likewise all the neighbors and kinsfolk. And they carry the body into the country with great rejoicings. And they have a great table in readiness, upon which the priests cut off the head, and then this is presented to the son. And the son and all the company raise a chant and make many prayers for the dead. Then the priests cut the whole of the body to pieces, and when they have done so they go up again to the city with the whole company, praying for him as they go. After this, the eagles and vultures come down from the mountains and everyone takes his morsel and carries it away. Then all the company shout aloud, saying, "Behold! The man is a saint! For the angels of God come and carry him to Paradise."

I suspect that Waldseemüller took his iconographic inspiration for the scene from an illustration of the "birds summoned" episode in Revelation 19:17–21, but I have not found such an image that is particularly close to Waldseemüller's. There is an illustration of essentially the same scene of Tibetans cutting up a cadaver and feeding it to birds in the 1501 Strasbourg edition of Mandeville's *Travels*, but Waldseemüller's scene is not particularly similar to that one.[148] As far as the image of Bathi, it is interesting that he is shown without a scepter or weapon; his open-handed gesture is perhaps to be interpreted as alluding to the giving of benefits mentioned in the legend.

4.23
BVRITHABETH PROV. hec diuisa est in .8. regna est execrabilis eciam morus in propria ista nam filii machant patres suos egrotos senio. caput occisi datur filio illud enim manducat et ex ossibus adornatum facit cyphum alie autem corporis partes vulturibus advolantibus administrantur. Corallus hic exponitur pro moneta.

[144]The kneeless race is described in Vincent of Beauvais, *Speculum historiale*, Book 32, Chap. 8; see Beazley, *The Texts and Versions* (see note 43 in Chap. 1 above), p. 81, and for Plano Carpini's text, p. 54. Plano Carpini's text is also supplied in Menestò, *Storia dei Mongoli* (see note 31), pp. 254–255, 352, and 431–432. For an English translation of the passage in Plano Carpini see Dawson, *The Mongol Mission* (see note 31), p. 20.
[145]On the idea that elephants had no knees see A. T. Hatto, "The Elephants in the Strassburg Alexander," *London Medieval Studies* 1 (1937–1939), pp. 399–429, reprinted in Peter Noble, Lucie Polak and Claire Isoz, eds., *The Medieval Alexander Legend and Romance Epic: Essays in Honour of David J. A. Ross* (Millwood, NY: Kraus International Publication, 1982), pp. 85–105, esp. 93–96. Sir Thomas Browne, *Pseudodoxia epidemica* (London: Tho. Harper, 1646), Book 3, Chap. 1, pp. 115–119, also discusses the myth. St. Augustine, *De civitate Dei* 16.8, says that *sciapods*, who shade themselves with their giant single foot, have no joints in their legs.
[146]See Chet Van Duzer, *The World for a King: Pierre Desceliers' Map of 1550* (London: British Library, 2015), pp. 102 and 187 note 315.
[147]See Yule, *Cathay and the Way Thither* (see note 61), vol. 2, pp. 250 (English), 327–328 (Latin), and 363–364 (Italian).
[148]The illustration is in John Mandeville, *Johannes Montevilla der wyffaren de Ritter* (Strasburg: Mathias Hupfuff, 1501), fol. M 2 r, and is reproduced in *La Gravure d'illustration en Alsace au XVIe siècle* (Strasbourg: Presses universitaires de Strasbourg, 1992–2000), vol. 2, p. 212. There is a very similar illustration in the 1481 Basel edition of Mandeville.

The province of Burithabeth [Tibet]. This is divided into eight kingdoms. There is an execrable custom in this very province: the sons kill their fathers when they are worn out with old age. The head of the slain man is given to his son, which he chews on and makes from the bones a very ornate cup, but the other parts of the body are offered to the vultures that fly to the spot. Coral is used as currency here.

This legend also comes from Odoric, and is a summary of the passage from Odoric cited in the commentary on the previous legend. There are illustrations of the son being given the father's head in London, British Library, Harley MS 3954, f. 67v (which illustration includes birds carrying away pieces of flesh); Paris, Bibliothèque nationale de France, MS 2810, f. 223r; Vatican City, Biblioteca Apostolica Vaticana, MS Urb. lat. 1013, f. 26r; and in the 1481 Augsburg edition of Mandeville's *Travels*.

4.24
CHANAN SIVE CHAIRAM PROVINCIA Hec prouincia diuisa est in septem regna subiecta magno Chaam in quibus duobus regnis regnant duo filii (si qui sunt) Chaam. Sunt ydolatre et adulteria vxorum non animaduertentes

The province of Chanan or Chairam. This province is divided into seven kingdoms which are subject to the Great Khan, in two of which kingdoms reign two sons (if there are two) of the Khan. The people are idolaters and pay no heed to their wives' adultery.

This legend comes from Marco Polo,[149] but for this legend Waldseemüller must have been using a version of Polo quite different from the Milan manuscript, as that manuscript does not include the parts about the second king and the adultery.

4.25
TERRA PIGMEORUM. Kakath ciuit[as] pigmeorum magna. Isti pigmei sunt longi tribus palmis qui operantur opera minona [for *minora*] de Sameto et meliora quam alii homines de mundo. Et si ibi magni homines filios generant tamen medietas ass[im]ilantur pigmeis

The land of the pygmies. Kaketie, the great city of the pygmies. These pygmies are three palms tall and make small works of samite, better than any other men in the world. And if large men there have sons, nonetheless half of them are the size of pygmies.

This legend comes from Odoric.[150] On this sheet there are pygmies both in India (see Legend 4.9) and in Tartaria;[151] the latter seem much more muscular than the former.

4.26
In his montibus reperiuntur adamantes et alii lapilli preciosi

In these mountains diamonds and other precious stones are found.

[149]See Marco Polo, *Marka Pavlova z Benátek, Milion* (see note 54) pp. 117–118; and *The Book of Ser Marco Polo* (see note 54) Book 2, Chaps. 48–49, vol. 2, pp. 64–84.
[150]See Yule, *Cathay and the Way Thither* (see note 61), vol. 2, pp. 207 (English) and 316 (Latin); the Italian text he edits does not include the material about the pygmies.
[151]For discussion of the pygmies in Tartaria see Terrien de Lacouperie, "The Negrito-Pygmies of Ancient China," *Babylonian and Oriental Record* 5 (1891), pp. 169–174 and 203–210.

This legend comes from Marco Polo[152]; this is an abbreviated version of a legend that appears on Waldseemüller's 1507 map[153] about the Valley of Diamonds,[154] so this represents one of the relatively few cases in which Waldseemüller retained information from that earlier map. The Valley of Diamonds is discussed in some detail in the chapter on Murfuli in Fries's *Uslegung*,[155] and this chapter is accompanied by a curious image in which the Indians have decorated their bodies with the gems.[156] In his manuscript globe of 1520, Johann Schöner drew many legends from Waldseemüller's *Carta marina*, but in this case rather than borrowing the brief legend from the 1516 map, he copied the much longer legend from his 1507 map.

4.27
Hec terra est fertilis in omnibus. rex huius contrate tributarius est regi de Cusch. habet enim multas vxores et concubinas etiam elephantes domesticos. Est quidem admira[t]ionis dignum quod fit in huius littore maris. Ut fertur: cum enim in illo mari sint varia genera piscium quelibet manerie: seruat naturaliter suum tempus et cum pisces solent se terrae suo inmitere tempore veniunt in magna quantitate et multitudine ad ripam et proiciunt se ad aridam et sic capiuntur ab hominibus et hoc fit in aliquas dies singulis annis.

This land is fertile in all things. The king of this country pays tribute to the king of Cusch. He has many wives and concubines and even domesticated elephants. In fact something worthy of admiration happens in the shore of this sea. It is said that in that sea there are many different types of fish; they naturally keep their time, and as fish are wont to cast themselves ashore at the appropriate time, they come to the shore in great numbers and multitudes and throw themselves on dry land, and so they are caught by men, and this happens on specific days each year.

This legend comes from Odoric.[157] Yule cites some later references to this phenomenon, which certainly involves a beach-spawning species of fish, but it is not clear which species.

4.28
Cathay sunt homines pagani habentes literam spetialem et veteris ac noui testamenti scripturam [emended to *scripturas*]. Vnum deum cristum venerantur et credunt vitam eternam sed non baptisantur Cristianos diligunt elimosinasque faciunt homines benigni satis Barbam non habent sed dispositione faciei cum mongalis concordant. Meliores artifices in mundo non reperiuntur. Terra eorum opulenta nimis in bladis vino auro serico et aliis rebus.

Cathay. The men are pagans and have their own alphabet, and the scriptures of the Old and New Testament [in their language]. They worship one god, Christ, and believe in eternal life, but are not baptized. They love Christians and do works of charity, and are fairly gentle. They have no beards, but their faces are similar to those of Mongols. No better artificers can be found in the world. Their land is very rich in grain, wine, silk, and other things.

The name Cathay was used by early modern Europeans to refer to northern China, and Mangi to designate southern China. The legend is from Plano Carpini.[158] The emendation by hand of *scripturam* to *scripturas* is in accordance with the list of corrections in Legend 9.2.

---

[152]See Marco Polo, *Marka Pavlova z Benátek, Milion* (see note 54), pp. 175–176, and *The Book of Ser Marco Polo* (see note 54), Book 3, Chap. 19, vol. 2, pp. 359–363.

[153]Waldseemüller's 1507 legend reads: *In quibusdam huius regni montibus inueniuntur preciosi lapides adamantes sed periculosum est propter sperpentes magnos quorum ibi est maxima multitudo eciam post pluuias uadunt homines ad flumina que ueniunt de montibus et deficiente aqua in arena reperiuntur adamantes*, "In some of the mountains in this kingdom precious stones, namely diamonds, are found, but it is dangerous because of the large serpents, of which there is a great number there. But after it rains men go to the rivers that flow from these mountains and when the waterflow decreases, diamonds are found in the sand.".

[154]For discussion of the Valley of Diamonds see Thomas Reimer, "Die Diamantenadler des Marco Polo," pp. 241–247, and in French "Marco Polo et les aigles chercheurs des diamants," pp. 491–496, in the commentary volume accompanying Marco Polo, *Das Buch der Wunder = Le livre des merveilles* (Lucerne: Faksimile Verlag, 1995). For additional bibliography see Chet Van Duzer, *Johann Schöner's Globe of 1515: Transcription and Study* (Philadelphia: American Philosophical Society, 2010), pp. 142–143, note 14.

[155]The Valley of Diamonds is discussed in Chap. 36 (i.e. 75) of Fries's *Uslegung*, and this chapter is translated into modern German by Petrzilka, *Die Karten des Laurent Fries* (see note 202 in Chap. 1 above), p. 148.

[156]For discussion of this image see Johnson, *Carta Marina* (see note 21 in Chap. 1 above), pp. 110–111.

[157]See Yule, *Cathay and the Way Thither* (see note 61), vol. 2, pp. 164–165 (English), 303–304 (Latin), and 348–349 (Italian).

[158]See Giovanni da Pian di Carpine, *Storia dei Mongoli* (see note 31), pp. 257–258 (Latin) and 353–354 (Italian); and Dawson, *The Mongol Mission* (see note 31), pp. 21–22.

4.29
CAMBALV METROPOLIS In hac ciuitate sedes est Imperialis magni Chaam de Chathayo Imperatoris omnium tartarorum generalis et maximi. habet enim hec in circuitu miliaria 40. italie habet et 12. portas distantes abinuicem .2. miliaria. In medio est palatium Iustitie seu habitationis domini Chaam cuius murus circuit miliaria 4. sunt enlm [for *enim*] in eo .4. columne de auro.

The great city of Cambalu. In this city is the imperial seat of the great Khan of China, the supreme overall Emperor of all the Tartars. This city is forty Italian miles in circumference and has twelve gates, spaced two miles apart. In the middle is the palace of justice or the residence of the lord Khan, the wall of which is four miles in circumference and inside there are four columns of gold.

Marco Polo describes both the palace and the city,[159] but Waldseemüller chose to use the description of Odoric instead.[160] Odoric says that there are twenty-four gold columns, while Waldseemüller says that there are four; this could be a mistake on Waldseemüller's part, or it could be a clue regarding which version of the text he was using.

4.30
Magnus Gog Chaam diuisit prouinciam mangi in novem regna que habent ferme 2000. magnarum ciuitatum, non est dicior prouincia mundo. habundat omnibus necesariis humane vite sunt ydolatre et magni artifices et mercatores etiam astrologie multum intendentes. sunt in ea maximi serpentes quas auide commedunt in conuiuiis. sunt et hic multa monasteria religiosorum ydolatrarum rigorosissimam vitam ducentium.

The Great Gog Khan divided the province of Mangi into nine kingdoms which have about 2000 large cities, and there is not a richer province in the world. It abounds in everything necessary to human life. The people are idolaters and great artisans and merchants, and they are great students of astrology. In the province there are huge serpents, which they eagerly eat at feasts. And here there are many monasteries of devout idolaters who follow a very scrupulous life.

The title "Gog Khan" was understood to mean "Emperor." Marco Polo describes the province of Mangi, roughly southern China, and Waldseemüller takes the detail about the division of the province into nine kingdoms from him, but Polo gives the number of cities in the province as 1200 rather than 2000,[161] and Waldseemüller took most of his description of the province from Odoric.[162] The detail about the people being great students of astrology comes from a different part of Odoric's narrative about the Khan and his realm.[163]

[159]See Marco Polo, *Marka Pavlova z Benátek, Milion* (see note 54), pp. 81–84; and *The Book of Ser Marco Polo* (see note 54), Book 2, Chaps. 10–11, vol. 1, pp. 362–366 and 374–375.
[160]See Yule, *Cathay and the Way Thither* (see note 61), vol. 2, pp. 217–222 (English), 318–320 (Latin), and 356–357 (Italian).
[161]See Marco Polo, *Marka Pavlova z Benátek, Milion* (see note 54), p. 144; and *The Book of Ser Marco Polo* (see note 54), Book 2, Chap. 76, vol. 2, pp. 185–193.
[162]See Yule, *Cathay and the Way Thither* (see note 61), vol. 2, pp. 179–182 (English), 309–310 (Latin), and 351 (Italian).
[163]See Yule, *Cathay and the Way Thither* (see note 61), vol. 2, pp. 238 (English), 325 (Latin), and 362 (Italian).

2.5 Sheet 5. Northern South America and the Caribbean (Plate 2.5)

The large crescent in the South American hinterland has been the subject of controversy: Waldseemüller copied it from Caverio's chart of c. 1503, but its meaning is not entirely clear. On Caverio's chart it is at the westernmost node of the network of rhumb lines, and was no doubt intended to indicate West, as the easternmost node of the rhumb lines marked by a sun, clearly indicating East. But on the *Carta marina* there is no connection between the crescent and the rhumb lines, and no sun in the eastern part of the map. In East Africa and Arabia (see sheet 7), Waldseemüller used a crescent to indicate Muslim control of a region, but it seems very unlikely indeed that Waldseemüller understood the symbol to indicate that the New World was Muslim. The image of the opossum is the earliest surviving European representation of the animal. The cannibals derive from the account of an early explorer, probably either Columbus or Vespucci, both of whom mention man-eaters in the lands they visited. Note the attempt at ethnographic verisimilitude in the representation of the cannibals' clothing as being made of feathers.

> 5.1
> Spagnolla que et Offira dicitur gignit aurum Masticem Aloen porcellanam etiam Canellam et zinzibrem. Latitudo Insule .440. miliariis longitudo .880. miliariis est: et inuenta per Cristoferum Columbum Januensem capitaneum regis Castile Anno domini .1492. Accole vescuntur serpentibus maximis et radicubus dulcibus saporem castanearum preferentibus (quas agres vocant) loco panis: carent quorum ferro sed pro eo vtuntur lapidibus acutis.
>
> Hispaniola, also called Ophir, produces gold, mastic, aloe, purslane [a kind of portulaca], in addition to cinnamon and ginger. The island is 440 miles wide, and 880 miles long: it was discovered in 1492 by the Genoese Christopher Columbus, Admiral of the King of Castile. Instead of bread, the natives eat large snakes and sweet roots that taste like chestnuts (which they call *agres*). They have no iron, but use sharp stones.

This legend is transcribed and translated in *Columbian Iconography*,[164] but I supply a revised translation here; Maria Teresa di Palma in her discussion of the legend in that book says that the measurements of the island come from the *Libretto de tutta la navigatione* of 1504, rather than later books into which the text of the *Libretto* was incorporated (the *Paesi novamente retrovati*, *Occeani decas*, *De orbe nove*), but this is not the case: in Chap. 13 of the *Libretto*, the island is said to be 340 miles wide, rather than 440 (the figure Waldseemüller cites),[165] and I do not see that it is possible to determine which book Waldseemüller was using here on the basis of these measurements, though we can be reasonably confident that he was not using the *Occeani decas* of 1511, which does not give a figure in miles for the width of the island, saying only that it is five degrees wide.[166] In the *Paesi novamente retrovati* the details about Hispaniola listed here may be found in chapters 89 and 96. The dimensions of the island are in Italian miles, and thus agree with the scale of Italian miles at the bottom of sheet 9. As indicated in the introduction, the identification of Hispaniola with Ophir[167] is part of the Columbian conception of the newly discovered lands adopted in the *Carta marina*, in contrast to the Vespuccian conception Waldseemüller had adopted in his 1507 map. As di Palma notes, this same legend appears in the 1522 (Strasbourg), 1525 (Strasbourg), 1535 (Lyon), and 1541 (Vienna) editions of Ptolemy's *Geography* and on Lorenz Fries's *Carta marina*;[168] the identification of Hispaniola with Ophir continued to be discussed until the end of the sixteenth century.[169]

> 5.2
> Terra parias Hic Margaritarum et Auri copia vescuntur testudinibus et radicibus loco panis vinum palmarum bibunt capras boues et oues non habent
>
> Land of Parias. Here there is an abundance of pearls and gold; the inhabitants eat turtles and roots instead of bread, and they drink palm wine. They have no goats, cows or sheep.

---

[164]See Gaetano Ferro et al., *Columbian Iconography* (see note 108 in Chap. 1 above), p. 527.

[165]See Thacher, *Christopher Columbus* (see note 64), vol. 2, pp. 467 (in the facsimile of the *Libretto*) and 496 (in Thacher's English translation).

[166]Pietro Martire d'Anghiera, *Opera: Legatio Babylonica; Occeani decas; Poemata; Epigrammata* (Seville: I. Corumberger, 1511), in Book 3 of the *Occeani decas*.

[167]Hispaniola is identified with Ophir in the *Libretto*, Chap. 13, see Thacher, *Christopher Columbus* (see note 64 in Chap. 1 above), vol. 2, pp. 467 (in the facsimile of the *Libretto*) and 495–496 (in Thacher's English translation). In the *Paesi novamente retrovati* the passage is in Chap. 96; in Pietro Martire d'Anghiera's *Occeani decas* of 1511 it is in Book 3.

[168]Incidentally Fries discusses Spagnola in Chap. 96 of the *Uslegung*; this is translated into modern German by Petrzilka, *Die Karten des Laurent Fries* (see note 202 in Chap. 1 above), p. 154.

[169]On the identification of Hispaniola with Ophir see the references cited in notes 85 and 86 in Chap. 1 above.

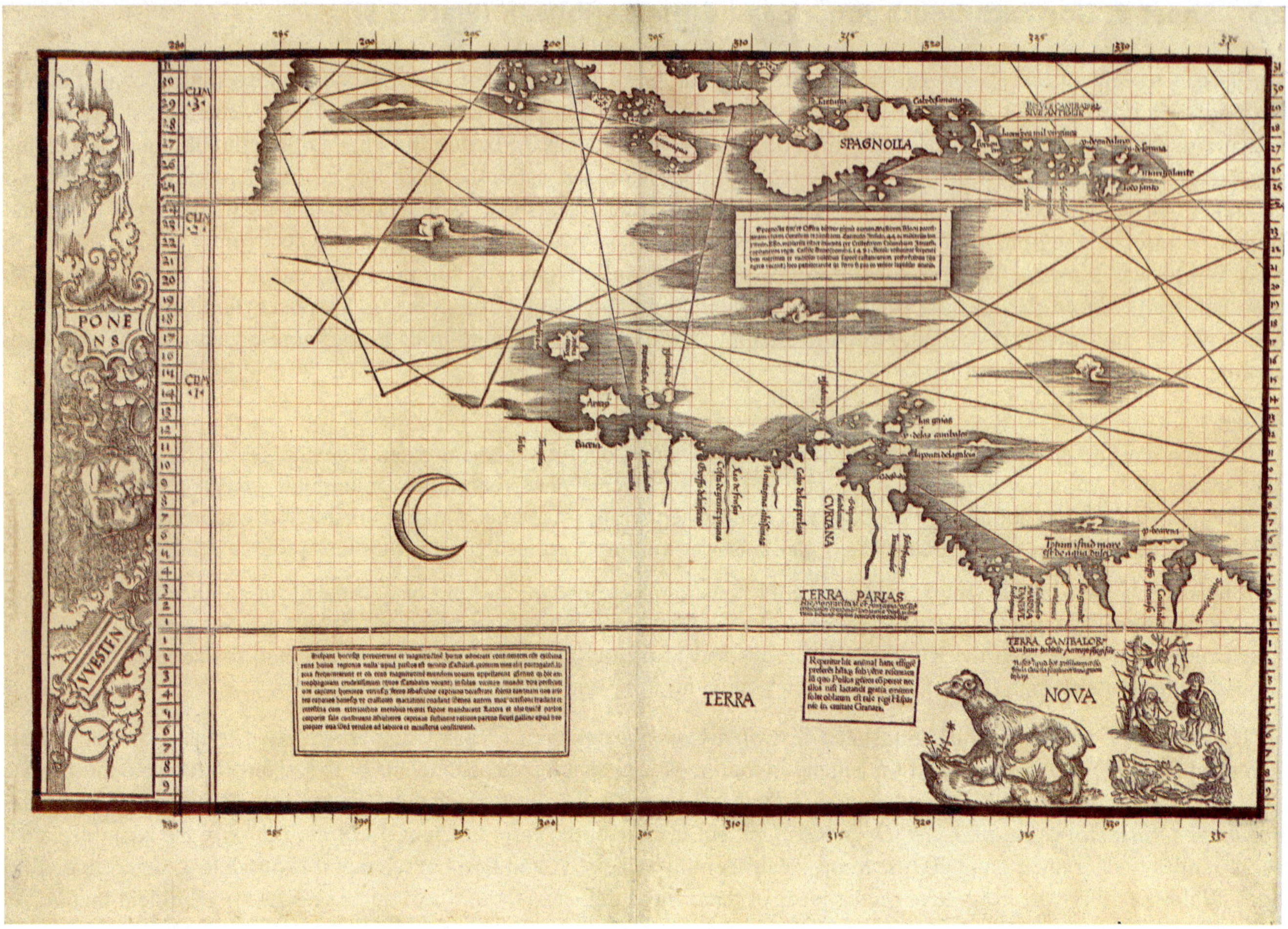

Plate 2.5 Sheet 5 of the *Carta marina*: Northern South America and the Caribbean. Courtesy of the Library of Congress

This legend is transcribed and translated in *Columbian Iconography*,[170] but I supply a slightly revised translation here. The name "Parias" comes from Columbus's account of his Third Voyage (1498–1500), and the name is given in the *Paesi novamente retrovati*, Chap. 105.[171] The wine of the Native Americans is mentioned in the same chapter, while the roots they eat are mentioned in Chap. 89, their eating of turtles (here called calandre) in Chap. 98, and their lack of goats, cows, or sheep is in Chap. 109. Johann Schöner has a legend very similar to this of Waldseemüller on his manuscript globe of 1520.

5.3
Hispani hucusque peruenerunt et magnitudinem huius admirati continentem esse existimarunt huius regionis nulla apud priscos est mentio Castilienses primum mox alii portogalenses litora frequentarunt et ob eius magnitudinem mundum nouum appellarunt Genus que hic antrophagorum crudelissimum (quos Canibales vocant) insulas vicinas inuadit dira persecutione capiunt homines vtriusque sexus. Masculos captiuos decastrare solent tanquam nos arietes capones bouesque vt crassiores mactationi euadant. Senes autem mox occisioni tradunt et intestina cum exterioribus membris recenti sapore manducant. Latera et alie quidem partes corporis sale conseruant. Mulieres captiuas sustinent ratione partus sicuti galline apud nos propter oua. Sed vetulas ad labores et ministeria constituunt.

The Spanish came to this point, and admiring the size of the land, judged that it must be a continent. There is no mention of this region by ancient authors. First the Castilians and then others, the Portuguese, frequented these

[170]See Gaetano Ferro et al., *Columbian Iconography* (see note 108 in Chap. 1 above), p. 547.

[171]For a fuller account of the name Parias see Henry Harrisse, *The Discovery of North America* (London: H. Stevens, 1892; Amsterdam: N. Israel, 1961), p. 318; and Christopher Columbus, *Select Letters of Christopher Columbus, with Other Original Documents, Relating to His Four Voyages to the New World*, ed. and transl. R. H. Major (London: Hakluyt Society, 1847), p. 120.

coasts, and because of its size, they called it a New World. The race that lives here are very cruel man-eaters (whom they call cannibals) who attack the neighboring islands and with terrible pursuit capture people of both sexes. They castrate their male captives, just as we do with rams, roosters and bulls, so that they are fatter for slaying, but they kill the old men directly, and they eat the intestines together with the exterior parts of the body while their taste is fresh. They preserve the sides and other body parts in salt. They keep the female captives alive to bear children, as we keep hens for eggs, but the old women they set to toil and servitude.

The derivation of this legend is interesting. The first part is borrowed from a legend on Johann Ruysch's world map of 1507–08, which appeared in some copies of the 1507 edition and in the 1508 edition of Ptolemy's *Geography* published in Rome.[172] On the Ruysch map there is a large banner covering the western coast of South America, and the text in it reads[173]:

Hucusque naute Hispani venerunt et hanc terram propter eius magnitudinem Mundum Novum appellarunt quai vero eam totaliter non viderunt nec usque in tempore hoc longius quam ad hunc terminum perlustrarunt ideo hic imperfecta relinquitur presertim cum nesciatur quo vergitur.

Spanish sailors came as far as this and called this land the New World because of its size, since indeed they did not see it all, nor up to this time have they surveyed further than this limit. Therefore here it is left unfinished, especially since it is not known in which direction it tends.

As far as I know, Waldseemüller's use of Ruysch here has not been pointed out before. The latter part of Waldseemüller's legend comes from the *Libretto*, the *Paesi novamente retrovati*, or the *Itinerarium Portugallensium*[174]—it does not seem possible to determine from which.[175] Johann Schöner has a legend very similar to Waldseemüller in the northern part of South America on his manuscript globe of 1520 at the Germanisches Nationalmuseum in Nuremberg.

5.4
Reperitur hic animal hanc effigiem proferens huiusque sub ventre reservaculum quo Pullos genitos comportat nec illos nisi lactandi gratia emittere solet oblatum est tale regi Hispaniae in ciuitate Granata

An animal that looks like this is found here; it has a bag under its belly where it carries its offspring, and it only allows them out for nursing. One such animal was given to the King of Spain in Granada.

This legend is transcribed and translated in *Columbian Iconography*; I give a slightly different translation here.[176] Waldseemüller could have taken this information from the *Paesi*, the *Itinerarium Portugallensium*, or the *Occeani decas*. Here is the passage in the *Itinerarium*, Chap. 113:

Conspexere etiamnum ibi animal quadrupes prodigiosum quidem: nam pars anterior vulpem: posterior vero simiam praesentabat, nisi quod pedes effingit humanos: aures autem habet noctuae, & infra consuetam alvum aliam habet instar crumenae, in qua delitescunt catuli ejus tantisper: donec tuto prodire queant: & absque parentis tutela cibatum quaerere: nec unquam exeunt crumenam: nisi cum sugunt: portentosum hoc animal cum catulis tribus sibiliam delatum est: & ex sibilia illiberim id est granatam: in gratiam regum: qui novis semper rebus oblectantur.

---

[172]On the Ruysch map see John Boyd Thacher, *The Continent of America: Its Discovery and Its Baptism* (New York: William Evarts Benjamin, 1896), pp. 209–219; Bradford F. Swan, "The Ruysch Map of the World (1507–1508)," *Papers of the Bibliographical Society of America* 45 (1951), pp. 219–236; Donald L. McGuirk, Jr., "Ruysch World Map: Census and Commentary," *Imago Mundi* 41 (1989), pp. 133–141; and Peter Meurer, "Der Maler und Katrograph Johann Ruysch," *Geschichte in Köln* 49 (2002), pp. 85–104. Ruysch's map is well reproduced in Nebenzahl, *Atlas of Columbus* (see note 217 in Chap. 1 above), pp. 48–49.

[173]The legend on Ruysch's map is translated by Thacher, *The Continent of America* (see note 172), p. 214; transcribed by Mauro Bini et al., *Alla scoperta del mondo: l'arte della cartografia da Tolomeo a Mercatore* (Modena: Il Bulino, 2001), p. 173; and transcribed and translated by Gaetano Ferro et al., *Columbian Iconography* (see note 108 in Chap. 1 above), p. 437; and John Thorley, *Documents in Medieval Latin* (Ann Arbor: University of Michigan Press, 1998), pp. 134 and 196.

[174]The relevant passage are the *Libretto*, Chap. 5 (see Thacher, *Christopher Columbus* (see note 64 in Chap. 1 above), vol. 2, pp. 458 and 487–488); and Chap. 88 in both the *Paesi* and the *Itinerarium Portugallensium*.

[175]Surekha Davies, *Renaissance Ethnography and the Invention of the Human: New Worlds, Maps and Monsters* (Cambridge and New York: Cambridge University Press, 2016), p. 91, seems to suggest that Waldseemüller's legend is closest to that in Pietro Martire d'Anghiera's *Occeani decas*, but Waldseemüller's legend does not include any of the extra phrases from the *Occeani decas* that do not appear in the other texts, such as *Mulieres comedere apud eos nefas est et obscenum*.

[176]See Gaetano Ferro et al., *Columbian Iconography* (see note 108 in Chap. 1 above), p. 547.

For they saw there a bizarre quadruped animal: its front section it is like that of a fox, while its rear section is like that of a monkey, except that it has feet that look human; it has ears like those of an owl, and below the usual belly it has another like a purse, in which its young hide until they can safely emerge, and seek food without the protection of their parent, and they never leave this pouch, except to suckle. This monstrous beast with three of its young was brought to Seville, and from Seville to Illiberis, or Granada, as a gift for the kings, who always delight in new things.

Thus there are a few different sources from which Waldseemüller might have obtained his description of the creature, but the image is a different matter. As I mentioned in the introduction (see pp. 22–23), this is the earliest surviving European image of the opossum, which was of great interest to European explorers as it was the first marsupial they had ever seen. Waldseemüller's image was very influential,[177] and while generally not very accurate, seems a little too good in some details (such as the creature's nipples) to have been created simply on the basis of the verbal description in the *Itinerarium* or other related sources. Thus it seems likely that Waldseemüller had access to an image of the creature, whether manuscript or printed, that no longer survives.

5.5
Terra cannibalorum qui hanc habitant Anthropophagi sunt

Land of cannibals. Those who inhabit this land are man-eaters.

This legend is transcribed and translated in *Columbian Iconography*; I give a slightly different translation here.[178] A very similar legend appears on Johann Schöner's manuscript globe of 1520, no doubt borrowed from the *Carta marina*. The images of cannibals near this legend on Waldseemüller's map—with one cannibal roasting human body parts on a spit over a fire, and a cannibal couple by a tree from whose branches human body parts are suspended (see Fig. 2.5)—are clearly part of the European textual and iconographic tradition of alleged New World cannibalism,[179] but it is not clear which precise iconographic source Waldseemüller was using. It seems likely that Waldseemüller was influenced by a woodcut of some New World natives, including cannibals, attributed to Johann Froschauer and made in about 1505,[180] but it is not possible to be sure of this influence.[181] The roasting of a man whole on a spit recalls the similar image on the Kunstmann II map of c. 1506,[182] but as rich as Waldseemüller's library was, it is difficult to believe that he had access to this map. Waldseemüller's

[177]For discussion of Waldseemüller's image of the opossum see Charles R. Eastman, "Early Portrayals of the Opossum," *The American Naturalist* 49.586 (1915), pp. 585–594, esp. 589; Wilma B. George, *Animals and Maps* (Berkeley: University of California Press, 1969), pp. 61–62; and Susan Scott Parrish, "The Female Opossum and the Nature of the New World," *The William and Mary Quarterly*, 3rd Series, 54.3 (1997), pp. 475–514, esp. 483–486.

[178]See Gaetano Ferro et al., *Columbian Iconography* (see note 108 in Chap. 1 above), p. 547.

[179]See Peter Hulme, "Columbus and the Cannibals: A Study of the Reports of Anthropophagy in the Journal of Christopher Columbus," *Ibero-Amerikanisches Archiv* 4 (1978), pp. 115–139; reprinted in his *Colonial Encounters: Europe and the Native Caribbean, 1492–1797* (London: Methuen, 1986), pp. 1–43; Samuel Roy Dunlap, "Among the Cannibals and Amazons: Early German Travel Literature on the New World," Ph.D. Dissertation, University of California, Berkeley, 1992; Philip P. Boucher, *Cannibal Encounters: Europeans and Island Caribs, 1492–1763* (Baltimore: The Johns Hopkins University Press, 1992); Yobenj Aucardo Chicangana, "El festín antropofágico de los indios Tupinambá en los grabados de Theodoro de Bry, 1592," *Fronteras de la historia* 10 (2005), pp. 19–82.

[180]Two copies of this woodcut survive, one in the New York Public Library and the other in the Bayerische Staatsbibliotek, Munich. For discussion and illustration of this woodcut see Wilberforce Eames, "Description of a Wood Engraving Illustrating the South American Indians (1505)," *Bulletin of the New York Public Library* 26.9 (1922), pp. 755–760; Rudolph Schuller, "The Oldest Known Illustration of South American Indians," *Journal de la Société des Américanistes* 16.1 (1924), pp. 111–118, reprinted in *Indian Notes* 7 (1930), pp. 484–497, and *Historical Records and Studies* 20 (1931), pp. 89–97; William C. Sturtevant, "First Visual Images of Native America," in Fredi Chiappelli, Michael J. B. Allen, and Robert Louis Benson, eds., *First Images of America: The Impact of the New World on the Old* (Berkeley, CA and London: University of California Press, 1976), vol. 1, pp. 417–454, at 420 with Fig. 2; and Samuel Roy Dunlap, "Among the Cannibals and Amazons: Early German Travel Literature on the New World," Ph.D. Dissertation, University of California at Berkeley, 1992, pp. 11–17.

[181]So Davies, *Renaissance Ethnography* (see note 175), pp. 88–91.

[182]The "Kunstmann II" map is Munich, Bayerische Staatsbibliothek, Cod. icon. 133; there is a tracing of the map in Konrad Kretschmer, *Die historischen Karten zur Entdeckung Amerikas: Atlas nach Konrad Kretschmer* (Frankfurt: Umschau, 1991), plate 8; it is reproduced at a large scale in black and white in Edward Luther Stevenson, *Maps Illustrating Early Discovery and Exploration in America 1502–1530* (New Brunswick, NJ, 1903); and it is illustrated and discussed in Hans Wolff, ed., *America: Early Maps of the New World* (Munich: Prestel-Verlag, 1992), pp. 134–136; and Ivan Kupčík, *Münchner Portolankarten: Kunstmann I–XIII und zehn weitere Portolankarten = Munich Portolan Charts: Kunstmann I–XIII and Ten Other Portolan Charts* (Munich: Deutscher Kunstverlag, 2000), pp. 28–34. The scene with the roasting on the spit is discussed by Davies, *Renaissance Ethnography* (see note 175), pp. 84–86.

Fig. 2.5 Detail of the cannibals in South America on Waldseemüller's *Carta marina* (sheet 5). Courtesy of the Library of Congress

image is also reminiscent of a woodcut by Jan van Doesborch printed to illustrate *Of the New Landes* (Antwerp, c. 1520),[183] but as the date of that woodcut is uncertain, it is possible that Waldseemüller's image inspired it, rather than vice versa.[184] The fact that Waldseemüller's natives are wearing feathers accords not only with the print of c. 1505 attributed to Froschauer, but also early reports about New World natives.[185]

---

[183]Robert Proctor, *Jan van Doesborgh, Printer at Antwerp: An Essay in Bibliography* (London: Printed for the Bibliographical Society, at the Chiswick Press, 1894), pp. 32–33.

[184]Again, so concludes Davies, *Renaissance Ethnography* (see note 175), p. 91.

[185]For discussion of Native American clothing made of feathers see Nicole Pellegrin, "Vêtements de peau(x) et de plumes: La nudité des Indiens et la diversité du monde au XVIe siècle," in Jean Céard and Jean-Claude Margolin, eds., *Voyager à la Renaissance. Actes du colloque de Tours 30 juin-13 juillet 1983* (Paris: Maisonneuve et Larose, 1987), pp. 509–529. Amy J. Buono, "Feathered Identities and Plumed Performances: Tupinamba Interculture in Early Modern Brazil and Europe," Ph.D. Dissertation, University of California, Santa Barbara, 2007, is excellent on surviving Native American feather capes, but devotes little attention to early textual and iconographic evidence for feather clothing.

5.6
Nascitur apud hos prisilicum cassia fistula Canella silvestris et varia genera bestiarum

Among these peoples grow brazilwood, cassia pods, wild cinnamon, and various types of animals.

This legend is transcribed and translated in *Columbian Iconography*; I give a slightly different translation here.[186] Waldseemüller took this information from one of the usual sources: the *Libretto*, the *Paesi*, or the *Itinerarium Portugallensium*.[187]

[186]See Gaetano Ferro et al., *Columbian Iconography* (see note 108 in Chap. 1 above), p. 547.
[187]The passage is at the very end of the *Libretto*, that is, at the end of Chap. 30: see Thacher, *Christopher Columbus* (see note 64 in Chap. 1 above), vol. 2, pp. 484 and 512; in the *Paesi* and *Itinerarium* the passage is in Chap. 113. For discussion of representations of brazilwood on maps see Yuri T. Rocha, Andrea Presotto and Felisberto Cavalheiro, "The Representation of *Caesalpinia echinata* (Brazilwood) in Sixteenth-and Seventeenth-Century Maps," *Anais da Academia Brasileira de Ciências* 79.4 (2007), pp. 751–765.

2.6 Sheet 6. Western Africa (Plate 2.6)

This sheet of the map is unusual: it is printed on a piece of paper that is smaller than the other eleven and has a different watermark, and also none of the corrections from the list of errata on sheet 9 have been entered on sheet 6, as they have been on the other sheets. Moreover, this sheet was merely laid into the Schöner Sammelband, rather than bound into it. The situation is made more interesting by the fact that Schöner made a manuscript copy of this sheet (see below on sheet 6A), which had been bound into the Sammelband. Schöner seems to have made this manuscript copy as part of a preliminary study for making his 1520 manuscript globe,[188] which draws on Waldseemüller's *Carta marina*, and also perhaps to replace the missing sheet of the map. The sequence of events was perhaps as follows. Schöner obtained a copy of the *Carta marina*, but sheet 6 was missing or damaged. He made a manuscript copy of that sheet, either from the damaged sheet in his possession, or from another copy to which he had access. Subsequently, another printed copy of the sheet was obtained and added to the Sammelband.

> 6.1
> CANARIE INSVLE X (QUE AB ANTIQVIS FORTUNATE DICEBANTVR QVARUM VII SVNT HABITATE SCILICET LANSOROTA FORTENENtura Grancanarina Teneriffa Gimera & ferra quarum quatuor a cristianis habitate scilicet lansarota Teneriffa Gimera et ferra que sunt sub dicione regis castilie oppida et ciui[tates] non habent sed pagos reliquam a saracenis po
>
> The ten Canary Islands, which the ancients called the Fortunate Islands, of which seven are inhabited, namely Lanzarote, Forteventura, Grand Canary, Tenerife, La Gomera, and El Hierro, of which four are inhabited by Christians, namely Lanzarote, Tenerife, La Gomera and El Hierro, which are under the dominion of the King of Castile. They have neither towns nor cities, just villages. The rest by the Saracens….

There is no passage about the Canary Islands in the *Libretto*, but this legend comes from Chap. 7 of the *Paesi*, the *Itinerarium Portugallensium*, or *Newe unbekanthe landte*, the German translation of the *Paesi* of 1508—from part of the description of Cadamosto's voyage.[189] The names of the islands are a mess in the *Itinerarium*, and there is some temptation therefore to think that Waldseemüller must have used the Italian or German versions of the text, but certainly he knew the names of the islands well enough from other sources that he could have seen beyond the strange spellings in the *Itinerarium*, if indeed he was using that book. Waldseemüller follows one of these texts in saying that seven of the islands are inhabited, but only lists six, leaving out La Palma, and he also makes a mistake in indicating which of the islands are inhabited by Christians: the Italian, Latin, and German versions of the text all include Forteventura on this list, but Waldseemüller lists Tenerife in its place. Incidentally there is a surprisingly long discussion of the Canary Islands in Fries' *Uslegung*, Chap. 60 bis.[190]

> 6.2
> A septentrionali parte fluminis de Senega habitant gentes (qui vocantur Asenegi) de primo regno anterioris [for *interioris*] Ethiopie sunt fusci coloris fallaces et loquaces portantes penniculum in facie ad tegendum nares et os que pudenda esse volunt. vivunt sine lege et rege temporali nec oppida pagos vel ciuitates habent sed degunt in desertis hinc inde vagantes et tuguria sua de loco in locum transfferentes [*sic*] iste fluvius de senega diuidit nigros Ethiopes (que zilofi dicuntur) a fulcis [for *fuscis*] scilicet senagis etiam segregat terram fertilem a sterili predicti deserti scinditur in mulos riuulos facitque regnum Seneg (eoltm [for *olim*] Experias dictum) cuius gentes pecunie vsum non habent sed rem pro alia venundant. habent regem proprium non tamen ciuitates sed pagos. mulieres crasse et magnis vberibus laudantur. nude vulgus incedit.
>
> On the northern side of the Senegal River live people called the Asenegi of the first realm of Interior Ethiopia. They are dusky in color, deceitful, and talkative, wearing a little cloth on their faces to cover their nose and mouth, which they consider shameful to behold. They live without law or temporal king, and have no towns, villages, or cities, but live in the desert, wandering here and there, moving their huts from place to place. The Senegal River divides the black Ethiopians, who are called Zilofi, from the dusky people [i.e. the Asenegi], and it also divides the fertile land from the aforementioned barren desert. The river is divided into many streams and creates the kingdom of Seneg

[188]On Schöner's hand-painted globe of 1520 see note 224 in Chap. 1 above.
[189]For an English translation see Gerald R. Crone, ed. and trans., *The Voyages of Cadamosto and Other Documents on Western Africa in the Second Half of the Fifteenth Century* (London: Printed for the Hakluyt Society, 1937), p. 11.
[190]The passage on the Canary Islands is translated into modern German by Petrzilka, *Die Karten des Laurent Fries* (see note 202 in Chap. 1 above), pp. 141–142.

Plate 2.6 Sheet 6 of the *Carta marina*: Western Africa. Courtesy of the Library of Congress

> (which was formerly called Experias), whose inhabitants do not use money, but trade one thing for another. They have their own king, but nonetheless do not have cities, only villages. Their women are fat, and are praised for their large breasts. The common people go naked.

This legend is assembled from several different passages in the *Paesi* or *Itinerarium*, all but one of which relate to the voyage of Cadamosto. The name Asenegi comes from the beginning of Chap. 14 (Crone p. 27)[191]; their dusky color is mentioned in chapters 9 and 14 (Crone pp. 17 and 28); the part about the people being deceitful and so on comes Chap. 17 (Crone pp. 32–33); the part about them covering their mouths is from the end of Chap. 9 (Crone p. 19); the part about them having no temporal king comes from the beginning of Chap. 13 (Crone p. 26); the fact that they wander in the desert is briefly mentioned in Chap. 9 (Crone p. 16); I do not see the lack of towns and villages mentioned in the *Paesi* or *Itinerarium*; the passage about the separation created by the Senegal river is from the beginning of Chap. 14 (Crone p. 27); the name Experias comes from the beginning of Chap. 125; the system of barter is described in Chap. 11 (Crone pp. 19–23) and also briefly in Chap. 12 (Crone p. 25); the part about the people having their own king, and having no cities, only villages, comes from Chap. 15 (Crone p. 29); the part about the women being praised for their large breasts is at the beginning of Chap. 11 (Crone p. 19); and the part about them going naked comes from Chap. 8 (Crone p. 13).

It certainly took some time to select and assemble the parts of this paragraph about these two peoples. In fact, although the paragraph seems to be about the area around the Senegal River, Waldseemüller has assembled sentences from chapters about several different areas, and if he had wanted to, he could have made a different assemblage of sentences from those same chapters that asserted the opposite characteristics: he could have supplied details about the natives' clothes, their towns, etc. By modern standards, Waldseemüller's mixing of sentences with insufficient regard for which precise area they applied to

[191]The references to Crone are to Gerald R. Crone, ed. and trans., *The Voyages of Cadamosto and Other Documents on Western Africa in the Second Half of the Fifteenth Century* (London: Printed for the Hakluyt Society, 1937).

would be considered cavalier, but it is not so unusual among Renaissance scholars. It is perhaps of greater interest to note, first, that Waldseemüller decided to create such a selection in the first place, rather than just supplying information about the region around the Senegal River: evidently the information just about the Senegal region was not interesting enough in his eyes. And second, that while he could have chosen details from this section of the *Paesi* or *Itinerarium* to convey a variety of impressions of the region, he chose details that tend to focus on the primitive and exotic.

It is remarkable that although Waldseemüller was content to make use of illustrations from the 1509 edition of Balthasar Springer's account of his voyage from Lisbon to India, he made no use of the excellent ethnographical material about the native peoples of West Africa in compiling this legend. The system of barter that Waldseemüller briefly cites from Chap. 11 of the *Paesi* or *Itinerarium* is known as the silent trade, and there is an ample bibliography on the subject.[192] Fries, incidentally, discusses the Senegal in Chap. 105 of the *Uslegung*.[193] Schöner on his manuscript globe of 1520 borrowed just part of the first sentence of this legend, beginning with the word *fallaces*.

6.3
HINC PARS SEPTENTRIONALIS ALEMNE CONVERtitur virtute Magnetis versus austrum & e conuerso peragrando equatorem.

From this point [south], the northern end of the compass needle is turned by magnetic force toward the south, and vice versa when crossing the equator [sailing north].[194]

This legend, which is located right at the equator, says that when a ship crosses the equator sailing to the south, the northern tip of the compass needle swings around to point south, and vice versa when it crosses back to the north. The legend, then, is simply false, but it is not clear where Waldseemüller would have obtained such an idea—neither Cadamosto nor Springer says anything similar, for example, and I find no trace of the idea in other sources. I do not think it is possible to read the legend as referring to magnetic declination, particularly as there is no dramatic change in magnetic declination in the waters off equatorial West Africa. Logically the word "alemne" can only mean "of the compass needle"; it is tempting to think that it might be a corruption of the Italian word for magnet, *calamita*, but the matter is not clear.

6.4
Hic nascitur corallus mire magnitudinis.

Here there grows coral of great size.

I do not know the source of this legend; it does not come from Cadamosto or Springer. It may come from one of the nautical charts that Waldseemüller had. Any information about sources of coral would be of great interest, as in the list of commodities available in Calicut supplied on sheet 12, coral is the most expensive. This is one of the legends that Fries did not have room to copy onto his *Carta marina*.

6.5
Hic sunt magne solitudines et deserte in quibus sunt leones pardi tigrides et elephantes

Here there are large deserts and wildernesses in which there are lions, leopards, tigers, and elephants.

Waldseemüller retained very few legends from his 1507 map on the *Carta marina*, but among those few legends are some in West Africa, including this one. The legends are in almost the same positions on the two maps, just a bit further north on the *Carta marina*: it is probably further south on the 1507 map just because Waldseemüller had to leave room for the Niger River. There is a short doublet of this legend just to the west on the *Carta marina* that reads *Magne solitudines hic sunt*, "Here there are large deserts."

---

[192]On the silent trade see P. J. H. Grierson, *The Silent Trade, a Contribution to the Early History of Human Intercourse* (Edinburgh: W. Green and Sons, 1903), pp. 41–68; E. W. Bovill, "The Silent Trade of Wangara," *Journal of the Royal African Society* 29.113 (1929), pp. 27–38; and P. F. de Moraes Farias, "Silent Trade: Myth and Historical Evidence," *History in Africa* 1 (1974), pp. 9–24.

[193]The passage on the Senegal is translated into modern German by Petrzilka, *Die Karten des Laurent Fries* (see note 202), p. 156.

[194]Osvaldo Baldacci, *Columbian Atlas of the Great Discovery*, trans. Lucio Bertolazzi and Luciano Farina (Rome: Istituto poligrafico e Zecca dello Stato, Libreria dello Stato, 1997), p. 124, mistakenly translates the legend "Here north of La Mina the magnetized needle turns toward Auster, and viceversa when crossing the equator," but the idea that "alemne" refers to La Mina is very difficult to accept, first because the legend is south of La Mina, and second because La Mina is not indicated on the map; further, this translation ignores the word "virtute," takes "Magnetis" as nominative when it is genitive, and introduces the word "needle" into the translation from nowhere, if "Alemne" refers to La Mina. "Alemne" must refer to the compass needle, since otherwise it is not clear what the "pars septentrionalis" is part of.

6.6
Hic reperiuntur leones Teopardi [*sic*] elephantes gyraffae et strues etiamque papapagalli [*sic*] et mire magnitudinis serpentes et formice.

Here there are lions, leopards, elephants, giraffes, ostriches, and also parrots, and huge serpents and ants.

This legend, which does not derive from the 1507 map, is another compilation from Cadamosto: the lions, leopards, and ostriches come from Chap. 9 of the *Paesi* or *Itinerarium* (Crone, *The Voyages of Cadamosto*, p. 17); the elephants and giraffes from Chap. 29 (Crone, pp. 46–47), the parrots from Chap. 30 (Crone, pp. 47–48), and the serpents and ants from Chap. 28 (Crone, p. 44). The error "Teopardi" is included in the list of errors in Legend 9.2, but like the other errors listed there for sheet 6, it is not corrected on the printed sheet 6 that has come down to us.

6.7
Hic videtur caballus piscis marinus.

Here is seen the aquatic horse-fish [hippopotamus].

This legend also comes from Cadamosto, from Chap. 44 in the *Paesi* or *Itinerarium* (Crone, *The Voyages of Cadamosto*, p. 73). The legend is located right by the Gambra River, exactly where Cadamosto says that hippopotamuses are found.

6.8
Hic reperitur zibetum et varia genera simearum

Here is found the civet cat and various types of apes.

This legend also comes from Cadamosto, from Chap. 42 in the *Paesi* or *Itinerarium* (Crone, *The Voyages of Cadamosto*, p. 69), where Cadamosto lists some items that natives brought to sell to the Europeans while they were staying with the Lord Batimaussa some sixty miles up the Gambra River.

6.9
Gens ydolatra et carnifagi sunt

This race is idolatrous and they eat meat.

This legend also comes from Cadamosto, from Chap. 43 in the *Paesi* or *Itinerarium* (Crone, *The Voyages of Cadamosto*, p. 70).

6.10
iste fluvius totam ethiopiam interiorem sua inundatione irrigat et fecundat equali condicione sicut Nilis Egiptum vnde opinatur esse pars seu brachi[u]m nilo fluvio.

This river irrigates all of Interior Ethiopia when it floods and makes it fertile, just as the Nile irrigates Egypt, whence it is thought that this river is a part or branch of the Nile.

The legend describes the Senegal River, and comes from Cadamosto, from the end of Chap. 14 in the *Paesi* or *Itinerarium* (Crone, *The Voyages of Cadamosto*, p. 28). One can imagine that Waldseemüller was excited about being able to include new information about a feature as important as the Nile. In the introduction I presented evidence that Waldseemüller had in his workshop a large nautical chart beside the Caverio chart, a chart similar to the Catalan Atlas of 1375, Mecia de Viladestes's chart of 1413, or the Catalan-Estense *mappamundi* of c. 1460, and on all of these maps, the Nile does have a branch that flows west to the Atlantic.[195] So this was a case where by using the most recent information available, Waldseemüller was actually going back to an earlier and incorrect geographical idea. Schöner paraphrases this legend on his manuscript globe of 1520.

[195] On this geography of the Nile see William L. Bevan and H. W. Phillott, *Medieval Geography: An Essay in Illustration of the Hereford Mappa Mundi* (London: E. Stanford, 1873; Amsterdam: Meridian, 1969), pp. 100–101; John K. Wright, *The Geographical Lore of the Time of the Crusades* (New York: American Geographical Society, 1925), pp. 304–306; and Brigitte Postl, *Die Bedeutung des Nil in der römischen Literatur* (Vienna: Verlag Notring, 1970). On the depiction of the western branch of the river on nautical charts see Sandra Sáenz-López Pérez, "Imagen y conocimiento del mundo en la Edad Media a través de la cartografía Hispana," Ph.D. Dissertation, Universidad Complutense de Madrid, 2007, vol. 1, pp. 378–388.

6.11
in hoc regno fiunt magne mercationes cum sale quod a Tagaza portatur nauigio usque ciuitatem Melli deinde baiulatur et ab hominibus portatur ad terras equinoctiales ubi tantus est ardor solis quod animalia vivere non possunt. victualia huius regionis sunt dactili (quarum magna est copia) et hor[d]eum milium et lac camelorum bibunt.

In this kingdom there is much commerce in salt, which is carried by ship from Taghaza to the city of Melli, from which it is borne and carried by men to the equatorial regions where the heat of the sun is so great that animals cannot live. The food stuffs of this region are dates (of which there is a great abundance), barley, millet, and they drink the milk of camels.

Taghaza, in what is now northern Mali, has salt mines that important through the end of the sixteenth century. This legend summaries the description of the salt trade given by Cadamosto, and related in Chap. 11 of the *Paesi* and *Itinerarium*. It is located just to the east of an image of the King of Melli and Guinea, and just southeast of the legend is the city of Cothia, which is mentioned by Cadamosto in Chap. 12 of the *Paesi* and *Itinerarium* (Crone p. 25) as a destination of part of the gold that is traded in Melli under the name Cochia, is now called Gao (in Mali). The part about the foods that the people eat comes from Chap. 9 of the *Paesi* or *Itinerarium*. Johann Schöner repeats much of this legend on his manuscript globe of 1520.

6.12
Hic reperiuntur Rinocerontes Tigrides et elephantes albi

Here are found rhinoceroses, tigers, and white elephants.

This legend is one of the few on the *Carta marina* that comes from Ptolemy, and also one of the few that Waldseemüller repeats from his 1507 map. In Book 4, Chap. 9 of the *Geography*, Ptolemy says that *...regio magna ethyopum est in qua elephantes albi sunt and rinocerontes and tigrides*,[196] and a closely related legend appears on the fourth map of Africa in the 1482 Ulm Ptolemy. This legend is clearly the source of Waldseemüller's legend. On the 1507 map this legend is further to the south than this iteration of it on the *Carta marina*, but there is also another iteration of it further southeast on the *Carta marina*, on sheet 7 (see Legend 7.4 below)—close to its original location on the 1507 map.

6.13
Hic monopedes sive scipodes

Here there are one-footed men or *sciapods*.

I have not been able to determine the source of this legend. It is logical that *sciapods*, who lie on their backs and use their oversized foot to shade themselves from the bright sun,[197] should be near the equator, and they are located in Ethiopia by Isidore, *Etymologiae* 11.3.23, and there are two *sciapods* in southern Africa on Martin Behaim's globe of 1492.[198] But I do not know of any evidence that Waldseemüller saw Behaim's globe.

6.14
Hic est multa habundantia aur[i]

Here there is great abundance of gold.

This legend is just above the toponym REGNVM MELLI, and thus the reference is no doubt to the gold used in Melli in the salt trade described by Cadamosto, and related in the *Paesi* and *Itinerarium*, Chap. 11.

6.15
Rinoceron seu Mononoceron [*sic*] animal

The animal rhinoceros or monoceros.

The image of the rhinoceros above the legend was discussed in the introduction (see pp. 26–27), where it was shown to be based on Burgkmair's 1515 print of the rhino rather than on Dürer's of the same year. Waldseemüller no doubt placed the

[196]I quote from the 1482 Ulm edition of Ptolemy; the passage in the 1513 Ptolemy is very similar: *...regio magna AEthiopum est: in qua Elephantes albi omnino gignuntur, & Rhinocerontes & Tigrides.*
[197]There is a good discussion of the history of the *sciapods* in Claude Lecouteux, "Herzog Ernst, les monstres dits 'sciapodes' et le problème des sources," *Études Germaniques* 34.1 (1979), pp. 1–21.
[198]For a study and facsimile of Behaim's globe see E. G. Ravenstein, *Martin Behaim, his Life and his Globe* (London: G. Philip & Son, Ltd., 1908).

image in Africa because he knew that Ptolemy spoke of rhinos in Africa (see Legend 6.12); as he was using Burgkmair's print, he may not have known that the rhino depicted by that artist was actually from India, but this information is included in the text on Dürer's print.

6.16
cicloped[es] siue monoculi homines sunt grandi et nigri horribil[es]

Cyclopes or one-eyed men—they are big and black and horrible.

I have not found a text that talks about black Cyclopes in Africa. But this legend was quite influential: in Africa near the coast of the Gulf of Guinea in the *Tab[ula] Mo[derna] Primae Partis Africae* in the 1522 (Strasbourg), 1525 (Strasbourg), 1535 (Lyon), and 1541 (Vienna) editions of Ptolemy, there is a legend that reads *Colopedes siue monoculi homines sunt grandes nigri et horribiles*. The word *colopedes* is a misreading of "cicloped[es]"; beside the legend is an illustration of one of the race, in this case a one-eyed *blemmyae*. A cyclops or monoculus is also depicted in this same part of Africa on Sebastian Münster's map of Africa (Africa XVIII) in his editions of Ptolemy (beginning in 1540), and in his *Cosmographia* (beginning in 1544), for example in the 1552 edition of the *Cosmographia* the map numbered 13, and is titled *Totius Africae tabula* and *descriptio uniuersalis etiam ultra Ptolemaei limites extensa*.[199] There is a one-eyed *blemmyae* in Africa on Pierre Desceiler's world map of 1550,[200] but somewhat further inland; and a *monoculos* in the much the same location as on the Ptolemaic maps in Guillaume Le Testu's *Cosmographie universelle* of 1556, ff. 4v and 18v.[201]

[199]Christopher Slogar, "*Polyphemus Africanus*: Mapping Cannibals in the History of the Cross River Region of Nigeria, ca. 1500–1985," *Terrae Incognitae* 37 (2005), pp. 16–27, discusses the monocolus on Münster's maps, but does not address the fact the creature appeared on Waldseemüller's *Carta marina* and in the 1522, 1525, 1535, and 1541 editions of Ptolemy.
[200]See Van Duzer, *The World for a King* (see note 146), pp. 112–113.
[201]Le Testu's *Cosmographie universelle* is in Paris, Bibliothèque du Service Historique de l'Armée de Terre (Château de Vincennes), MS D.L. Z.14.

2.7 Sheet 6A. Western Africa (Plate 2.6A)

As I mentioned above in my general remarks on sheet 6, this hand-drawn copy of the printed sheet 6 of the *Carta marina* was made by Johann Schöner. Schöner made this copy of the sheet as some type of preparation for creating his hand-painted globe of 1520. There is clear evidence of this in the presence of a legend in the Gulf of Guinea on sheet 6A that does not appear on sheet 6, but does in the same position on Schöner's 1520 globe, which I list just below. As the same time, the 1520 globe is not at all merely a copy of the *Carta marina*: for example, it does not include images of the West African sovereigns or the rhinoceros. Schöner may not have completely drawn these elements on the copy of sheet 6 simply because he considered these graphic elements less important, or he may have left them out because he knew that his globe would not include them.

6A.1

Insule hec inuente sunt Anno 1484. per portugalenses & per eos inhabitate fuerunt enim ante deserte

These islands were discovered in 1484 by the Portuguese, and were inhabited by them—before they had been uninhabited.

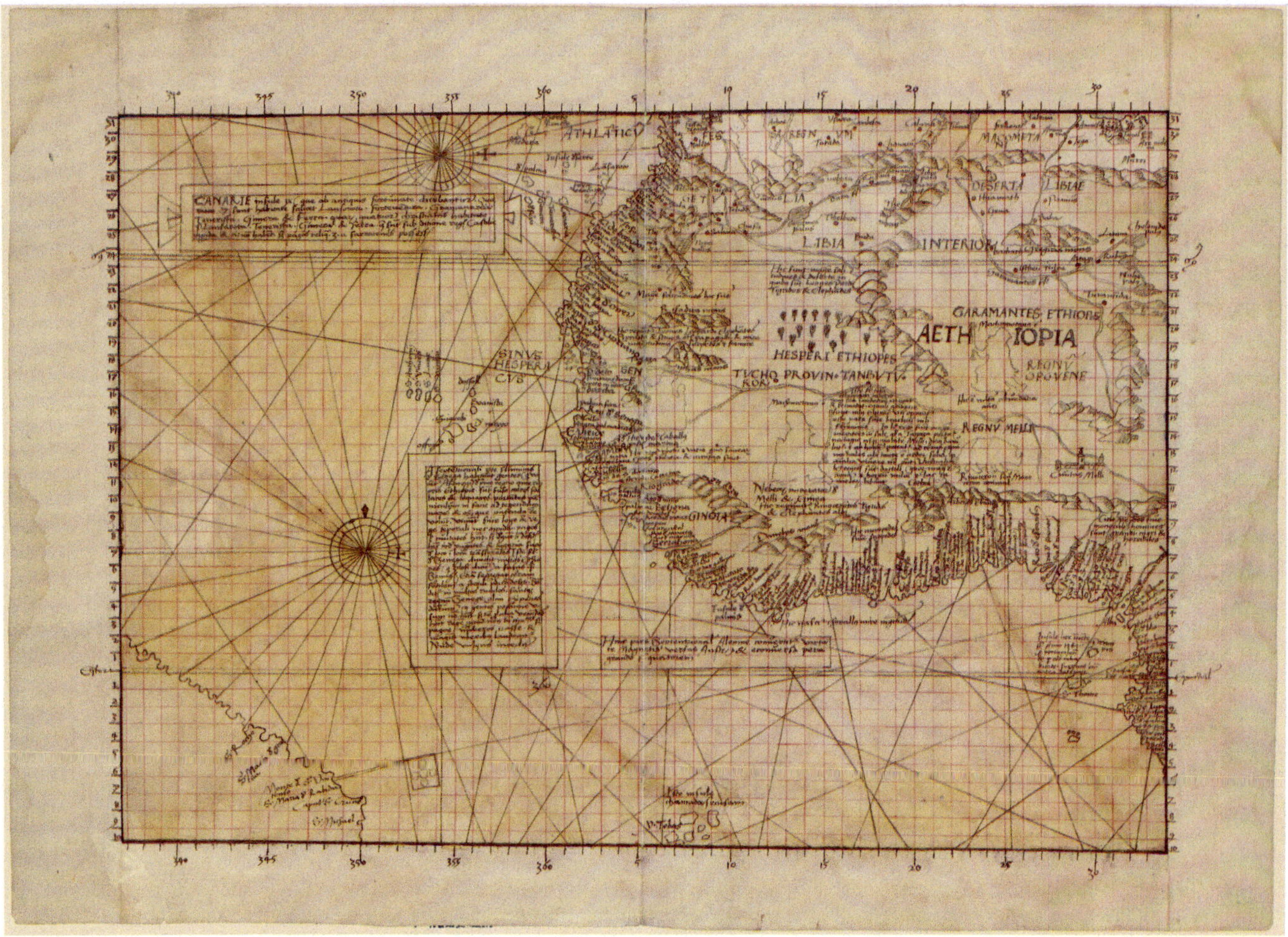

Plate 2.6A Sheet 6A of the *Carta marina*: Western Africa. Courtesy of the Library of Congress

This legend is one that was added to the manuscript sheet by Schöner, that is, it does not appear on the printed sheet 6. Very similar information appears in Schöner's *Luculentissima quaedam terrae totius descriptio* (Nürnberg: Ioannis Stuchssen, 1515), which he wrote to accompany his globe of the same year, on f. 40r[202]:

> Hae tres insulae sunt Australes a nobis: & inuentae Anno domini 1484. In his nihil aliud fuit repertum quam desertum sylvarum: & opaca nemorum. Ibi nullum animal: neque hominem: neque quadrupedum. Sed solum aues diuersi generis: nunc vero continue inhabitantur vtriusque sexus hominibus per Portugalenses.

> These three islands are to the south of us, and were discovered in 1484. Nothing was found in them except a desert of woods and dense forests: there was no animal, no human, no quadruped, just birds of various types. Now, however, it is continuously inhabited by Portuguese people, both men and women.

The discovery of the islands is usually dated to 1470,[203] but Schöner is perhaps following a legend on Martin Behaim's globe of 1492, which ascribes the discovery to 1484, on a voyage in which Behaim claimed to have participated.[204]

[202]For discussion of the *Luculentissima* see Henry Harrisse, *Bibliotheca americana vetustissima: A Description of Works Relating to America, Published between the Years 1492 and 1551* (New York: G. P. Philes, 1866), nos. 80–81, pp. 140–143. A PDF of the *Luculentissima* is available for download from the Munich Digitisation Centre of the Bayerische Staatsbibliothek, at http://www.digital-collections.de.

[203]Robert Garfield, "A History of São Tomé Island, 1470–1655," Ph.D. Dissertation, Northwestern University, 1971, p. 1: "The exact date of discovery is open to question, but the most commonly accepted one is 21 December 1470—St. Thomas' Day—after whom, in keeping with Portuguese tradition, the island was named. The discoverers, Pero Escobar and João de Santarem, also found on the same voyage the islands of Annobon (1 January 1471) and Príncipe (17 January 1471)." Also see Viriato Campos, "Os dias de descobrimento das ilhas de S. Tomé e Príncipe," *Anais do Clube Militar Naval* 7.9 (1970), pp. 454–483, reprinted in *Elementos de história da ilha de S. Tomé (em comemoração do V centenário do descobrimento)* (Lisbon: Centro de Estudos de Marinha, 1971), pp. 45–69; and Armando Cortesão, *Descobrimento e cartografia das ilhas de S. Tomé e Príncipe* (Coimbra: Junta de Investigações do Ultramar, 1971) = Agrupamento de Estudos de Cartografia Antiga, Série separata, 62 (18 pp. and 6 plates).

[204]Behaim's legend, translated, runs: "These islands were found with the ships which the King of Portugal sent out to these parts of the country of the Moors in 1484. There was a perfect wilderness then, and we found no men there, only forests and birds. But at present the king sends there people who have been condemned to death, men as well as women, and he affords them the means of cultivating the land and of multiplying, so that this country may be inhabited by Portuguese." See E. G. Ravenstein, *Martin Behaim, His Life and His Globe* (London: G. Philip & Son, Ltd., 1908), p. 101, and for discussion of his claims, see (in the same book) "Behaim's African Voyage, 1484–85," pp. 20–30, esp. 24.

2.8 Sheet 7. East Africa, the Red Sea, Arabia, and the Western Indian Ocean (Plate 2.7)

The shape of this part of Africa is much different on the *Carta marina* than on Waldseemüller's 1507 map, particularly the orientation of the Red Sea. On the 1507 map the sea runs close to north and south, while here it is closer to running east and west—in accordance with the Caverio chart. The large king near the center of the sheet is Prester John, the essentially mythical Christian monarch of great power and wealth. On his 1507 map, Waldseemüller had placed Prester John in Asia, but the location of this mythical king had long been uncertain. The longest legend on this sheet, at the right edge in the Indian Ocean, describes the important trading center of Calicut (Kozhikode, India), which is represented on sheet 8. The three Portuguese flags from the Horn of Africa northward mark recent conquests by navigators of that country.

> 7.1
> Hic omnes sunt Cristiani iacobini scismatici Circumcisionem observant more saracenorum, cremantur ferro in faciem quo originali peccato mundari credunt nec mutuo confitentur peccata sed deo, profitentur quod unicam naturam in christo tantum
>
> Here all of the people are Jacobite Christian schismatics who practice circumcision according to the custom of the Saracens. They are burned with iron in the face, which they believe is the way to be cleansed of original sin, nor do they confess their sins to other men, but to god, and they profess that in Christ there is only a single nature.

On the Jacobite Christians see above on Legend 3.14. Much of the information here comes from Bernhard von Breydenbach, *Peregrinatio in terram sanctam* (Mainz: Peter Schöffer the Elder, 1486), f. [82v], "De Abbasinis sive Indianis et eorum cerimoniis," who writes[205]:

> Nam circumcisionem carnalem servant more sarcenorum et jacobitarum pavulos suos circumcidentes... Adurunt quoque infantes suos in frontibus ferreo calamo in modum crucis... credentes eos per huiusmodi adustionem ab originali peccato mundari.
>
> For they practice physical circumcision according to the custom of the Saracens and Jacobites, circumcising their baby boys... They also burn their infants on the forehead with an iron rod in the form of a cross... believing that though this type of burning they are cleansed of original sin.

Thus we can be confident that Waldseemüller had a copy of Breydenbach's *Peregrinatio* as part of his extensive library—or at least had access to the book. Schöner copied this legend on his 1520 globe, and much of the information also appears in his *Opusculum geographicum* ([Nuremberg]: [Johann Petrejus], [1533]), part 2, Chap. 12.

> 7.2
> Hic absorbitur per tria miliaria
>
> Here [the river] is absorbed [and flows underground] for three miles.

[205]See Jaynes, *Christianity beyond Christendom* (see note 58), p. 287.

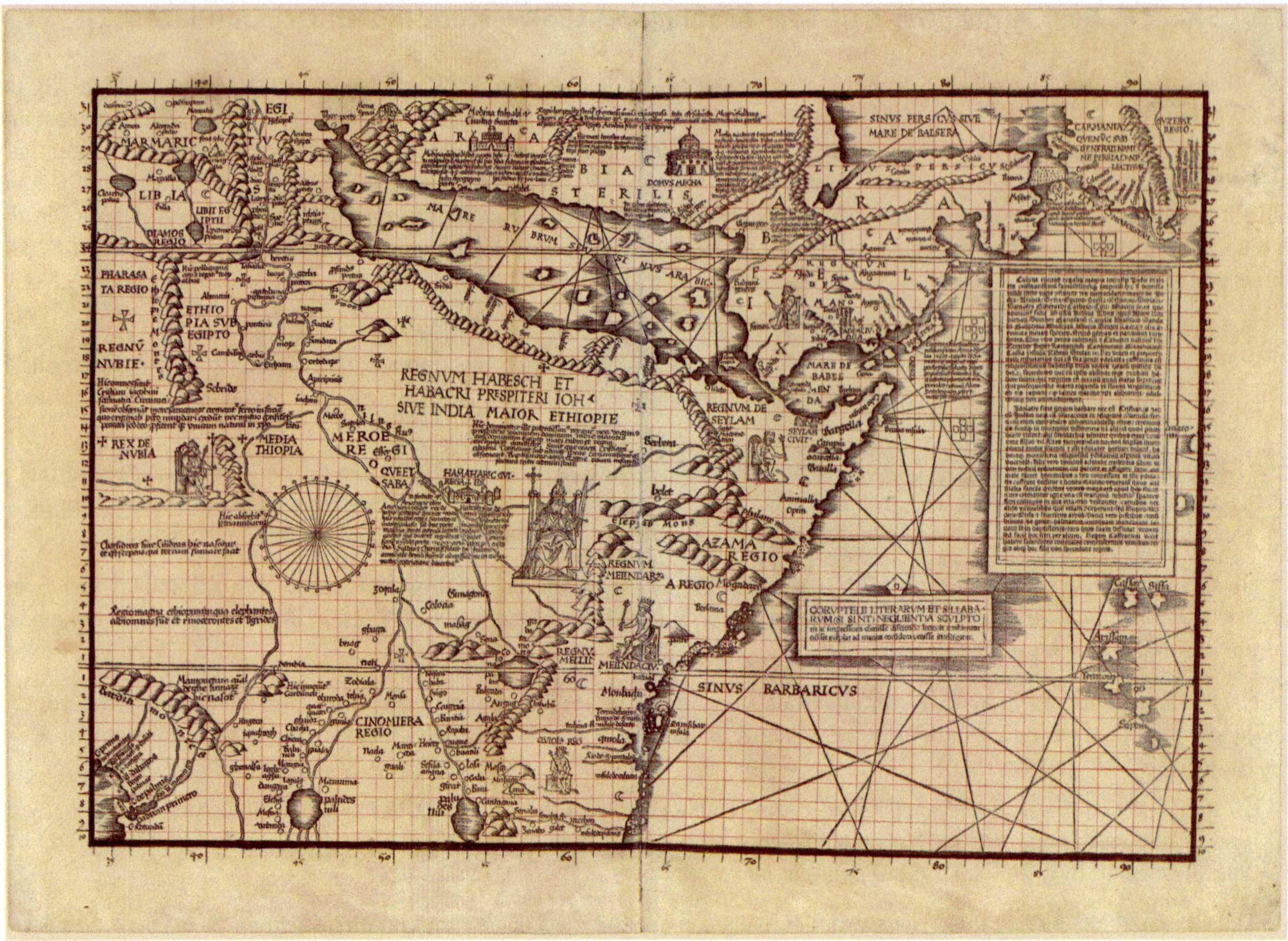

Plate 2.7 Sheet 7 of the *Carta marina*: East Africa, the Red Sea, Arabia, and the Western Indian Ocean. Courtesy of the Library of Congress

Both Honorius Augustodunensis and Gervase of Tilbury speak of the Nile being absorbed and flowing underground,[206] but the reference to a distance of three miles indicates the real source of the legend: a map of the *Egyptus novelo* family, a *tabula moderna* of Africa that survives in three manuscripts of Ptolemy's *Geography* painted between 1469 and about 1480 by Pietro del Massaio in Florence. These maps combine a Ptolemaic depiction of Egypt and the northern Nile system with a

[206]Honorius Augustodunensis, *Imago mundi*, ed. Valerie Flint in *Archives d'histoire doctrinale du Moyen Age* 57 (1982), pp. 48–93, at 52, (Sect. 1.9): *Geon qui et Nilus iuxta montem Athlantem surgens, mox a terra absorbetur, per quam occulto meatu currens, in littore rubri maris denuo funditur, Aethiopiam circumiens per Aegyptum labitur, in septem ostia divisus, magnum mare iuxta Alexandriam ingreditur*; Gervase of Tilbury, *Otia imperialia: Recreation for an Emperor*, ed. and trans. S. E. Banks and J. W. Binns (Oxford: Clarendon Press, 2002), 2.3, pp. 182–183: *Gion, qui et Nilus, iuxta montem Athlantem fluens, mox a terra absorbetur, per quam occulto meatu currens in littore Rubri maris denuo refunditur, et uersus orientem surgens et Ethiopiam circuiens, et sic per Egyptum labitur, ac in septem ostia divisus, Magnum mare iuxta Alexandriam ingreditur*, that is, "The Gihon, also called the Nile, rises near Atlas, but is soon absorbed back into the earth, and runs through a hidden channel underground until it re-emerges again on the shore of the Red Sea; it then heads eastwards and flows round Ethiopia, and so it comes to glide through Egypt until, dividing into seven channels, it enters the Great Sea near Alexandria." On the fact that the river is absorbed into the earth and reappears again also see 1.12, pp. 72–73.

map of Ethiopia (i.e. Africa south of Egypt) that was based on a new, post-Ptolemaic description of the regions.[207] The three manuscripts of Ptolemy's *Geography* that include this *tabula moderna* or modern map are[208]:

1. Vatican City, Biblioteca Apostolica Vaticana, MS Vat. lat. 5699, f. 125r, dated 1469, where it the map is titled *Aegyptus cum Ethiopia moderna*;
2. Vatican City, Biblioteca Apostolica Vaticana, MS Urb. lat. 277, ff. 128v–129r, dated 1472, where the map is titled *Descriptio Egypti Nova*[209]; and
3. Paris, Bibliothèque nationale de France, MS lat. 4802, ff. 130v–131r, made c. 1475–80,[210] where the map is titled *Egyptus Novelo*.

The influence of an *Egyptus novelo* map on Waldseemüller's depiction of southern Africa in the *Carta marina* is visible mainly in the toponyms and hydrography,[211] rather than in the long legends, but nonetheless the map was an important source—probably indirectly by way of a world map by Henricus Martellus—for the *Carta marina*. On Vat. lat. 5699, f. 125r, the legend in question reads *Absorbitur hic fluvius per tria milia*; on Urb. lat. 277, ff. 128v–129r, it reads *Labitur occulte per 3 milia passus*; and on BnF Lat. 4802, ff. 130v–131r, *Per 3 milia sub terra labitur hic fluvius*. Thus we see that Waldseemüller's legend is closest to that on Vat. lat. 5699, f. 125r. Recent multispectral images of the world map by Henricus Martellus at Yale, which shows many signs of the influence of *Egyptus novelo* cartography, and of which Waldseemüller made abundant use in creating his 1507 map, reveal that it has a closely related legend that runs *hic absorbetur per tres leucas*, "Here the river is absorbed for three leagues."[212]

7.3
Chersidras sive Cilidras hic nascitur et est serpens qui terram fumare facit

---

[207]For brief general discussions of the *Egyptus Novelo* map see Laura Mannoni, *Una carta italiana del bacino del Nilo e dell'Etiopia del secolo XV* (Rome: s.n., 1932) = Pubblicazioni dell'Istituto di geografia della R. Università di Roma, Ser. B, num. 1; Bertrand Hirsch, "Cartographie et itinéraires: Figures occidentales du nord de l'Ethiopie aux XVe et XVIe siècles," *Abbay* 13 (1986) pp. 91–122; and Gianfranco Fiaccadori, "Egyptus Novelo," in Siegbert Uhlig, ed., *Encyclopaedia Aethiopica* (Wiesbaden: Harrassowitz, 2003–), vol. 2, pp. 246–247. Also see O. G. S. Crawford, "Some Medieval Theories about the Nile," *Geographical Journal* 114 (1949), pp. 6–23, esp. 8–19; and W. G. L. Randles, "South-East Africa as Shown on Selected Printed Maps of the Sixteenth Century," *Imago Mundi* 13 (1956), pp. 69–88, esp. pp. 73–76 and the reproduction between pp. 69 and 70.

[208]All three manuscripts are described by Germaine Aujac, "Le peintre florentin Piero del Massaio et la 'Cosmographia' de Ptolémée," *Geographia antiqua* 3–4 (1994–1995), pp. 187–204, and Louis Duval-Arnould, "Les manuscrits de la *Géographie* de Ptolémée issus de l'atelier de Piero del Massaio (Florence, 1469—vers 1478)," in Didier Marcotte, ed., *Humanisme et culture géographique à l'époque du Concile de Constance autour de Guillaume Fillastre: Actes du Colloque de l'Université de Reims, 18–19 novembre 1999* (Turnhout: Brepols, 2002), pp. 227–244.

[209]Biblioteca Apostolica Vaticana MS Urb. lat. 277 has been published in facsimile as *Codex Urb. Lat. 277: La cosmografía de Claudius Ptolomeus: transcrito entre los años 1472 y 1473* (Madrid: Encuentro, 1983), with a commentary volume by Arthur Dürst. The *Descriptio Egypti Nova* map in this manuscript is briefly described by Dürst on p. 68 of his commentary.

[210]In earlier literature a date of 1456 was often assigned to BnF MS lat. 4802, but this dating is corrected by Bertrand Hirsch, "Cartographie et itinéraires: Figures occidentales du nord de l'Ethiopie aux XVe et XVIe siècles," *Abbay* 13 (1986) pp. 91–122, at 98; Aujac, "Le peintre florentin Piero del Massaio" (see note 208), and Duval-Arnould, "Les manuscrits de la *Géographie* de Ptolémée" (see note 208). The maps in BnF MS lat. 4802 have been reproduced in facsimile in *Géographie de Ptolémée: traduction latine de Jacopo d'Angiolo de Florence; reproduction réduite des cartes et plans du manuscrit latin 4802 de la Bibliothèque nationale*, ed. Henri Auguste Omont (Paris: Catala frères, 1926), where the *Egyptus novelo* map is on plates 64–65.

[211]Bertrand Hirsch, "Les sources de la cartographie occidentale de l'Éthiopie (1450–1550): les régions du lac Tana," *Bulletin des Études Africaines de l'INALCO* 7.13–14 (1987), pp. 203–236, at 233–235, provides a table showing correspondences among some of the toponyms on the three surviving *Egyptus novelo* maps and the *Carta marina*, but this table is both incomplete and inaccurate; a much superior table is available in Bertrand Hirsch, "Connaissance et figures de l'Éthiopie dans la cartographie occidentale du XIVe siècle au XVIe siècle," Thèse de Doctorat, Université de Paris I, 1990, "Transcription de la Tabula moderna de l'Ethiopia," vol. 2, pp. 17–26.

[212]See Van Duzer, *Henricus Martellus's World Map at Yale* (see note 14), pp. 116–117.

Chersydras or Cilidras is born here, and it is a serpent that causes the ground to smoke.

This is another of the few legends that Waldseemüller retained from his 1507 map.[213] This serpent is described by Isidore,[214] Albertus Magnus in his *De animalibus*,[215] in the *Experimentator*,[216] and also in the *Hortus sanitatis*,[217] an illustrated encyclopedia first published in 1491.[218] In this case, however, we do know Waldseemüller's immediate source, which was the Yale Martellus map (or one very similar to it), which has an almost identical legend in the same location,[219] and this text, like many other texts about animals on the Yale Martellus map, came from the *Hortus sanitatis*.

7.4
Regio magna ethiopum in qua elephantes albi omnes sunt et rinocerontes et tigrides

The great land of the Ethiopians in which all of the elephants are white, and there are rhinoceroses and tigers.

This legend is another of the few that Waldseemüller retained from his 1507 map; for discussion see above on Legend 6.12.

7.5
Mamonetum animal de genere simiarum hic nascitur – Hic carbunc[u]li inveniuntur.

The marmoset, an animal from the race of apes, is born here. Here precious stones are found.

The texts are on opposite sides of a mountain. The text about the marmoset is also from Waldseemüller's 1507 map, but there it is somewhat longer: *Mamonetum est animal de genere simearum quod ligatur in ventre propter colli grossiciem huius caudam pilosam*. This animal is described by Albertus Magnus, *De animalibus* 22.76,[220] Thomas of Cantimpré, *Liber de natura rerum* 4.74,[221] and in the *Hortus sanitatis*, "De animalibus," Chap. 94.[222] Waldseemüller borrowed may of the descriptive texts about animals on his 1507 map from a map by Henricus Martellus similar to his world map at Yale, and Martellus used the *Hortus sanitatis* for many of his texts about animals, but I have not found a legend about the marmoset on the Yale Martellus map. I have not found the source of the text about the precious stones.

---

[213]The legend on the 1507 map is almost identical: *Cherisidal sive cilidras hic nascitur et est serpens qui terram fumare facit.*

[214]Isidore 12.4.24, translated into English in Isidore of Seville, *The Etymologies of Isidore of Seville*, trans. Stephen A. Barney et al. (Cambridge, UK and New York: Cambridge University Press, 2006), p. 256: "The *chelydros* is a snake that is also known as the *chersydros*, as it if were cerim, because it dwells both in the water and on land; for the Greeks call land χέρσος and water ὕδωρ. These make the earth on which they move smoke, as Macer thus describes it (fr. 8): 'Whether their backs froth out poison, or it smokes on the earth, where the hideous snake crawls.' And Lucan (*Civil War* 9.711): 'And the chelydri drawn along with their smoking trails.' But it always proceeds in a straight line, for if it turns when it moves, it immediately makes a sharp noise.".

[215]The passage is in Albertus's *De animalibus* 25.21; for the Latin text see Hermann Stadler, ed., *Albertus Magnus, De animalibus libri XXVI, nach der Cölner Urschrift* (Münster: Aschendorff, 1916) = Beiträge zur Geschichte der Philosophie des Mittelalters, vol. 15–16, vol. 2, p. 1564; for an English translation see Albertus Magnus, *On Animals: A Medieval Summa Zoologica*, trans. Kenneth F. Kitchell, Jr., and Irven Michael Resnick (Baltimore: Johns Hopkins University Press, 1999), vol. 2, pp. 1723–1724.

[216]For the text in the *Experimentator* see Janine Deus, "Der 'Experimentator'—eine anonyme lateinische Naturenzyklopädie des frühen 13. Jahrhunderts," Dissertation, Universität Hamburg, 1998, p. 268.

[217]The relevant chapter is number 36 in the section "De animalibus" of the *Hortus sanitatis*.

[218]The *Hortus sanitatis* "major," which is the work that interests us here, is to be distinguished from the *Hortus sanitatis* "minor," which is a Latin translation of the German herbal often titled *Gart der Gesundheit*, first published by P. Schoeffer, Mainz, 1485. The herbal published in 1485 has 435 chapters, while the *Hortus Sanitatis* "major" of 1491 has 1066 chapters. For discussion of the early editions of the *Hortus Sanitatis* see Arnold C. Klebs, "Herbals of 15th Century," *Papers of the Bibliographical Society of America* 11 (1917), pp. 75–92, and 12 (1918), pp. 41–57, esp. 48–51 and 54–57. There is a more detailed discussion in Joseph Frank Payne, "On the 'Herbarius' and 'Hortus sanitatis'," *Transactions of the Bibliographical Society* 6.1 (1901), pp. 63–126, esp. 105–124. The first edition of the work was published in Mainz by Jacob Meydenbach, June 23 1491.

[219]See Van Duzer, *Henricus Martellus's World Map at Yale* (see note 14), p. 114.

[220]For the Latin text see Hermann Stadler, ed., *Albertus Magnus, De animalibus libri XXVI, nach der Cölner Urschrift* (Münster: Aschendorff, 1916) = Beiträge zur Geschichte der Philosophie des Mittelalters, vol. 15–16, vol. 2, p. 1413; for an English translation see Albertus Magnus, *On Animals: A Medieval Summa Zoologica*, trans. Kenneth F. Kitchell Jr., and Irven Michael Resnick (Baltimore: Johns Hopkins University Press, 1999), vol. 2, p. 1522.

[221]Thomas of Cantimpré, *Liber de natura rerum*, ed. H. Boese (Berlin and New York: DeGruyter, 1973), pp. 150–151.

[222]Both Thomas of Cantimpré and the anonymous author of the *Hortus sanitatis* indicate that their information comes from Isidore, but in fact it comes from Albertus Magnus.

7.6
Regio hec penitus sterilis et arenosa inuia et inaquosa tota est subiecta Magno Soldano Chauri et Babilonie Item paucas habet villas rarissime autem ciuitat[es]. victualia portantur ad hanc de Egipto Arabia felice et de Ethiopia.

This region is totally sterile and sandy, impassable and without water. All of it is subject to the great Sultan of Cairo and Babylon [i.e. Cairo]. It has few villages and extremely few cities; food stuffs are brought here from Egypt, Arabia Felix, and Ethiopia.

This legend is at the top of the sheet, in Arabia. The source is Varthema's chapter on Jeddah, the port of Mecca.[223]

7.7
Hic mare terrestre et arenosum in quo reperitur mumia peragratur hoc directione Magnetis et instrumentorum sicut pelages

Here there is a terrestrial and sandy sea in which mummy is found; it is crossed with the aid of a compass and instruments, just like seas.

This legend comes from Varthema,[224] who says of crossing the sea of sand, "and the pilots go in advance with their compasses as they do at sea." For discussion of mummy see the commentary on Legend 3.18 above.

7.8
Medina ciuitas distat a portu Jubo .3. dietarum itineris in ea sepultus est Macomet cum sua filia fatonia et pseudosius [for *pseudosiis*] appostolis et complicibus scilicet Buhichar Aumar Halii et Otman in templo sunt .45. codices quibus lex et historia eorundem est commendata hec Medina habet .300. lares

The city of Medina is a three days' journey from the port of Jubo. In the city Mahommed is buried with his daughter Fatonia and his false apostles and accomplices, namely Babicher, Aumar, Haly, and Othman. In the temple there are forty-five codices in which the law and history of the Muslims are recorded. This Medina has three hundred houses.

This legend also comes from Varthema.[225] As mentioned in the introduction (see p. 29), Waldseemüller used the image of Medina in the 1515 edition of Varthema to illustrate Mecca on his map; for Medina he used a somewhat elaborated version of his standard image for a large and important city, with no attempt to follow the description of the mosque of Medina in Varthema. The moon symbol to the right of the city indicates that the area is under Muslim control. Incidentally Fries discusses Medina in Chap. 71 bis of the *Uslegung*.[226] Johann Schöner has a very heavily abbreviated version of this legend on his manuscript globe of 1520.

7.9
Mecha ciuitas et Emporium celeberr[ima] in sterili Arabia sita distat a medina talnabi plus quam .100. et a portu Zida .40. miliariis Rex tributarius est Soldano de Chayro. Hec a mercatoribus et Saracenis omnibus frequentatur sicut Roma a cristianis gratia indulgenciarum et mercature in ea summum est templum Machometi

The very famous city and trading center of Mecca is located in the desert of Arabia, and is more than a hundred miles distant from Medina of the Prophet, and forty miles from the port of Jeddah. The king is a tributary of the Sultan of Cairo. The city is frequented by merchants and by all Saracens, just as Rome is by Christians, because of the indulgences and trade. In the city is the highest temple of Mohammed.

This legend is assembled from a few different passages in Varthema,[227] except for the part about Mecca and Medina being more than a hundred miles apart, whose source I do not know. It should be mentioned that according to Waldseemüller's scale of German miles, the cities are close to 150 miles apart. Fries discusses Mecca in Chap. 72 bis of the *Uslegung*.[228] Schöner borrows the latter part of this legend on his 1520 globe.

---

[223]Varthema, *The Travels of Ludovico di Varthema* (see note 103), p. 53.
[224]Varthema, *The Travels of Ludovico di Varthema* (see note 103), p. 33.
[225]See Varthema, *The Travels of Ludovico di Varthema* (see note 103), pp. 26–28.
[226]The passage on Medina is translated into modern German by Petrzilka, *Die Karten des Laurent Fries* (see note 202 in Chap. 1 above), p. 146.
[227]The material on the trade and pardons available in Mecca come from Varthema, *The Travels of Ludovico di Varthema* (see note 103), pp. 38–41; the fact that the city is subject to the Sultan of Cairo is on p. 36; the fact that the port is forty miles from the city is on pp. 37 and 51.
[228]The passage on Mecca is translated into modern German by Petrzilka, *Die Karten des Laurent Fries* (see note 202 in Chap. 1 above), pp. 146–147.

7.10
In Zidam ciuitatem non intromittuntur Cristiani neque iudei

Neither Christians nor Jews are allowed to enter the city of Jeddah.

Jeddah is the port city of Mecca; the legend comes from Varthema.[229]

7.11
nascuntur hic serpentes[?] minores Medicinal[es] estque habunda[n]cia omnium victualium gentes sunt Macometani annum .150. etatis atingentes

There are born small serpents which are used in medicine, and there is an abundance of all food stuffs. The people are followers of Mohammed and live to be 150 years old.

This legend is near the port of Gezan. The parts about the abundance of food stuffs and the people being followers of Mohammad comes from Varthema's chapter on that port.[230] I have not found the source of the claim that serpents are used in medicine or that the people live to be 150 years old.

7.12
Adem metropolis et regia fortis habens 6000 lares

Aden, a metropolis and a mighty royal city with 6000 houses.

This legend comes from Varthema,[231] and it offers a good object lesson in Waldseemüller's preferences for more recent sources. Marco Polo has a detailed and interesting account of Aden,[232] but the cartographer chose to use Varthema instead.[233]

7.13
REGNUM HABESCH ET HABACCI PRESBITERI JOH. SIVE INDIA MAIOR ETHIOPIE. Hic dominatur ille potentissumus rex quem nos vocamus presbiterem johannem dominum indie maioris cuius dominiam Egypto mari rubro et regno melindarum clauditur. Suntque omnes cristiani abasini Conficiunt sub utraque specie circumcisionemque obseruant. Baptisantur aqua et igne eciam infantibus sacramenta administrant.

The Kingdom of Abyssinia of Prester John, or Greater India in Ethiopia. Here rules that most powerful king whom we call Prester John, lord of Greater India, whose dominion is bounded by Egypt, the Red Sea, and the kingdom of Melinde. They are all Abyssinian Christians. They take communion under both forms [i.e. bread and wine] and practice circumcision. They are baptized by water and fire and even administer the sacraments to infants.

Part of this legend seems to come from or have been inspired by Giolamo Sernigi's "Second Letter to a Gentleman of Florence," specifically a passage published in Chap. 60 of the *Paesi* and *Itinerarium*, which in English runs[234]:

I do not understand that there are any Christians there [i.e. in Calicut] to be taken into account, excepting those of Prester John, whose country is far from Calicut, on this [i.e., the western] side of the Gulf of Arabia, and borders upon the country of the King of Melinde, and, far in the interior, upon the Ethiopians, that is the black people of Guinea, as also upon Egypt, that is the country of the Sultan of Babylon [Cairo]. This Prester John has priests who offer sacrifices, and respect the Gospels and the Laws of the Church, much as is done by other Christians.

---

[229]See Varthema, *The Travels of Ludovico di Varthema* (see note 103), p. 52.
[230]See Varthema, *The Travels of Ludovico di Varthema* (see note 103), pp. 55–56.
[231]See Varthema, *The Travels of Ludovico di Varthema* (see note 103), p. 59.
[232]See Marco Polo, *Marka Pavlova z Benátek, Milion* (see note 55), pp. 192–194; for an English translation see *The Book of Ser Marco Polo* (see note 55), Book 3, Chap. 36, vol. 2, pp. 438–439.
[233]On Aden in the late medieval period see Robert Bertram Serjeant, "The Ports of Aden and Shihr," *Recueils de la Société Jean Bodin* 32 (1974), pp. 207–224 = *Les grandes escales: colloque organisé en collaboration avec la Commission internationale d'histoire maritime (10e Colloque d'histoire maritime)*, vol. 1.
[234]The translation is from Ravenstein, *A Journal of the First Voyage* (see note 57 in Chap. 1 above), pp. 137–141, at 138. On Sernigi see Ravenstein's introduction to the two letters, pp. 119–123; and Carmen M. Radulet, "Girolamo Sernigi e a importância económica do Oriente," *Revista da Universidade de Coimbra* 32 (1985), pp. 67–77.

This letter contains material from Caspar the Jew of India, also called Gaspar de Gama, whom Waldseemüller cites as a source in the text block on sheet 9 of the *Carta marina* (see Legend 9.3). As I indicated in the introduction, Waldseemüller says that he had access to a travel narrative by Caspar that was sent to the King of Portugal, which we may imagine was more detailed than the account published as Sernigi's second letter in the *Paesi* and *Itinerarium*, and it seems likely that this legend derives from that more detailed version of Caspar's account, which unfortunately is lost. Incidentally Fries discusses Prester John in Chap. 90 of the *Uslegung*.[235]

The use that Johann Schöner made of this legend on his manuscript globe of 1520 is quite interesting. Schöner repeats much of Waldseemüller's legend in the same spot in Africa, except that speaks of an anonymous king, omitting Waldseemüller's two identifications of the king as Prester John in the beginning of the legend. At the same time, in Asia he copies the legend about Prester John in that region on Waldseemüller's 1507 map, but adds a more sonorous beginning. Thus while Waldseemüller transferred Prester John from Asia on his 1507 map to Africa on his 1516 map, Schöner opted to keep him in Asia, copying Waldseemüller's 1507 legend about him in Asia, and also his 1516 legend about him in Africa, but omitting the name of the monarch. Other sources, including Waldseemüller, reflect a shift of Prester John from Asia to Africa,[236] but Schöner elects to keep him in Asia on his 1520 globe.

7.14
In ciuitate Amaharic olim Auxuma residet rex de Habesch sive de India Ethiopie quem nos vocamus presputerum iohannes habens multos reges Saracenos tributarios in cuius palacio sunt due porte in quibus Leones et maximi canes catenis ligati ne quis ignotus alienus sine ductore nigrediat [for *ingrediat*] et habet eciam semper in qualibet porta .1000. armatos pro custodia. Suldanus Chari ei desoluit tributum annuale de nili fluminis admissione ne in catharactis restrictione diuertatur

In the city of Amaharic, formerly Axum, resides the king of Abyssinia or of India in Ethiopia, whom we call Prester John. He has many Saracen kings who pay him tribute, and in his palace there are two gates at which lions and huge dogs are tied up with chains, lest some unknown stranger should enter without a guide, and further, he always has a thousand armed guards at each gate. The Sultan of Cairo pays him an annual tribute for permitting the flow of the Nile, lest it be diverted by restriction at the cataracts.

I have not been able to find the source of much of this legend, and am inclined to suspect that it comes from the narrative that Caspar sent to the King of Portugal, as did the previous legend. It should be noted that the description of Prester John's palace here is entirely different from that in the (spurious) letter of Prester John, through which he was known in Europe[237]: in the description of his palace in that letter (Sects. 56–63), the emphasis is on precious construction materials. The illustration of Prester John in the 1525 edition of Fries's *Uslegung* (f. 24r) includes the lions and dogs mentioned in this legend.

[235]The passage on Prester John is translated into modern German by Petrzilka, *Die Karten des Laurent Fries* (see note 202 in Chap. 1 above), pp. 153.
[236]On the transfer of Prester John from Asia to Africa see J. Richard, "L'Extrême–Orient Légendaire au Moyen Âge: Roi David et Prêtre Jean," *Annales d'Éthiopie* 12 (1957), pp. 225–244; Jeremy Lawrance, "The Middle Indies: Damião de Góis on Prester John and the Ethiopians," *Renaissance Studies* 6.3–4 (1992), pp. 306–324; Francesc Relaño, *The Shaping of Africa: Cosmographic Discourse and Cartographic Science in Late Medieval and Early Modern Europe* (Burlington, VT, and Aldershot, UK: Ashgate, 2001), pp. 51–72; and Manuel João Ramos, *Essays in Christian Mythology: The Metamorphosis of Prester John* (Lanham, MD: University Press of America, 2006), pp. 106–116. In fact there was a mid-fifteenth-century nautical chart that is lost but whose legends are preserved, which offers an account of Prester John's transfer from Asia to Africa: see Jacques Paviot, "Une mappemonde génoise disparue de la fin du XIVe siècle," in Gaston Duchet-Suchaux, ed., *L'Iconographie: études sur les rapports entre textes et images dans l'Occident médiéval* (Paris: Le Léopard d'Or, 2001), pp. 69–97, at 87–88 and 96. Paviot claims that the lost chart was from the late fourteenth century, but as one of its legends cites Antoniotto Usodimare, it must be from 1455 or later.
[237]The text of Prester John's letter is edited by F. Zarncke, "Der Brief des Presters Johannes an den byzantinischen Kaiser Emanuel," *Abhandlungen der königlich sächsischen Gesellschaft der Wissenschaften, phil.-hist. Klasse* 7 (1879), pp. 873–934, which is reprinted in Charles F. Beckingham and Bernard Hamilton, eds., *Prester John, the Mongols, and the Ten Lost Tribes* (Aldershot, Hampshire; and Brookfield, VT: Variorum, 1996), pp. 40–102; it is translated into English in Michael Uebel, *Ecstatic Transformation: On the Uses of Alterity in the Middle Ages* (New York: Palgrave Macmillan, 2005), pp. 155–160.

Perhaps the most interesting and distinctive part of the legend is the indication that the Sultan of Cairo pays Prester John an annual tribute not to divert the Nile, that is, to allow the water to come to Cairo. The myth that the emperor of Ethiopia exacted this tribute from the Sultan of Cairo has a rich history that goes back to the thirteenth century,[238] and moreover, there are authors who say that it was Prester John who had the power to control the river and exacted the tribute.[239] So the myth about Prester John controlling the Nile should not be entirely unexpected. It seems quite possible that Caspar recounted this myth in his narrative. There is another source that Waldseemüller consulted that contains a reference to the myth, though not enough details (it omits mention of the tribute) to have been Waldseemüller's only source on this subject, namely a letter from King Manuel to Pope Leo X dated June 1513 regarding the conquest of Malacca.[240] In a passage about Prester John, King Manuel writes[241]:

> Haud exiguum adorandae & verae Crucis lignum ad nos mittit, viros vafros et industrios poscens, quorum ingenio & artificio a Sulcani territorio et Regione, Nilum deflecti aliqua [for *atque*] diverti posse existimat.
>
> He sent us a sizeable piece of the wood of the true and worshipful Cross, and asked for clever and industrious men, by whose ingenuity and skill he thinks that the Nile can be turned away and diverted from the territory and country of the Sultan.

It would be interesting to know the basis for this part of King Manuel's letter.

7.15
CAMBEIA Regnum Gentes sunt ydolatre et ciuitas emporial magna et ditissima

The kingdom of Cambay [Khambhat]; the people are idolaters and the city is a center of trade, large and very rich.

Possible sources for this legend include Marco Polo, Varthema, and Joseph the Indian (the last from the *Paesi* and *Itinerarium*). The account in the Naples manuscript of Polo is very brief and does not include even the few details in this legend.[242] But it would have been possible to assemble these details from Varthema,[243] and easy to do so from the account of Joseph the Indian.[244]

7.16
Ormus insula emporcialis ditissima apud hanc piscantur perle habet regem Macometanum terra est sterilis

The island of Hormuz is a very rich center of trade. They fish for pearls here; the king follows Mohammed, and the land is barren.

[238]See Yule, *Cathay and the Way Thither* (see note 61), vol. 2, pp. 348–350; Elisabeth-Dorothea Hecht, "Ethiopia Threatens to Block the Nile," *Azania* 23.1 (1988), pp. 1–10; and Richard Pankhurst, "Ethiopia's Alleged Control of the Nile," in Haggai Erlich and Israel Gershoni, eds., *The Nile: Histories, Cultures, Myths* (Boulder, CO: L. Rienner, 2000), pp. 25–37, esp. 31–32.

[239]See Georges Lengherand, *Voyage de Georges Lengherand, mayeur de Mons en Haynaut, a Venise, Rome, Jérusalem, Mont Sinaï & Le Kayre (1485–1486)* (Mons: Masquillier & Dequesne, 1861), p. 185, reprinted in Enrico Cerulli, *Etiopi in Palestina* (Rome: Libreria dello Stato, 1943–47), p. 286; C. Schefer, ed., *Le voyage d'Outremer de Bertrand de la Brocquière* (Paris: Ernest Leroux, 1892; Farnborough: Gregg, 1972), pp. 145–146, with an English translation in Bertrandon de La Broquière, *The Voyage d'Outremer*, trans. and ed. Galen R. Kline (New York: P. Lang, 1988), pp. 92–93. Fra Mauro on his world map of c. 1455 depicts the gates used to control the Nile and alludes to this control in a brief legend: see Falchetta, *Fra Mauro's World Map* (see note 143 in Chap. 1 above), pp. 266–267, *403.

[240]The letter was published in several editions, the first being *Epistola potentissimi ac inuictissimi Emanuelis Regis Portugaliae & Algarbiorum &c. De victoriis habitis in India & Malacha: ad S. in Christo Patrem & D[omi]n[u]m nostrum D[omi]n[u]m Leonem X. Pont. Maximum* (Rome: Impressa per Iacobum Mazochium, 1513). For evidence that Waldseemüller used the letter and also for additional references on it see the commentary on Legend 12.5.

[241]The passage is on signature Aiii^{v} of the pamphlet, and is transcribed by William Roscoe, *The Life and Pontificate of Leo the Tenth*, revised by Thomas Roscoe (London: Chatto and Windus, 1876), vol. 1, p. 523.

[242]See Marco Polo, *Marka Pavlova z Benátek, Milion* (see note 54), p. 182; for Yule's translation see *The Book of Ser Marco Polo* (see note 54), Book 3, Chap. 28, vol. 2, pp. 397–398.

[243]See Varthema, *The Travels of Ludovico di Varthema* (see note 103), pp. 105–110.

[244]The relevant chapter in the *Paesi* or *Itinerarium* is 141. For an English translation of this chapter see Greenlee, "The Account of Priest Joseph" (see note 70 in Chap. 1 above), pp. 111–112.

This legend comes from Varthema;[245] Waldseemüller might have used the account of the island given by Joseph the Indian (Priest Joseph) in the *Paesi* or *Itinerarium*, specifically Chap. 140,[246] but he did not, for this chapter does not include the detail about the island being barren. The presence of a Portuguese flag by Hormuz on the *Carta marina* is interesting. Varthema was in Hormuz in 1504, before the Portuguese conquered it, which took them two attempts, in 1507 and 1515.[247] So in addition to Varthema, Waldseemüller had another source later than 1507, perhaps a nautical chart with a Portuguese flag on Hormuz.[248] Incidentally Fries discusses Hormuz in Chap. 81 of the *Uslegung*.[249]

7.17
Cacotora siue Scutora insula cristiana Anno domini .1507. erepta de Saracenorum imperio et regi portugalie subacta. diues populosa et fecunda. Habet ciuitates preclaras et emporiales optimum Aloen secuteinum [for *socotorinum* or *socotranum*] profert incole sunt cismatici.

Sacotora or Socotra, a Christian island, was torn from the control of the Saracens in 1507 and brought under the power of the King of Portugal. It is rich and populous and fertile. It has cities that are famous trading centers. It grows excellent Socotran aloe. The inhabitants are schismatics.

All of the printed sources that Waldseemüller cites in the long text block on sheet 9 (Legend 9.3) were published in at least one edition in 1507 or before, so at least some of this legend comes from a more recent source that he does not cite there. The parts about Socotra being a Christian island, rich, and involved in trade could be from Marco Polo,[250] but Polo does not mention any cities, so this seems doubtful. Varthema was in Socotra before 1507, and in any case does not describe the island. So the source of most of this legend is not clear, though I searched in accounts of the island's history for sources that might have been available in Europe by 1516.[251] The aloe of Socotra was famous for centuries.[252]

On Waldseemüller's 1507 map he depicts two islands, Discordis and Scoyra, which in fact are both to be identified with Socotra, but on the *Carta marina* he has remedied this error. However, his depiction of the island on the *Carta marina* has a unique feature: it is depicted as two islands joined by a bridge. I have not seen any text that mentions such a division of the island or such a bridge, and there is no similar representation of the island on any other map, not even Fries's re-edition of the *Carta marina*. It is tempting to speculate that this feature is a visual acknowledgement by Waldseemüller that what he had thought were two islands were in fact one. But there is no way to confirm this hypothesis.

7.18
Calicut ciuitas regalis magna tociusque Indie maioris celebratissimus famosissimusque emporium. Hic ex diuersis mundi partibus varie confluunt res mercandarie, maxime de India, Arabia, Syria, Egipto, Persia, Ethiopia, Seylam, Samotra, Guzerath, Cathaio, Cini, Macini etc. ea que aportantur sunt Muscum Ambra Thus lignum Aloes

---

[245]See Varthema, *The Travels of Ludovico di Varthema* (see note 103), pp. 94–95.

[246]For an English translation of this chapter see Greenlee, "The Account of Priest Joseph" (see note 70 in Chap. 1 above), pp. 110–111.

[247]For an account of the first Portuguese attempt to conquer Hormuz see Afonso de Albuquerque, *Being the Portuguese Text of an Unpublished Letter of the Biblioteca Geral da Universidade de Coimbra Relating the Portuguese Conquest of Ormuz in 1507*, trans. Ronald Bishop Smith (Bethesda, MD: Decatur Press, 1972); Jean Aubin, "Cojeatar et Albuquerque," *Mare Luso-Indicum* 2 (1971), pp. 99–134; and the description of Gaspar Correia, *Lendas da India*, ed. Rodrigo José de Lima Felner (Coimbra: Por Ordem da Universidade, 1921), Chaps. 6–9, pp. 828–869. For an account of both the 1507 and 1515 attempts see António Dias Farinha, "A dupla conquista de Ormuz por Afonso de Albuquerque," *Studia* 48 (1989), pp. 445–472; and the same author's "Os Portugueses no Golfo Pérsico (1507–1538): Contribuição documental e crítica para a sua história," *Mare Liberum* 3 (1991), pp. 1–159, esp. 35–38 on the 1515 conquest.

[248]On the sixteenth-century Portuguese cartography of this area in general see Zoltán Biedermann, "Ormuz et sa région dans les cartes portugaises du XVIe siècle," in Dejanirah Couto and Rui Manuel Loureiro, eds., *Revisiting Hormuz. Portuguese Interactions in the Persian Gulf Region in the Early Modern Period* (Wiesbaden: Harrassowitz and Calouste Gulbenkian Foundation, 2008), pp. 121–133.

[249]The passage on Hormuz is translated into modern German by Petrzilka, *Die Karten des Laurent Fries* (see note 202 in Chap. 1 above), pp. 150.

[250]See Marco Polo, *Marka Pavlova z Benátek, Milion* (see note 54), pp. 183–184; for Yule's translation see *The Book of Ser Marco Polo* (see note 54), Book 3, Chap. 32, vol. 2, pp. 406–410.

[251]Pereira da Costa, José, "Socotorá e o domínio português no oriente," *Revista da Universidade de Coimbra* 23 (1973), pp. 323–371; Zoltán Biedermann, "Nas pegadas do apóstolo: Socotorá nas fontes europeias dos séculos XVI–XVII," *Anais de História de Além-Mar* 1 (2000), pp. 287–386.

[252]Zoltán Biedermann, "Uma erva de muitas virtudes: o aloés de Socotorá na mira de viajantes e botanistas desde a Antiguidade até à Época Moderna," *Revista de Cultura* (Macau) 21 (2007), pp. 28–46.

Reubarbarum Bombex Gariofanum Canella Brasilicum Sandalum Ganphora Muscatum Macis Benzui Lacca et alia aromata minarum specierum. Nec non gemarum et pannorum varie sortes. Que vero penes calicutium et Cananor nascuntur sunt Zinziber Piper Lamarindi Cardimonum Mirabolanum Cassia Fistula Zibetum Storax etc. Rex diues et prepotens nulli tributarius qui cum sua gente ydolatra caffranus est religiosissimus habensque duas vxores quarum quelibet habet.x. sacerdotes qui cum ipsis absente rege concubitum habere solent. hic virgines cum octavum annum etatis superauerint prosequuntur mares rogando vt florem virginitatis ab eis capiant, quod tameu [for *tamen*] maxime viri abhorrent: adulteria[m]que non animaduertunt.

Idolatre sunt gentes barbare nec cum Cristianis nec cum Judeis neque cum Saracenis in religione conueniunt, seruant enim execrabiles abhominabilesque ritus et ceremonias, suntque in multiplici differentia. nam alii id quod primus mane vident, alii simulachra adorant quibus etiam deuouent filios vel filias victimandas, necnon seipsos interemunt amore illorum et isti ydolatre barbari dicuntur. habentque monasteria religiosorum ydolatrarum asperimam vitam ducentium. Alii vero demonem adorant credentes illum diuini iudicii executorem cui parcere et affligere bene aut male facere hominibus a deo concessum et isti ydolatre caffrani dicuntur et boues maxime venerantur quos animalia sancta credunt vxores magnat[or]um apud hos solemniter cremantur igne vna cum maritorum funeribus sperantes fieri coniuges in alia vita. non vescunter carnibus nec aliis victualibus que vitam conceperunt, sed Butiro, saccaro, Orisa, et fructibus arborum, pauci vero piscibus. vinum bibunt de genere palmarum; antequam manducant. lauant se in baptisteriis, raro quis suam deflorat vxorem sed facit hoc fieri per alium. Regem Caffranum decet habere sacerdotes corriuales commiscentes vxoribus regis obque hoc filii non succedunt regnis.

The great regal city of Calicut is the most busy and famous trading center of all of Greater India. Here from different parts of the world various goods flow, particularly from India, Arabia, Syria, Egypt, Persia, Ethiopia, Ceylon, Sumatra, Gujarat, Cathay, China, Macini [i.e. India Beyond the Ganges], etc. The goods which are brought there are musk, amber, frankincense, aloe wood, rhubarb, silk, clove, cinnamon, brasil wood, sandalwood, camphor, nutmeg, mace, benzoin, gum lacca, and other minor spices, as well as various types of gems and cloths. Those that grow near Calicut and Cannanore are ginger, pepper, tamarind, cardamom, myrobalan, cassia fistula, civet, storax, etc. The king is rich and powerful, paying tribute to no one, and together with his people is a highly devout Caffranus idolater. He has two wives, each of whom has ten priests who when the king is absent lie with them. Here when virgins turn nine years old they pursue the men, asking them to take the flower of their virginity, but the men are very averse to doing so. They do not punish adultery.

Idolaters are barbarous people who do not agree with Christians, Jews, or Saracens in their religion, for they practice execrable and abominable rites and ceremonies, and they are greatly diverse. For some of them worship whatever they see first in the morning, while others worship idols to whom they even offer up their own sons and daughters for sacrifice, and even kill themselves for love of the idols, and are known as barbarous idolaters. They have monasteries of idolatrous monks who lead a very harsh life. Others worship a demon, believing him to be the minister of divine justice, to whom god gave the right to spare or punish men who do good or evil. They are known as Caffrani idolaters, and principally worship cattle, which they believe are sacred animals. Among these people, the wives of nobles are solemnly cremated at the funerals of their husbands, hoping to become their wives in another life. They do not eat meat or other foods which were alive, but butter, sugar, rice[?], and fruit from trees, but a few of them eat fish. They drink wine from a species of palm. Before they eat they wash themselves in bathtubs. It is rare that a man deflowers his own wife, but rather he has this done by another man. It is thought good for the king, who is a Caffranus, to have rival priests who lie with his own wives, and because of this, sons do not succeed their fathers as rulers.

This long legend is a complex compilation, a couple of whose sources I have not been able to identify. Certainly Waldseemüller was faced with a difficult task in assembling this text, as he had material about Calicut from Varthema,[253] the so-called Anonymous Narrative, Joseph the Indian, and Caspar the Jew, some parts of which were contradictory. The early

[253]For discussion of Varthema's account of Calicut see Joan-Pau Rubiés, "Ludovico de Varthema: The Curious Traveller at the Time of Vasco da Gama and Columbus," in his *Travel and Ethnology in the Renaissance: South India through European eyes, 1250–1625* (Cambridge and New York: Cambridge University Press, 2000), pp. 125–163, at 155–162.

part of the legend, which describes the sources of the spices available at Calicut, lists both some of the spices and other goods available, and also indicates which ones grow near Calicut, are summarized from the longer lists in the *Paesi* or *Itinerarium*, Chaps. 82 and 83.[254] I do not know of a source that says that the king of Calicut pays tribute to no one. The author of the Anonymous Narrative says that the king has two wives, each of whom has ten priests who lie with her: see the *Paesi* or the *Itinerarium*, Chap. 75.[255] The behavior of the young virgins is described earlier in the same chapter.[256]

The first sentence of the second paragraph about the Caffrani idolaters seems to be Waldseemüller's composition; on the Caffrani see Legend 3.37 above. Later in Chap. 75 of the *Paesi* and *Itinerarium* the idolatry of the Guzerates, who trade in Calicut, is mentioned (they are said to worship the sun, moon, and cattle), but this is certainly not the source of the material about idolatry, which comes rather from Varthema's chapter on Java, in the context of a comparison of Java and Calicut.[257] I do not know the source of the material about the sacrifices of sons and daughters and the suicides, which is puzzling since Joseph the Indian said that he knew little of the sacrifices of the people of Caranganor (end of Chap. 131 of the *Paesi* and *Itinerarium*[258])—and the customs in Caranganor and Calicut are compared at the beginning of Chap. 139 of the *Paesi* and *Itinerarium*.[259] The monasteries of idolatrous monks come from Joseph the Indian in the *Paesi* or *Itinerarium*, Chap. 134.[260] The alleged demon-worship in Calicut comes from Varthema,[261] while the worship of cattle is mentioned both by Varthema and in the so-called first letter of Girolamo Sernigi.[262] The burning of wives upon the death of their husbands, that is, the practice of *suttee* or *sati*, is mentioned by Joseph the Indian and also by Varthema (as discussed in the introduction above, and as illustrated on sheet 4 of the *Carta marina*).[263] Their avoidance of meat is mentioned in the so-called first letter of Girolamo Sernigi, in a passage that appears in the *Paesi* and *Itinerarium*, Chap. 57.[264] The palm wine is briefly mentioned in the same passage in the first Sernigi letter (*Paesi* and *Itinerarium* Chap. 57),[265] and in more detail by Joseph the Indian (*Paesi* and *Itinerarium* Chap. 138).[266] The frequent bathing of the inhabitants of Calicut is mentioned by both Joseph the Indian (Chap. 132 of the *Paesi* or *Itinerarium*)[267] and Varthema.[268] The passage about having another man deflower one's wife is somewhat similar to the passage about the behavior of the young virgins at the end of the first paragraph, but seems in fact to derive from Varthema's account of the king of Calicut,[269] even though Varthema says that this practice was only that of the king, while Waldseemüller says that it was a general usage. With regard to the priests who lie with the king's wives, the author of the Anonymous Narrative says that the king has two wives, each of whom has ten priests who lie with her: see

[254]For an English translation of these chapters see Greenlee, "The Anonymous Narrative" (see note 67 in Chap. 1 above), pp. 91–94. There is also a chapter on the spices that grow near Calicut in Varthema, *The Travels of Ludovico di Varthema* (see note 103), pp. 157–158.

[255]See Greenlee, "The Anonymous Narrative" (see note 67 in Chap. 1 above), p. 80. Incidentally Joseph the Indian says that he has many wives in Chap. 132 of the *Paesi* or *Itinerarium*; for a translation see Greenlee, p. 101. There is a similar passage in a letter sent from the King of Portugal to the King of Castile, and published in Rome in 1505: see *Copy of a Letter of the King of Portugal Sent to the King of Castile Concerning the Voyage and Success of India*, trans. Sergio J. Pacifici (Minneapolis: University of Minnesota Press, 1955), p. 18, but in this case the number of priests per wife is not specified.

[256]See Greenlee, "The Anonymous Narrative" (see note 67 in Chap. 1 above), p. 79.

[257]See Varthema, *The Travels of Ludovico di Varthema* (see note 103), pp. 251–252: "Their faith is this: some adore idols as they do in Calicut, and there are some who worship the sun, others the moon; many worship the ox; a great many the first thing they meet in the morning; and others worship the devil in the manner I have already told you.".

[258]For an English translation see Greenlee, "The Account of Priest Joseph" (see note 70 in Chap. 1 above), pp. 100–101.

[259]See Greenlee, "The Account of Priest Joseph" (see note 70 in Chap. 1 above), pp. 108–109.

[260]For an English translation see Greenlee, "The Account of Priest Joseph" (see note 70 in Chap. 1 above), p. 104.

[261]On the alleged worship of demons in Calicut see *The Travels of Ludovico di Varthema* (see note 103), pp. 136–138; for discussion of this misinterpretation of Hindu religion see note 128.

[262]For the passage in Varthema see note 485; for the passage in the first Sernigi letter see the *Paesi* or *Itinerarium*, Chap. 57, and Ravenstein, *A Journal of the First Voyage* (see note 57 in Chap. 1 above), p. 132.

[263]For the description by Joseph see the *Paesi* or *Itinerarium*, Chap. 132, and Greenlee, "The Account of Priest Joseph" (see note 70 in Chap. 1 above), pp. 101 and 110. For the passage in Varthema see, *The Travels of Ludovico di Varthema* (see note 103), pp. 206–208.

[264]For an English translation see Ravenstein, *A Journal of the First Voyage* (see note 57 in Chap. 1 above), p. 132.

[265]See Ravenstein, *A Journal of the First Voyage* (see note 57 in Chap. 1 above), p. 133.

[266]See Greenlee, "The Account of Priest Joseph" (see note 70 in Chap. 1 above), pp. 107–108.

[267]For an English translation see Greenlee, "The Account of Priest Joseph" (see note 70 in Chap. 1 above), p. 101.

[268]See Varthema, *The Travels of Ludovico di Varthema* (see note 103), p. 149. Curiously in Sernigi's first letter it is said that the people of Calicut bathe very infrequently, specifically three times a year: see the *Paesi* or *Itinerarium*, Chap. 53, and for an English translation, Ravenstein, *A Journal of the First Voyage* (see note 57 in Chap. 1 above), p. 126.

[269]See Varthema, *The Travels of Ludovico di Varthema* (see note 103), p. 141.

the *Paesi* or the *Itinerarium*, Chap. 75.[270] The detail about the sons not inheriting the throne comes from a different source, however, namely Joseph the Indian in Chap. 132 of the *Paesi* or *Itinerarium*.[271]

The increase in the attention given to Calicut[272] on the *Carta marina* versus Waldseemüller's 1507 map is remarkable: on the 1507 map there is just one brief legend,[273] and in the 1513 Ptolemy the legend on the modern map of India is very similar to that on the 1507 map. On the 1516 map, in addition to the long legend just studied, there is a long legend about the sources and prices of the spices available in Calicut on sheet 12[274]: clearly the city was regarded as one of the most important in the world. But this eminence did not last long: the fortunes of Calicut as a trading center were already in decline in the early seventeenth century.[275]

On Johann Schöner's globe of 1520, much of the geography of Asia comes from Waldseemüller's 1507 map, but his treatment of Calicut is of interest as it seems to involve material from the *Carta marina*. On his 1507 map Waldseemüller has a brief legend about Calicut to the east of Taprobana (just mentioned, and transcribed in a footnote). On Schöner's 1520 globe the place name *Calicut* is on the small, westward jutting peninsula just south of the delta of the Indus River, and there is a very faded and all-but-illegible legend about Calicut in the Indian Ocean on the other side of Taprobana, west of the island. This legend is much longer than that on Waldseemüller's 1507 map, so it seems likely that Schöner borrowed from the *Carta marina*, as he did in various other places on his 1520 globe, but a determination of Schöner's source here will have to await multispectral imaging of the globe.

7.19
CORUPTELE LITERARUM ET SILLABARUM (SI SINT) NEGLIGENTIA SCULPTOris ac impressoris commisse discretione lectoris emendetur non sic exemplar ad manus eorundem venisse intelligatur

Mistakes (if there are any) in letters or words that were made through the negligence of the engraver or printer may be corrected at the reader's discretion. Let it be understood that the original did not reach their hands with such errors.

This legend is transcribed and translated by Baldacci, and also translated by Harris; the translation here is my own.[276] The legend does not explicitly refer to the list of corrections printed on sheet 9 of the map,[277] and thus seems best taken as a piece of general advice that was on the map from the beginning, rather than a response to problems in the printing of this map—though it certainly may be a response to problems that had occurred in other projects in which Waldseemüller had been involved. Waldseemüller's expression of concern here is the first such to appear on a map, and was followed a few years later (as Baldacci notes) by Giovanni Vespucci's curt *Errata si quid excussoris culpa*, "If there are errors, they are the

[270]See Greenlee, "The Anonymous Narrative" (see note 67 in Chap. 1 above), p. 80.
[271]For an English translation see Greenlee, "The Account of Priest Joseph" (see note 70 in Chap. 1 above), p. 101.
[272]For references on Calicut in the sixteenth century see note 94.
[273]The legend on Waldseemüller's 1507 map: *Calliqut prouincia nobilis in ea sunt multa genera minerarum pimenta et alia genera mercatorum que veniunt de multis partibus canella zinamomum zinziber gariofolus sandalum et de omnibus specibus: hec est inuenta per regem portugallie*, that is, "The noble province of Calicut; in it there are many types of minerals and other types of merchandise which come from many parts: cinnamon, ginger, clove, sandalwood, and a bit of every spice: the region was discovered by the King of Portugal." Waldseemüller's legend depends on that on the Caverio chart, which is translated by Stevenson, *Marine World Chart* (see note 32), p. 112.
[274]Incidentally Calicut is described in Chap. 26 of Fries's *Uslegung*, and this passage is translated into modern German by Petrzilka, *Die Karten des Laurent Fries* (see note 202 in Chap. 1 above), pp. 132–133. Accompanying the text in the 1525 edition of the *Uslegung* is a large two-page illustration of the city; the image is briefly discussed by Johnson, *Carta marina* (see note 21 in Chap. 1 above), pp. 111–112.
[275]On the decline in Calicut's fortunes see Lewes Roberts, *The Merchants Mappe of Commerce* (London: Printed by R. O., 1638), Chap. 92, p. 188.
[276]See Baldacci, *Columbian Atlas of the Great Discovery* (see note 194), p. 125; and Harris, "A Typographic Appraisal" (see note 227), p. 48.
[277]For references on the list of corrections on the *Carta marina* see note 226.

engraver's fault," on his world map of c. 1524.[278] Vespucci's indication of where the blame for errors should go was evidently well warranted, as the map has a large number of mis-spellings.[279] Hildegard Binder Johnson points to Martin Luther's *Admonition to Printers* of September 1525 as another early sixteenth-century expression of frustration with printing errors.[280]

[278]Vespucci's map, titled *Totius orbis descriptio: tam veterum quam recentium geographorum*, survives in two copies, one at Harvard, and the other in Wolfenbüttel, shelfmark 15 Astron. 2°. For discussion of the map see Henry Harrisse, *The Discovery of North America: A Critical, Documentary and Historic Investigation, with an Essay on the Early Cartography of the New World* (London: Henry Stevens and Son; Paris: H. Welter, 1892), nos. 147 (p. 533) and 148 (pp. 533–534); Rodney W. Shirley, *The Mapping of the World: Early Printed World Maps 1472–1700* (London: Holland Press, 1983), pp. 58–59; and Gaetano Ferro et al., *Columbian Iconography* (see note 108 in Chap. 1 above), pp. 145–148 (with a very good reproduction). The Wolfenbüttel copy is discussed in Christian Heitzmann, "Wem gehören die Molukken? Eine unbekannte Weltkarte aus der Frühzeit der Entdeckungen," *Zeitschrift für Ideengeschichte* 1.2 (Summer, 2007), pp. 101–110.

[279]See Juan Gil, "Loores de la crítica textual," in Maurilio Pérez González, ed., *Actas: III Congreso Hispánico de Latín Medieval (León, 26–29 de septiembre de 2001)* (León: Universidad de León, Secretariado de Publicaciones y Medios Audiovisuales, 2002), vol. 1, pp. 17–30, at 24.

[280]Johnson, *Carta marina* (see note 21 in Chap. 1 above), p. 123. Luther's "An Admonition to the Printers in Nürnberg" is translated in *The Letters of Martin Luther*, selected and translated by Margaret A. Currie (London: Macmillan and Co., Limited, 1908), pp. 144–145.

2.9 Sheet 8. Southern India and Southeast Asia (Plate 2.8)

The two large peninsulas that dominate this sheet are copied from Caverio's chart of c. 1503, but Waldseemüller provides much more detail about the hinterlands, and the density of information in the left peninsula, which is southern India, clearly indicates that this was an area of great interest. While northern India on sheet 4 is the abode of the monstrous races of men, southern India is a realm of Eastern potentates and exotic goods of trade. In the other, larger peninsula, which is labeled MACINI REGIO (India Beyond the Ganges), Waldseemüller shows the most interest in Pegu (in modern Burma, or Myanmar), which was a rich trading city. The lesser density of information near the eastern edge of Asia is interesting, and contrasts with the fairly dense Marco Polo-derived information in this area on the 1507 map. This difference must be interpreted as a decrease in Waldseemüller's confidence in Polo, as must his decision not to include Japan on the *Carta marina*, and indeed to show less of the eastern Indian Ocean and the eastern coast of Asia than the map he was using as a model, the Caverio chart, did.

8.1
Rex Cambaie tributarius est regi de Guzerat

The king of Cambay pays tribute to the king of Gujarat.

I have not been able to determine the source of this legend. It does not come from Varthema's discussion of Cambay, nor from that of Joseph the Indian.[281] The manuscript of Marco Polo that seemed to be closest to what Waldseemüller was using does not mention the king of Cambay, and Yule's more inclusive synthetic text actually says that the people of Cambay pay tribute to no one.[282]

8.2
REGNUM ORIZA Sunt ydolatre caffrani. Rex huius regni nulli est tributarius inuadit hostes .100. elephantibus et .100000. pensionariis equitum et peditum. Hic est habundancia in prisilto [for *brasilico*] serico porcelana et bladis

The kingdom of Orissa. They are Caffrani idolaters. The king of this realm pays tribute to no one. He attacks his enemies with a hundred elephants and a hundred thousand mercenary horsemen and foot soldiers. There is an abundance of brasil wood, silk, porcelain, and grain.

I have not found the source of this legend. Franz Hümmerich has suggested that this legend comes from the lost work of Gaspar of India that had been sent to the king of Portugal, to which Waldseemüller says that he had access in the long text block on sheet 9.[283] This is a plausible suggestion. On the Caffrani idolaters see Legend 3.37 above.

8.3
NARSINGA CIUITAS Hic dominatur Rex Narsinge omnium regum Indie potentiss[imus] cuius imperii circumferentia extenditur plus quam ad .3000. miliar[ia] qui cum sua gente Cafranus ydolatra est tum non minus habet multos cristianos sub mandato eorum que et praesertim portugallensium amicus. Habet et reges tributarios, hostes inuadit .600. elephant[ibus] et equitum et peditum sine numero. Rex habet .200. vxores quae omnes rege mortuo comburentur.

The city of Narsinga. Here rules the king of Narsinga, the most powerful of all of the kings of India, the bounds of whose power extend more than 3000 miles. Together with his people he is a Caffranus idolater, but nonetheless he has many Christians under their control and is a particular friend of the Portuguese. He has kings who pay him tribute, and he attacks his enemies with six hundred elephants and cavalry and foot soldiers without number. The king has two hundred wives who will all be burned when the king dies.

---

[281]Varthema discusses Cambay in *The Travels of Ludovico di Varthema* (see note 103), pp. 105–111; for Joseph the Indian's discussion see the *Paesi* or *Itinerarium*, Chap. 141, and for an English translation of this chapter see Greenlee, "The Account of Priest Joseph" (see note 70 in Chap. 1 above), pp. 111–112.

[282]For the text of the manuscript of Marco Polo that generally seems close to what Waldseemüller was using see Marco Polo, *Marka Pavlova z Benátek, Milion* (see note 54), p. 182; for Yule's translation which includes the indication that the people of Cambay pay tribute to no one see *The Book of Ser Marco Polo* (see note 54), Book 3, Chap. 28, vol. 2, pp. 397–398.

[283]Franz Hümmerich, "Studien zum 'Roteiro' der Entdeckungsfahrt Vascos da Gama 1497–1499," *Revista da Universidade de Coimbra* 10 (1927), pp. 53–302, at 125–126.

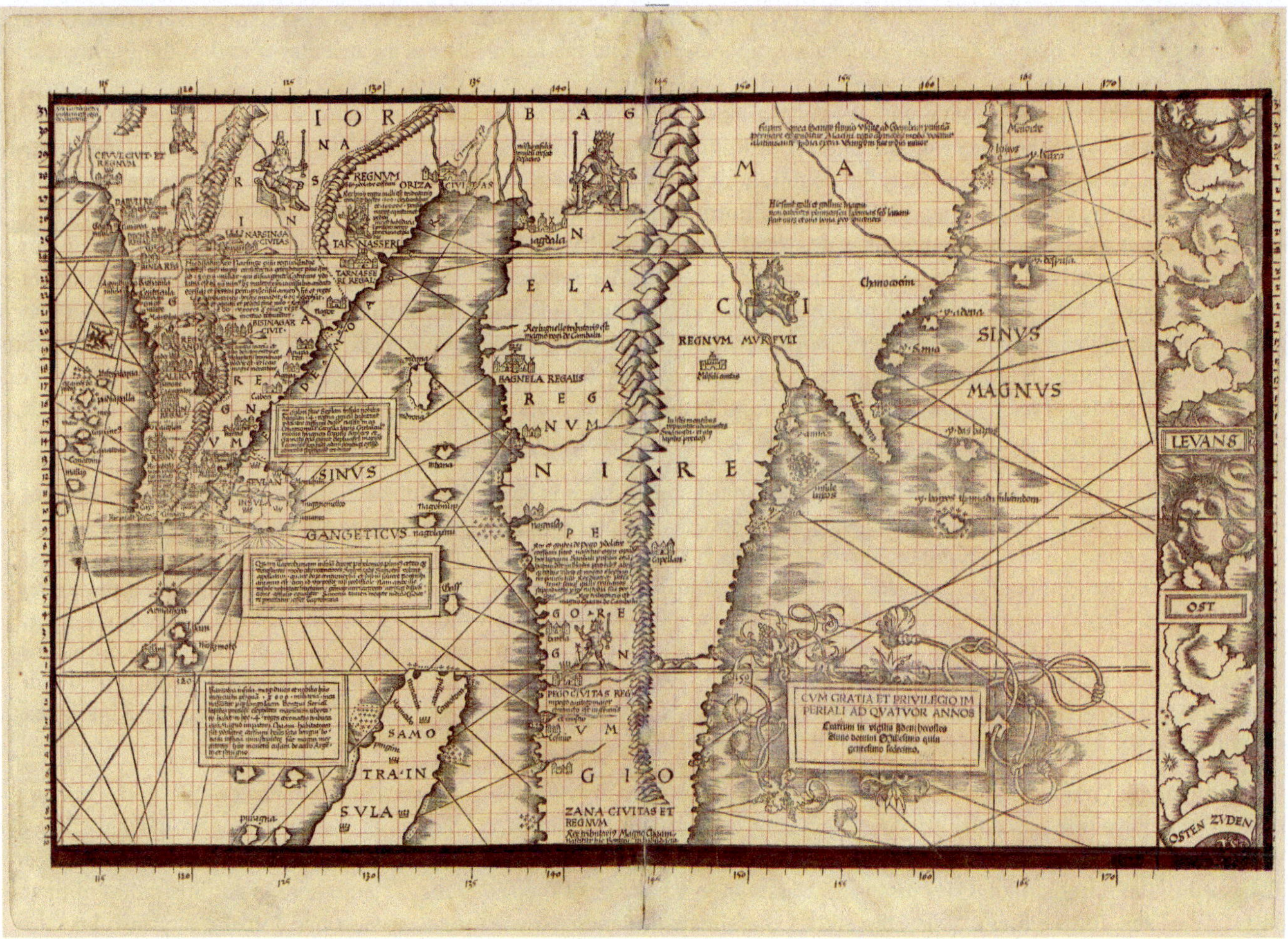

Plate 2.8 Sheet 8 of the *Carta marina*: Southern India and Southeast Asia. Courtesy of the Library of Congress

Varthema describes Narsinga and says that the king has 40,000 horses, but no details match between Waldseemüller's description and Varthema's.[284] Some of the details Waldseemüller supplies come from Joseph the Indian, who says that the King of Narsinga is very powerful, that his kingdom is 3000 miles around, and that when he attacks his enemies, "he takes with him eight hundred elephants, four thousand horses, and innumerable foot soldiers."[285] The number of elephants cited is different, but clearly Waldseemüller was using Joseph the Indian as a source. Other details come from a passage in the Anonymous Narrative about a king named Naremega who lives in the mountains near Calicut—which Waldseemüller correctly understood as a reference to Narsinga. In that passage we are told that the king "has two or three hundred wives.

[284]See Varthema, *The Travels of Ludovico di Varthema* (see note 103), p. 126; for discussion of Varthema's account of Narsinga see Joan-Pau Rubiés, "Ludovico de Varthema: The Curious Traveller at the Time of Vasco da Gama and Columbus," in his *Travel and Ethnology in the Renaissance: South India through European Eyes, 1250–1625* (Cambridge and New York: Cambridge University Press, 2000), pp. 125–163, at 147–154.

[285]See the *Paesi* or *Itinerarium*, Chap. 142, and for an English translation of this chapter see Greenlee, "The Account of Priest Joseph" (see note 70 in Chap. 1 above), p. 113.

The day he dies they burn him and all of his wives with him."[286] But Waldseemüller was also using a very recent source. The details about the many Christians under the king's control and the fact that he is a particular friend of the Portuguese do not appear in any of the sources that Waldseemüller cites in the long text block on sheet 9.

There were intermittent negotiations between different kings of Narsinga and the Portuguese in 1505, 1510, and 1514. The negotiations were motivated by the interest of both parties in having an ally against local enemies, but as the strategic situation in India was in constant flux, the negotiations never amounted to much.[287] An optimistic pronouncement that the king of Narsinga was a special friend of the king of Portugal might have been made at several points during this period, but it is perhaps more likely the pronouncement was made in 1514 or 1515, as an agreement between the king of Narsinga and Afonso de Albuquerque was reached in 1514. In a letter of December 4, 1513, Albuquerque wrote to King Manuel of Portugal assuring him that the kings of both Narsinga and Deccan would want to conclude a treaty with Manuel as a way to gain victory over the other[288]; in a letter of October, 25, 1514, Albuquerque wrote to King Manuel of Portugal, "In India there are several things to be done. The first is to conclude a treaty with the King of Narsinga, which cannot fail to be of great benefit to your Majesty."[289] In a letter sent from Cananor about a month later, on November 27, 1514, Albuquerque mentions that the treaty with the King of Narsinga had been concluded[290]:

> On the 8th day of November, as I was on the point of starting from Goa to Cochin, the ambassadors from the King of Narsinga arrived, bringing me some bracelets and jewels, which I now send to your Majesty. Their instructions were to conclude, on behalf of the King of Narsinga, a treaty of peace and friendship with your Majesty; to wage war against the Turks in the kingdom of the Deccan, and arrange about the free importation of horses into their ports from Arabia and Persia. The first thing we talked about was the war with the Turks, in which I agreed to help the King of Narsinga; and as the King of Onor was a tributary of Narsinga, and was at war against Melique Az (captain of the Adil Khan) then at Cintacora, I wrote to the Adil Khan requesting him to instruct his captains to cease hostilities, which he did at once. As regards the question of horses, I could not agree to their proposals, and they at last returned to the King laden with presents from us.

This treaty did not last long, but this period seems the most likely time when the source of Waldseemüller's report that the king of Narsinga is a particular friend of the king of Portugal originated. Thus this is a source that would have reached Waldseemüller rather close to the 1516 completion date of the map, like the 1515 Burgkmair print of the rhinoceros discussed in the introduction. Fries discusses Narsinga in Chap. 79 of the *Uslegung*,[291] and there is an illustration of the city on f. 21v of the 1525 edition.[292] By the end of the sixteenth century a war of succession had left the city in ruins.[293]

The density of legends and coastal cities in southern India from Narsinga southward is impressive when compared with that of the large peninsula (*Macini regio*) just to its east.

> 8.4
> BISINGAR CIVIT. Hec ciuitas maxima et omnium delicamentorum et voluptatum humanarum alitrix est, et locus magne mercature.

---

[286]The passage is in the *Paesi* and *Itinerarium*, Chap. 76; for an English translation see Greenlee, "The Anonymous Narrative" (see note 67 in Chap. 1 above), p. 82.

[287]See H. Heras, "Early Relations Between Vijayanagara and Portugal," *Quarterly Journal of the Mythic Society* 16 (1925), pp. 63–74; B. S. Shastry, "The First Decade of Portuguese-Vijayanagara Relations," *Journal of Indian History* 52 (1974), pp. 147–155, reprinted in his *Studies in Indo-Portuguese History* (Bangalore, 1981), pp. 80–91; Jorge Manuel dos Santos Alves, "A cruz, os diamantes e os cavalos: Frei Luis do Salvador, primeiro missionário e embaixador português em Vijayanagara (1500–10)," *Mare Liberum* 5 (1993), pp. 9–20; and Joan-Pau Rubiés, "The Portuguese and Vijayanagara: Politics, Religion and Classification," in his *Travel and Ethnology in the Renaissance: South India through European eyes, 1250–1625* (Cambridge and New York: Cambridge University Press, 2000), pp. 164–200.

[288]Afonso de Albuquerque, *Cartas de Affonso de Albuquerque, seguidas de documentos que as elucidam*, ed. Raymundo Antonia de Bulhão Pato (Lisbon: Academia real das sciencias de Lisboa, 1884–1935), vol. 1, letter 41, p. 199.

[289]Albuquerque, *Cartas de Affonso de Albuquerque* (see note 288), vol. 1, letter 76, p. 327; partially translated in Frederick Charles Danvers, *The Portuguese in India: Being a History of the Rise and Decline of their Eastern Empire* (London: W.H. Allen & Co., Limited, 1894), vol. 1, pp. 306–307.

[290]Albuquerque, *Cartas de Affonso de Albuquerque* (see note 288), vol. 1, letter 87, p. 340; the translation here is from Danvers, *The Portuguese in India* (see note 289). vol. 1, pp. 307–308.

[291]The chapter on Narsinga is translated into modern German by Petrzilka, *Die Karten des Laurent Fries* (see note 202 in Chap. 1 above), pp. 149–150.

[292]The illustration of Narsinga in the *Uslegung* is briefly described by Johnson, *Carta marina* (see note 21 in Chap. 1 above), p. 110.

[293]See Robert Sewell, "Destruction of Vijayanagar (A.D. 1565)," in his *A Forgotten Empire: Vijayanagar: A Contribution to the History of India* (New Delhi: National Book Trust, India, 1970), pp. 189–205; and Vasundhara Filliozat, ed., *Vijayanagar: As Seen by Domingos Paes and Fernao Nuniz (16th Century Portuguese Chroniclers and Others)* (New Delhi: National Book Trust, India, 1999), pp. 329–330.

The city of Vijayanagar. This city is the greatest nourisher of all human delicacies and pleasures, and a center of much trade.

This legend comes from Varthema.[294]

8.5
Zaylon siue Seylan insula nobilis diuisa in .4. regna. qui eam habitant ydolatre caffrani dicuntur. nascitur in ea Cinamomum sive Canella lapis Carbunculus rubinus hiacinctus Topasius Saphirus et Granatus eciam gignit elephantes maximos. in ea mons est super quem adam penitentiam egisse ab incolis supersticiose creditur.

The noble island of Ceylon which is divided into four kingdoms. Those who inhabit the island are Caffrani idolaters. In it grows cinnamon or canella, carbuncle, ruby, hyacinth, topaz, sapphire, and garnet; it even bears huge elephants. On the island there is a mountain on the peak of which the locals superstitiously believe Adam did penance.

Marco Polo, Odoric, and Varthema all discuss Ceylon,[295] but most of the legend comes from Varthema. The one detail in Waldseemüller's legend that is not in Varthema is the carbuncles on the island, and the cartographer may have added these to the list of precious stones from Odoric. Of course Waldseemüller's use of Varthema represents a change and updating from his 1507 map, where his legend on Ceylon comes from Marco Polo.[296] The words *diuisa in* near the beginning of the legend are the result of a correction by hand of *diuisam* in accordance with the list of corrections in Legend 9.2. On the Caffrani idolaters see Legend 3.37 above.

8.6
Hic sepultus est S. Thomas

Here St. Thomas is buried.

This legend may have come from Marco Polo, Odoric, Varthema, or Priest Joseph.[297] Interestingly, there is a closely related legend on the 1507 map, *his occisus est s. thomas*, "Here St. Thomas was killed," but Waldseemüller seems to have used one of his textual sources for this legend rather than copying from his own earlier map.

8.7
COLON REGNUM Hic habitant multi cristiani sub mandato ydolatrarum

The kingdom of Quilon. Here live many Christians under the control of idolaters.

This is one of the few cases on the *Carta marina* where information was available from Varthema, but Waldseemüller followed Marco Polo instead,[298] and also stayed with a legend very similar to that on his 1507 map.[299]

---

[294]See Varthema, *The Travels of Ludovico di Varthema* (see note 103), p. 126: "It is a place of great merchandise, is extremely fertile, and is endowed with all possible kinds of delicacies. It occupies the most beautiful site, and possesses the best air that were ever seen: with certain very beautiful places for hunting and the same for fowling, so that it appears to me to be a second paradise."

[295]See Marco Polo, *Marka Pavlova z Benátek, Milion* (see note 54), pp. 165–166; for an English translation see *The Book of Ser Marco Polo* (see note 54), Book 3, Chap. 14, vol. 2, pp. 314–330; Yule, *Cathay and the Way Thither* (see note 61), vol. 2, pp. 170–173 (English), 305–307 (Latin), and 347–348 (Italian); and Varthema, *The Travels of Ludovico di Varthema* (see note 103), pp. 188–191.

[296]The legend on the 1507 map reads: *SEYLAM Hec insula est vna de maioribus et melioribus mundi habens In circuitu miliaria duo milia et .xl. insula hec habet regem ditissimum qui nulli tributarius est homines insule hu*[*ius*] *ydolatre sunt omnes nudi ambulant nullum bladum habent excepto riso habent lapides preci*[*osi*], that is, "Ceylon: this island is one of the largest and best of the world, having a circumference of 2040 miles. This island has a very rich king who is tributary to no one. The men of this island are idolaters and go about nude; they have no grain except rice, and they have precious stones."

[297]See Marco Polo, *Marka Pavlova z Benátek, Milion* (see note 54), p. 189; for an English translation see *The Book of Ser Marco Polo* (see note 54), Book 3, Chap. 18, vol. 2, pp. 353–359; Yule, *Cathay and the Way Thither* (see note 61), vol. 2, pp. 141–143 (English), 297–298 (Latin), and 343–344 (Italian); Varthema, *The Travels of Ludovico di Varthema* (see note 103), p. 187; the *Paesi* or *Itinerarium*, Chap. 142, with an English translation in Greenlee, "The Account of Priest Joseph" (see note 70), p. 113.

[298]See Varthema, *The Travels of Ludovico di Varthema* (see note 103), pp. 182–184; and Marco Polo, *Marka Pavlova z Benátek, Milion* (see note 54), pp. 177–178; for an English translation see *The Book of Ser Marco Polo*, Book 3 (see note 54), Chap. 22, vol. 2, pp. 375–382. On Coilum see further Paul Pelliot, *Notes on Marco Polo* (Paris: Impr. nationale, 1959–73), vol. 1, pp. 399–402.

[299]The legend on the 1507 map runs: *Coilum ciuitas. Hic habitant christiani et Judei et idolatr*[*i*] *habent linguam propriam rex nulli tributarius habent omnium genera specierum*, "The city of Quilon. Here live Christians and Jews and idolaters; they have their own language. Their king pays tribute to no one, and they have all types of spicess."

8.8
Hic crescit piper in magna copia

Here pepper grows in great abundance.

This legend comes from the same passage in Marco Polo as the previous one.[300]

8.9
Hic piscantur perle.

Here they gather pearls.

This legend comes from Varthema.[301]

8.10
Quam Taprobanam insulam dixere ptholomeus plinius ceterique vetustiores modo alii recenciores Sailon, alii Samotram volunt appelatam. quare eorum controuersiam et dubium soluere non mihi animus est, cum id vix posset nisi perdifficile. Nam ambe ille insule nobilitate fructuum, gemmarum auri argenti aerisque disposicione affluunt equaliter. Samotra tamen meopte indicio (si dicere permittatur) esset Taprobana.

The island that Ptolemy, Pliny, and all of the other ancient authors called Taprobana, other more recent authors call Ceylon, and others Sumatra. I have no interest in resolving their controversy and doubt, as it would be very difficult or impossible. For those two islands are famed for their products and are equally endowed with supplies of gems, gold, silver, and copper. Nonetheless Sumatra, according to my opinion (if it is permitted to say so) is to be identified with Taprobana.

There was considerable confusion in the Renaissance about whether the island of Taprobana described by classical authors should be identified with Ceylon or with Sumatra.[302] In identifying it with Sumatra, Waldseemüller is following both Varthema and Joseph the Indian.[303] This legend seems to be one that Waldseemüller composed, rather than copying most of it from a specific source. A look at the island of Taprobana in Waldseemüller's world maps offers a good overview of the evolution of his cartographic thought. The image of the island in the 1507 map is purely Ptolemaic; in the modern map of India in the 1513 Ptolemy it retains its Ptolemaic name (*Taprobana Insula*), but has lost its Ptolemaic shape and location, and Waldseemüller instead uses the shape of the island from Caverio's chart, and the legend describing the island is adapted from Caverio's[304]; while on the *Carta marina* he retains Caverio's shape for the island, but calls it *Samotra insula*, and places between it and *Seylan insula* (Ceylon) this legend about the identity of Taprobana that derives from Varthema and/or Joseph the Indian. In the legend he says that he has no interest in resolving the controversy, and is hesitant to express his own opinion

[300]For a good discussion of the history of the trade in and cultivation of pepper see Luís Filipe F. R. Thomaz, "Pequena história da pimenta," in his *A questão da pimenta em meados do século XVI: um debate político do governo de D. João de Castro* (Lisbon: Centro de Estudos dos Povos e Culturas de Expressão Portuguesa, Universidade Católica Portuguesa, 1998), pp. 9–48; also published in Artur Teodoro de Matos and Luís Filipe F. R. Thomaz, eds., *A carreira da India e as rotas dos estreitos: actas do VIII Seminário Internacional de História Indo–Portuguesa (Angra do Heroísmo, 7 a 11 de junho de 1996)* (Angra do Heroísmo: O Seminario, 1998), pp. 37–306, with the history of pepper on pp. 45–84.

[301]See Varthema, *The Travels of Ludovico di Varthema* (see note 103), p. 185; for discussion of the pearl fishery in this area see C. R. de Silva, "The Portuguese and Pearl Fishing off South India and Sri Lanka," *South Asia*, new series, 1 (1978), pp. 14–28.

[302]On the confusion about the identification of Taprobana see Ananda Abeydeera, "Taprobane, Ceylan ou Sumatra? Une confusion féconde," *Archipel* 47 (1994), pp. 87–123; and the same author's "Giovanni Battista Ramusio y voit Sumatra et Immanuel Kant Madagascar," *Archipel* 56 (1998), pp. 199–230. For another complaint that Ptolemy had located Taprobana incorrectly see Donald William Ferguson, "The Discovery of Ceylon by the Portuguese in 1506," *Journal of the Ceylon Asiatic Society* 19.59 (1907), pp. 284–400, at 376, who quotes a letter from Andrea Corsali to Juliano de Medici dated January 6, 1515: "This island was not located by Ptolemy, whom I find deficient in many particulars. He placed Traprobana wrongly, as can be judged by Y. H. from the sailing chart that Don Michele the king's orator brought to Rome." Don Michele is Dom Miguel da Silva, and Ferguson speculates that the chart in question was a tracing sent by Albuquerque to King Manuel of a large Javanese chart which was lost in the wreck of the *Flor de la mar* in 1511.

[303]See Varthema, *The Travels of Ludovico di Varthema* (see note 103), pp. 228–232; and the *Paesi* or *Itinerarium*, Chap. 142, translated into English in Greenlee, "The Account of Priest Joseph" (see note 70 in Chap. 1 above), p. 113; and also see in Antony Vallavanthara, *India in 1500 A. D.: The Narratives of Joseph the Indian* (Mannanam: Research Institute for Studies in History, 1984), pp. 216–217.

[304]The legend on the modern map of India in the 1513 Ptolemy runs: *In hac insula que satis magna est reperitur aurum suntque ibi berilli zinziber et ciuiuslibet alterius generius aromata. Est autem gens ydolatrie dedita et cum quibusdam aliis negotiatur ita ut pro eis que efferuntur alie res in ipsorum insulam reportant.* For the legend on the Caverio chart with an English translation see Stevenson, *Marine World Chart* (see note 32), pp. 114–115.

(which is borrowed from Varthema and Joseph the Indian), but he does express his opinion very clearly by giving the name *Samotra insula* to the island that on Caverio's map represented Taprobana. Of course this legend is part of Waldseemüller's broader rejection of Ptolemy and other ancient authorities on the *Carta marina* in favor of more recent sources.

8.11
Samotra insula mag. diues et nobilis habens in circuitu plus quam 3000 miliaria in ea nascuntur piper longus lacca Bentzui sericum lapides preciosi elephantes maximum ubertate[m] habet in hec 4 reges coronatos tributarios Magno imperatori Chaam habitatores sunt ydolatre caffrani homines satis benigni bonam iusticiam ministrantes sunt magni mercatores habent monetam cusam de auro Argento et stangno.

The large island of Sumatra is rich and noble, having more than 3000 miles in circumference. Here are found long pepper, lacca, benzoin, silk, precious stones, and elephants. The land is very fertile. In the island there are four crowned kings who pay tribute to the Great Khan. The people are Caffrani idolaters and are reasonably kind, administer good justice, and are great merchants. They have stamped coins of gold, silver, and tin.

Some of this legend comes from Varthema,[305] but there is also influence from another source that I have not been able to identify. Varthema says that Sumatra is 4500 miles in circumference rather than 3000, and also that there are three crowned kings rather than four, and says nothing about them being tributary to the Khan. Varthema does talk about the careful administration of justice on Sumatra, about the coins of different metals, and about the elephants, but his list of the exotic goods available on the island is different: he mentions long pepper, benzoin, and silk, but not lacca or precious stones. On the Caffrani idolaters see Legend 3.37 above. Fries discusses Sumatra in Chap. 102 of the *Uslegung*.[306]

8.12
in istis insulis multum crescit de sacaro

In these islands much sugar grows.

I have not been able to locate the source of this legend.

8.13
BAGNELA REGALIS Rex bagnelle tributarius est magno regi de Cambalu

The royal city of Bengal. The king of Bengal pays tribute to the great king of Cambalu.

This is another of the relatively few cases in which Waldseemüller chose to follow Marco Polo rather than Varthema. Varthema describes the city of Bengal,[307] but this legend does not derive from his description: he speaks of the king of Bengal having a huge army and being at war with the king of Narsinga, but not a word about him paying tribute to the king of Cambalu. Marco Polo says that when he was at the court of the Great Khan, the Khan's armies had gone to conquer Bengal, but had not yet done so.[308] Thus it seems that Waldseemüller assumed that they must have succeeded. But it is not clear why he chose to use Marco Polo e.

8.14
PEGO REGNVM Rex et gentes de pego ydolatre caffrani sunt. nascitur enim apud hos lighum Sandali prisilia eciam habundat in Bladis pecoribus aliisque bestiis zibeto et muscio Elephantes paucos habent. Rex diues et potens tenet semper mille cristianos stipendiatos pro custodia sue persone. Rex tributarius est magno Chaam de Cambalu.

The kingdom of Pego. The king and people of Pego are Caffrani idolaters. Among these people grows sandalwood and brasil wood, and the region abounds in grain, cattle and other beasts, civet, and musk. They have few elephants.

[305]See Varthema, *The Travels of Ludovico di Varthema* (see note 103), pp. 228–234, and he discusses other aspects of the island through p. 243.
[306]Fries's chapter on Sumatra is translated into modern German by Petrzilka, *Die Karten des Laurent Fries* (see note 202 in Chap. 1 above), pp. 155–156. For a description of trade on Sumatra a bit more than a hundred years later see Roberts, *The Merchants Mappe* (see note 275), Chap. 105, pp. 213–215.
[307]See Varthema, *The Travels of Ludovico di Varthema* (see note 103), pp. 210–212.
[308]See Marco Polo, *Marka Pavlova z Benátek, Milion* (see note 54), pp. 126–127; for an English translation see *The Book of Ser Marco Polo* (see note 54), Book 2, Chap. 55, vol. 2, pp. 114–115.

The king is rich and powerful, and always has a thousand Christian mercenaries to protect his person. He pays tribute to the great Khan of Cambalu.

Most of this legend comes from Varthema,[309] but Varthema does not mention the sandalwood or brasil wood, or the idea that the king of Pego pays tribute to the great Khan—indeed Varthema says that the king is very powerful. I have not been able to determine the source of these additions: the suggestion that the king pays tribute to the Great Khan makes one think of Marco Polo, but Polo does not discuss Pego. On the Caffrani idolaters see Legend 3.37 above. Fries discusses Pego in Chap. 92 of the *Uslegung*.[310]

8.15
PEGO CIVITAS REG in pego ciuitate maior contractus est in gemmis et mustio

The royal city of Pego. In the city of Pego there is a great market in gems and musk.

This legend comes from the same passage in Varthema cited in the discussion of the previous legend: Varthema does not say that there is a market of musk in Pego, but he does give the price of civet cats. Also Pego is mentioned as a source of benzoin in the list of the sources of spices and other exotic goods in the *Paesi* and *Itinerarium*, Chap. 83.[311]

8.16
ZANA CIVITAS ET REGNUM Rex tributarius Magno Chaam nascitur hic Bentziu in habundancia

The city and kingdom of Siam. The king pays tribute to the Great Khan; here benzoin grows in abundance.

I do not know where the idea that the king of Siam pays tribute to the Great Khan comes from, but Zana (Siam) is mentioned as a source of benzoin in the list of the sources of spices and other exotic goods in the *Paesi* and *Itinerarium*, Chap. 83.[312]

8.17
MACINI REGIO Ea pars que a Gange fluuio vsque ad cyambaru prouinciam per mare extenditur Macini regio ab incolis modo vocatur a latinis autem India excra Gangem siue india minor

The Macini region. The part that extends from the Ganges River to the province of Cyambaru along the sea is called the Macini region by its inhabitants, but by the Latins it is called India Beyond the Ganges or Lesser India.

It has been suggested that the name Macini is another form of Mangi,[313] but this seems questionable; etymologically it is related to Machin, i.e. Burma, but was applied to China and Indo-China, as Waldseemüller does. The spelling Macini is used by Varthema once, but he does not define it,[314] and it is not used on the Caverio map. The definition in Waldseemüller's legend is the most precise I know, and is perhaps his own.

8.18
Hic sunt galli et galline magni non habentes plumos seu pennas sed lanam sicut oues et oua bona producentes

Here there are big roosters and hens that do not have plumes or feathers, but wool like sheep, and they produce good eggs.

[309]See Varthema, *The Travels of Ludovico di Varthema* (see note 103), pp. 215–219.
[310]Fries's chapter on Pego is translated into modern German by Petrzilka, *Die Karten des Laurent Fries* (see note 202 in Chap. 1 above), p. 153.
[311]For an English translation see Greenlee, "The Anonymous Narrative" (see note 67 in Chap. 1 above), p. 93.
[312]For an English translation see Greenlee, "The Anonymous Narrative" (see note 67 in Chap. 1 above), p. 93.
[313]Ivar Hallberg, *L'Extrême Orient dans la littérature et la cartographie de l'Occident des XIIIe, XIVe, et XVe siècles; étude sur l'histoire de la géographie* (Göteborg: W. Zachrissons boktryckeri a.-b., 1907), pp. 324–325.
[314]See Varthema, *The Travels of Ludovico di Varthema* (see note 103), p. 236.

Material about these chickens appears in both Marco Polo and Odoric,[315] and in this case Waldseemüller made use of Polo, for Odoric does not mention the birds' eggs. This legend is copied by Johann Schöner on his manuscript globe of 1520.

8.19
in istis montibus reperiuntur adamantes Smaragdi et alii lapides preciosi

In these mountains are found diamonds, emeralds and other precious stones.

This legend seems to be a duplicate of Legend 4.26 on the Valley of Diamonds mentioned by Marco Polo.

8.20
CVM GRATIA ET PRIVILEGIO IMPERIALI AD QUATUOR ANNOS Exartum [for *Exactum*] in vigilia Penthecostes Anno domini Millesimo quingentesimo sedecimo

By imperial favor and privilege [protected] for four years; published at the Pentecost Vigil [May 11], 1516.

It is unfortunate that the printer's name is not indicated. The elaborate decorations around the cartouche help fill what would otherwise be a large area of empty ocean. Waldseemüller's uneasiness about leaving a large empty space here is palpable, particularly when the *Carta marina* is contrasted with Caverio's chart.

---

[315]See Marco Polo, *Marka Pavlova z Benátek, Milion* (see note 54), p. 149; for an English translation see *The Book of Ser Marco Polo* (see note 54), Book 2, Chap. 80, vol. 2, p. 229; and Yule, *Cathay and the Way Thither* (see note 61), vol. 2, p. 186 (English), 311 (Latin), and 352 (Italian).

2.10 Sheet 9. Cartouches in the Map's Southwest Corner (Plate 2.9)

The cartouche in the upper left, which is in the shape of a shield, is surrounded by text that says the chart is dedicated to Hugo de Hassard (Hugues des Hazards), Bishop of Toul, a town in what is now eastern France about 60 miles northwest of Saint-Dié. The cartouche in the lower left is covered with a glued-on sheet of paper, but beneath that sheet is a list of printing errors, all of which Johann Schöner had corrected by hand on the surviving copy of the map, except for those on sheet 6. The large cartouche on the right contains Waldseemüller's introduction to the map: he describes his 1507 map as representing the world as it was known to the ancients, while his new map, the *Carta marina*, aims to show the world as it is known in his time. He also lists his most important sources in making the map.

9.1
Hugonis. de. Hassardis. ecclesie. Tullensis. episcopi. digniss. munus.

The most worthy gift of Hugues des Hazards, bishop of the church of Toul.

This text surrounds the shield-shaped cartouche in the upper left part of the sheet, and indicates that Hugues des Hazards had contributed financially to the production of the map, presumably by helping to defray the costs of its printing. Hugues was Bishop of Toul from 1506 to 1517.[316] Gautier (or Walter) Lud inscribed the dedicatory letter in Matthias Ringmann's *Grammatica figurata* of 1509 to the bishop, so there had been some previous association between Waldseemüller's circle and Hugues.[317] Waldseemüller's intention was certainly not that Hugues's coat of arms be printed in the cartouche,[318] but rather that the coat of arms of the purchaser of the map be added in the cartouche by hand.

9.2
Errata emendentur

In porta Norbogie partis prope oceanum deponatur nomen Groneland: In oceano Germanico deponatur nomen insule Islanda conuenit enim alteri insule magis septentrionali, Sub fluuio de Senega legatur seu brachium pro seu brachi. Prope oceanum hespericum legatur leopardi pro Teopardi. Prope baldac legatur soluit tributem imperatori Chaam: in tabula magna prope prasiliam legatur portogalenses pro portogalensis: et in eadem Januensis pro tanuensis. In Arabia deserta legatur que charoanam mercatorum inuadunt Braua insula pro brana prope Bismagar legatur rege mortuo comburentur pro comburetur in tabula zaylon legatur diuisa in .4. pro diuisam .4. prope Camul legatur sunt ydolatre sub dominio tartarorum. Penes Gangem fluuium legatur contra regem Narsinge pro contra rege Penes lineam circumscribentem Tartariam legatur Quod extra ambitur pro quod intra ambitur in tablua Tartarie Mongal legatur alendis pecoribus apta pro alendis pedoribus: in tabula Cathay prouincie legatur veteris ac noui testameti scripturas pro scripturraram [*sic*]: in eade legatur: sed dispositione faciei pro dispositioni in tabula sub ciuitate Cambalu legatur Imperatoris pro imperatorum et in eadem habitatonis [*sic*] domini Chaam pro babitatoni et in eadem legatur sunt in eo .4. columne de auro. Penes imperatorem Noy legatur imperator super .600000. armatorum. In tabula Mangi legatur diuisit prouinciam Mangi pro prouinciam magui: vbicumque ponitur prisilicum legatur brasilicum: cetera lectoris discreitioni comitantur.

Please Correct These Errors.
In the entrance of the part of Norway near the ocean [sheet 2] please add the name "Groneland"; in the German Ocean add the name of the island "Islanda," for it belongs to another island further to the north. Beneath the Senegal River read "seu brachium" for "seu brachi" [in Legend 6.10]. Near the western ocean read "leopardi" for "Teopardi" [Legend 6.6]. Near Baghdad read "soluit tributem imperatori Chaam" [Legend 3.26]. On the big sheet near Brazil read "portogalenses" for "portogalensis" [Legend 10.2], and on the same sheet "Januensis" for "tanuensis" [same legend]. In Arabia deserta read "que charoanam mercatorum inuadunt" [Legend 3.18]; also "Braua insula" for "brana" [off the tip of East Africa]; near Bismagar read "rege mortuo comburentur" for "comburetur" [Legend 8.3]; on the sheet for Ceylon read "diuisa in .4." for "diuisam .4." [Legend 8.5]; near Camul read "sunt ydolatre sub

[316]On Hugues des Hazards see Pierre-Étienne Guillaume, *Notice sur le bourg de Blénod-lès-Toul: précédée d'un éloge historique de monseigneur Hugues des Hazards, 72e évêque et comte de Toul* (Nancy: Grimblot, Raybois et Cie, 1843; Paris: le Livre d'histoire, Impr. Lorisse, 2004); Georges Viard, "Hugues des Hazards, évêque de la pré-réforme lorraine," *Annales de l'Est*, Ser. 6, vol. 55.2 (2005), pp. 9–20; and Dominique Notter, "Les évêques de Toul au temps de Hugues des Hazards," *Etudes Touloises* 134.1 (2010), pp. 3–18, esp. 11–16.
[317]See Albert Ronsin, "Hugues des Hazards et le gymnase vosgien de Saint-Dié," *Annales de l'Est*, Ser. 6, vol. 55.2 (2005), pp. 151–166.
[318]On Hugues's coat of arms see Notter, "Les évêques de Toul au temps de Hugues des Hazards" (see note 316), pp. 11–12; and Jean-Christophe Blanchard, "L'emblématique de Hugues des Hazards," *Annales de l'Est*, Ser. 6, vol. 55.2 (2005), pp. 127–150.

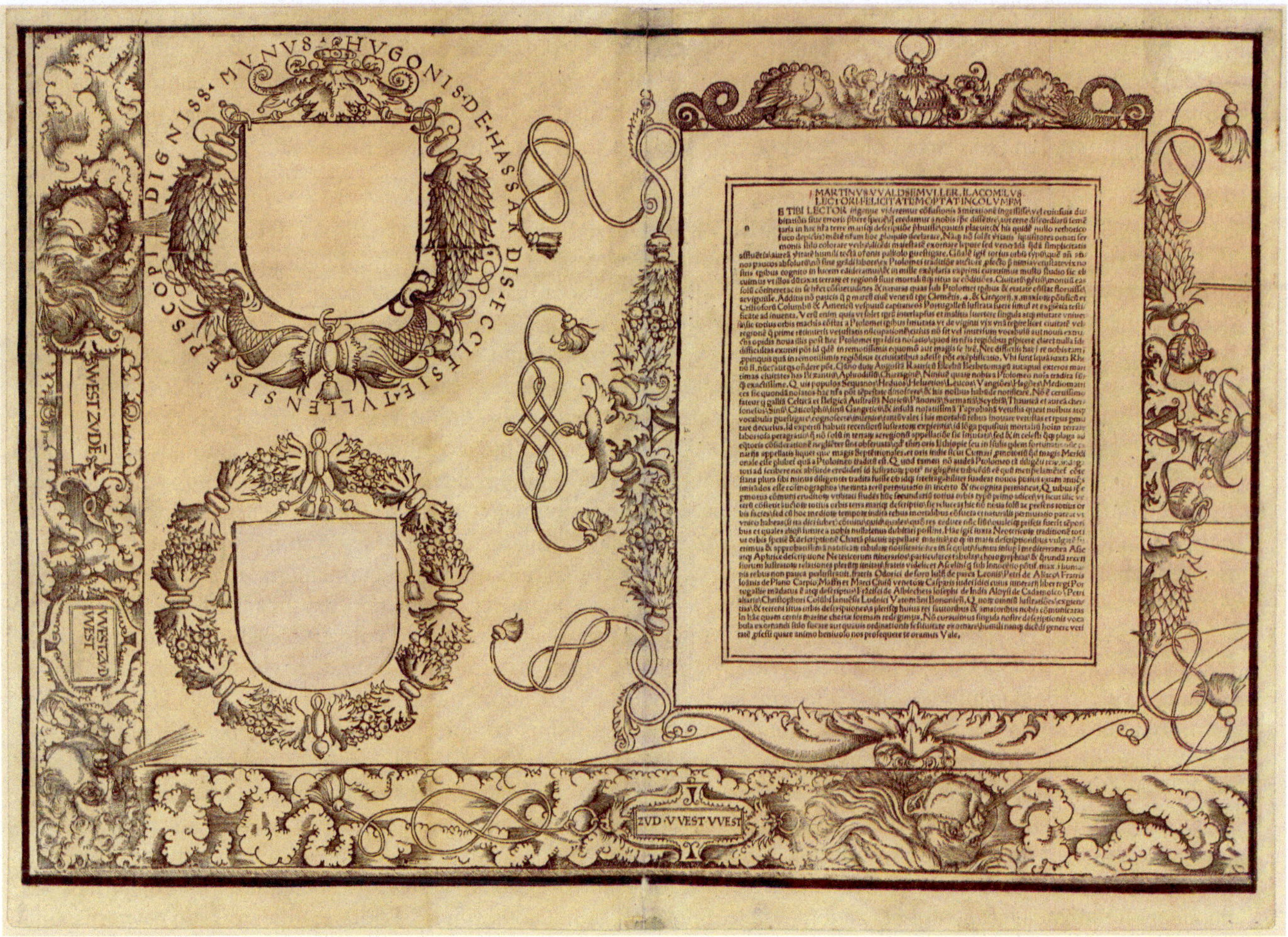

Plate 2.9 Sheet 9 of the *Carta marina*: Cartouches in the Map's Southwest Corner. Courtesy of the Library of Congress

dominio tartarorum" [Legend 4.20]. Near the Ganges River read "contra regem Narsinge" for "contra rege" [Legend 4.15]; near the line that circumscribes Tartaria read "Quod extra ambitur" for "quod intra ambitur" [Legend 4.1]. On the sheet of Mongol Tartaria read "alendis pecoribus apta" for "alendis pedoribus" [Legend 4.16]; on the sheet of the province of Cathay read "veteris ac noui testameti scripturas" for "scripturraram" [*sic*] [Legend 4.28]; on the same sheet read "sed dispositione faciei" for "dispositioni" [Legend 4.28]; on this sheet beneath the city of Cambalu read "Imperatoris" for "imperatorum" [Legend 4.29], and on the same sheet "habitatonis [*sic*] domini Chaam" for "babitatoni" [same legend], and on the same sheet read "sunt in eo .4. columne de auro" [same legend]. Near the emperor Noy read "imperator super .600000. armatorum" [Legend 3.8]. On the sheet of Mangi read "diuisit prouinciam Mangi" for "prouinciam magui" [Legend 4.30], and everywhere that "prisilicum" appears read "brasilicum." All other errors are entrusted to the reader's discretion.

This list of corrections was printed on the map but is covered by a pasted-on piece of paper that was on the map when it was discovered, and thus the printed text is not visible without using special methods to see through the pastedown. It is tempting to think that the piece was added after the corrections were entered by hand on the map, but there is no way to be certain. All of the corrections are entered on the map except the first one about the name "Groneland" and the one for sheet 6 (see the introductory paragraph for that sheet). The corrections have been transcribed twice before, in the first case by Fischer and von Wieser by shining a bright light through the map, and in the second by John Hessler and colleagues at the Library of Congress through hyperspectral imaging.[319] Both transcriptions contain errors, the abbreviations were not expanded, and no

[319]The corrections are transcribed by Joseph Fischer and Franz Ritter von Wieser, *Die älteste Karte mit dem Namen Amerika* (see note 9 in Chap. 1 above), pp. 20–21; and John W. Hessler, "Correcting Waldseemüller: Analysis of Hyperspectral Images of the Pastedown Shield on the 1516 Carta Marina," written in 2008 and available at http://www.kislakfoundation.org/download/Hessler-Analysis_of_Hyperspectral_Images.pdf.

translation was provided. The fact that there is a list of errata printed on the map obviously indicates that proof sheets had already been printed, and that the map was ready for publication. Most of the errors listed are on sheets 3 and 4, which is not unexpected, as those sheets have particularly large numbers of legends. The instruction regarding the name "Islanda" near the beginning of the list does not make good sense. Evidently no effort was made to put the corrections in any particular order. The possibility that there would be printing errors on the map is mentioned in Legend 7.19.

9.3
MARTINVS WALDSEEMVLLER ILACOMILVS LECTORI FELICITATEM OPTAT INCOLVMEM NE TIBI LECTOR ingenue videremur confusionis admirationem ingessisse, vel cuiusvis dubitationis sive erroris praebere spetiem, que credamur a nobis ipse dissentire, aut certe discordiarum sementaria in hac nostra terre marisque descriptione praebuisse, paucis placuit (& his quidem nullo rethorico fuco depictis) mentem nostram hoc proloquio declarare. Namque non solent veritatis inquisitores ornati sermonis stilo colorare verba, dicendi maiestatem exornare lepore, sed veneranda quedam simplicitatis affluentia, auream veritatem humili tectam orationis palliolo prouestigare. Generalem igitur totius orbis typum, quem ante annos paucos absolutum, non sine grandi labore, ex Ptolomei traditione, auctore profecto prae nimia vetustate vix nostris temporibus cognito, in lucem edideramus, & in mille exemplaria exprimi curauimus multo studio sic elicuimus vt illos dumtaxat terrarum et regionum situs mortaliumque ritus ac conditiones, Civitatum, gentium, montium eas solum contineret ac in se haberet consuetudines & naturas quas sub Ptolomei temporibus et etatate [for *etate*] constat floruisse, ac viguisse. Additis non paucis quae per marcum civem venetum tempore Clementis .4. & Gregorii.x. maximorum pontificum et Cristoforum Columbum & Americum vesputium capitaneos Portugallenses lustrata fuere simul et experientia testificante adinventa. Verum enim quia vt solet temporum interlapsus et malitia inuertere singula atque mutare vniuer[s]a, sic totius orbis machina constat a Ptolomei temporibus immutata vt de viginti vix vnam reperire licet ciuitatem vel regionem que prime retinuerit vetustatis nuncupationem, cuius non sit vel inuersum vocabulum aut nouis extructis opidis noua illis post hec Ptolomei tempora indita nominatio, quod in nostris regionibus perspicere claret. nulla inde difficultas exoriri potest id quidem in remotissimis equo modo aut magis se habere. Nec difficilis hac in re nobis tam in propinquis quam in remotissimis regionibus et ciuitatibus adesse potest exemplificatio. Vbi sunt inquam iuxta Rhenum flumen nuncupata aut quis ostendere potest Gannodurum Augustam Rauricum Elcebum Berbetomagum aut apud exteros maritimas ciuitates has Bizantium Aphrodisium Chartaginem Niniuem quarum nobis a Ptolomeo nomina tradita sunt quam exactissime. Quis populos Sequanos, Heduos, Heluetios, Leucos, Vangiones, Hagones, Mediomatrices sic quondam nominatos hac nostra potest tempestate dinoscere, & his nominibus habunde notificare. Non est certissimo fateor que galliam Celticam et Belgicam Austrasiam Noricum, Pannoniam, Sarmatiam, scythiam, Thauricam et auream chersonesos, sinum, Canticolphum, sinum Gengeticum, & insulam nominatissimam Taprobanam vetustis queat nominibus atque vocabulis prouestigare, cognoscere, inuenire, tantum valet in his mortalium rebus innouare vetustas et temporibus permutare decursus. Id expertum habuit recensiorum [for *recentiorum*] lustratorum experientia, id longa perquisiuit mortalium hominum terrarum laboriosa peragratio, que non solum in terrarum ac regionum appellatione sic immutata, sed & in celesti quoque plaga ad equatoris considerationem neglienter sint obseruata, quod tamen in oris Ethiopie seu in insulis quidem fortunatis nunc canariis appellatis liquet que magis Septentrionales, et oris indie sicut Cumari promontorium quod magis Meridionale esse perhibetur quam a Ptolomeo traditum est. Quod tamen non audeam Ptolomeo tan diligenti rerum indagatori adscribere nec absurde crediderim id lustratorum potius negligentie tribuendum esse quum met ipse lamentetur contestans plura sibi minus diligenter tradita fuisse ob idque irrefragabiliter suadeat nouos potius quam antiquos imitandos esse cosmographos, ne tanta rerum permutatio in incerto & incognita permaneat, Quibus ipse permotus communi eruditorum vtilitati studens hunc secundarium totius orbis typum primo adieci, vt sicut illic veterum constetit auctorum totius orbis terra marique descriptio, sic reluceat hic non noua solum ac presens totius orbis facies, sed cum hoc mediorum temporum indita rebus mortalibus consueta et naturalis permutatio pateat vt vnico habeas (si ita dici iubet) contuitu quid, quales, quomodo res caduce nunc fiunt, qualesque priscis fuerint temporibus et quales aliquin future a nobis nullatenus dubitari possint. Hanc igitur iuxta Neotericorum traditionem totius orbis spetiem & descriptionem Chartam placuit appellare marinam, eo que in maris descriptionibus vulgarem fuerimus & approbatissimam nauticarum tabularum notificationes insequuti, sumus insuper in mediterranea Asie atque Aphrice descriptione Ne[o]tericorum itinerarios, particulares tabulas, chorographias, & quorundam recensiorum [read *recentiorum*] lustratorum relationes plerumque imitati, fratris videlicit Ascelini qui sub Innocento pontifice maximo in humanis rebus non pauca prelustravit, fratris Odorici de foro Iulii de parca Leonis, Petri de Aliaco, Fratris Ioannis de Plano Carpio, Maffii et Marci Civium venetorum Casparis iudei indici cuius itinerarii liber regi Portugallie mandatus est atque descriptus, Francisci de Albiecheta Iosephi de India Aloysi de Cadamosco, Petri aliaris, Christophori Colubmi Ianuensis Ludo[v]ici

Vatomanni Bononiensis. Quorum omnium lustrationes, experientias, & terreni situs orbis descriptiones, a plerisque huius rei fautoribus & amatoribus nobis communicatas, In hanc quam cernis marine chartae formam redegimus. Non curavimus singula nostre descriptionis vocabula exornandi stilo fucare aut quauis ordinationis festiuitate exornare, humili nanque dicendi genere veritatem professi quare animo beniuolo nos prosequere te oramus. Vale.

Martin Waldseemüller (Ilacomilus) wishes the reader unblemished happiness.
Lest I should seem, noble reader, to be trying to dazzle you with confusion, or to give the appearance of any kind of doubt or error, in which I might seem to disagree with myself, or to supply the grounds for disagreements in this map of the land and sea, it seemed a good idea to declare my mind in this introduction, in a few words free of rhetorical coloring. For those who seek the truth do not adorn their words with stylistic flourishes, or embellish the dignity of their speech with a superficial charm, but use a venerable abundance of simplicity to cover the golden truth with a little cloak of speech. And so: a few years ago I finished a general map of the whole world, not without great labor, according to the tradition of Ptolemy, an author barely known today because of his great antiquity. I published it, and had a thousand copies printed with great care. I made it so that it would contain only those lands and regions, those religions and conditions of people, and those customs and natures of cities, peoples, and mountains—that it would only have in it those customs and characteristics that are known to have been extant or in use in Ptolemy's time. Many things were added that were discovered and confirmed through the testimony of experience by the Venetian citizen Marco during the papacies of Clement IV and Gregory X, and by the Portuguese captains Christopher Columbus and Amerigo Vespucci. For indeed, as the passage of time is accustomed to overturn a specific bad thing and to change wickedness as a whole, so it is well known that the whole machine of the world has changed since Ptolomey's time, so that out of twenty cities or regions hardly one can be found that retains its old name–whose name has not been changed, or new names given to new cities built after Ptolemy's time. This can even be seen in our own regions, but any difficulty in this regard would certainly be equal or greater in distant regions. And it is not difficult for me to provide examples of this, both in nearby cities and regions and in very distant ones. Where are those cities that were said to be by the Rhine, or who can point out Gannodurum [Konstanz], Augusta Rauricum [Augst, Aargau, Switzerland], Elcebus [Ell, near Strasbourg], and Berbetomagum [Worms], or among foreign maritime cities, Byzantium, Aphrodisias, Carthage, Ninive, whose names have been handed down to us by Ptolemy with great precision? The people once known as the Sequani, the Hedui, the Helvetians, the Leuci, the Vangioni, the Hagoni, the Mediomatrices—who can recognize them today, and who can easily make them known to us by these names? I acknowledge that one might not be able to locate, find, or recognize according to their ancient names Celtic Gaul and Belgian Gaul, Austrasia, Noricum, Pannonia, Sarmatia, Scythia, Thaurica, the Golden Chersonese, the Bay of Canticolphus, the Bay of the Ganges, and the very well-known island of Taprobane: so much does it prevail in these human things to alter ancient practices and to change course with the times. This is confirmed by the experience of recent explorers, and was found out by the long and laborious reconnaissance of the world: not only those things which in the naming of lands and regions have been thus altered, but even those which in the celestial regions and the study of the equatorial regions have been inaccurately recorded, so that it is clear that the shores of Ethiopia and the Fortunate Islands, now called the Canaries, should be more to the north, and in the shores of India, the Promontory of Cumari [Cape Comorin] is held to be further to the south than Ptolemy said. Which errors however I would not dare to ascribe to Ptolemy, so diligent an investigator, nor would I absurdly believe that the errors are to be attributed instead to the negligence of explorers, when he himself complains, asserting that many things were inaccurately passed down to him, and because of this, indiputably persuades us to follow the recent rather than the ancient cosmographers, lest so great a change in things remain uncertain or unknown. Moved by these considerations, and in the interest of the common utility of scholars, I have added this second image of the world to my first, so that just as in the first the image of the whold world, land and sea, agreed with that of the ancient authors, so in this one, not only may the new and present face of the world shine forth, but together with that, the customary and natural change introduced into worldly affairs in the intervening times, so that you can see (if I may say so) at a single glance why, of what kind, and how transitory things have come to be now, what they were like in former times, and how they will be in the future, without a doubt. Therefore, it seemed good to call this image and description of the whole world, made in accordance with the tradition of modern authors, a Carta marina, and for that reason, as far as the depiction of the oceans, I have followed the commonly used and the most approved nautical charts and their indications, while in the depiction of the Mediterranean, Asia and Africa I have made ample use of recent authors' travel narratives, regional maps, descriptions of countries, and the accounts of some recent explorers, such as that of Ascelinus the monk, who during the papacy of Innocent made extensive explorations in human affairs; of friar Odorico of Pordenone in Friuli, Pierre d'Ailly, friar John of Plano Carpini; Matteo and Marco [Polo],

citizens of Venice; Caspar the Jew of India, whose travel narrative was inscribed and sent to the King of Portugal; those by Francesco [de Almeida], Joseph the Indian, Alvise de Cadamosto; Pedro Alvares [Cabral]; the Genoese Christopher Columbus, and the Bolognese Ludovico de Varthema. All of whose voyages, experiences, and descriptions (or maps) of the world, which were sent to me by many who promote and love this type of work, I have reduced into the form of the marine chart that you now see. I was not concerned with embellishing each word of my description with ornate style or placing them in a striking order, for I prefer a plain style to express the truth. For that reason I beg you to follow me with a well-disposed spirit. Farewell.

The first and last thirds of this legend were transcribed and translated by M. T. di Palma[320]; the latter part where Waldseemüller discusses his sources has been translated by Baldacci[321]; and the whole legend was translated by John Hessler.[322] The transcription and translation here are my own. It has been questioned whether Waldseemüller's reference to an earlier world map in the early part of the legend is in fact about his 1507 map, but it is not at all clear to what other map Waldseemüller might be referring. Doubt has also been cast on the cartographer's claim that 1000 copies of the 1507 map were printed, particularly as only one copy of the map survives. While there is no way to be certain in this matter, it is worth keeping in mind that the 1507 map was a very ambitious project, and thus an ample print-run would be expected. Although there is little data regarding print runs for books in the early sixteenth century, and none for maps, there are some indications that a print run of 1000 books would not have been unusual.[323] With regard to the fact that only one copy of the 1507 map (and of the *Carta marina* for that matter) survives, that is not at all unusual: in the Incunabula Short-Title Catalogue (ISTC), there are thousands of editions of books for which only one copy survives.[324] Maps printed on large sheets are *prima facie* much more perishable than books, as they lack the built-in protection of a book's covers, and it is very easy to find examples of sixteenth-century maps that survive in only one or two exemplars.[325]

As mentioned briefly in the introduction above, the section of this legend on the *Carta marina* that addresses changes in place names is very similar to a passage in the introduction to the second part of Waldseemüller's 1513 edition of Ptolemy, and in fact some of the place names he uses as examples are the same in the two passages. Here is the passage from the 1513 Ptolemy[326]:

Verum quia temporis lapsus multa quidem labilitate quoque sua indies mutat: plaerisque visus est auctor notabilius a modernioribus deuiasse. Id quod cernere licet in utraque Pannonia, quae nunc Hungaria & Austria vocatur. Et quae regio dum floruit unica appellatione Sarmatia, siue Sauromatia dicebatur: nunc diuisim Poloniam, Russiam, Prussiam, Moscouiam & Lituaniam nominamus. Populorum denique usui placuit transmutatio vocabulorum. Quos enim vetustas Eluetios & Sequanos, nunc vulgo Burgundiones Suitensesque vocamus. Quaedam & ciuitates primitiuis nominibus orbati sunt. Quis enim iuxta Rhenum Flauium, Canodurum, Augustam rauricum, Elcebum & Berthomagum urbes a Ptolaemo commemoratas digito mostrabit?

But since the course of time changes many things from day to day as it passes, it has become generally evident that the author deviates notably from those more modern, as may be seen in the two Pannonias, which are now called Hungary and Austria; and the region which was called while it flourished, by the sole appellation of Sarmatia or Sauromatia, we now name in its divisions, Poland, Russia, Prussia, Muscovy and Lithuania. Change in the names of nations has also come into use. For those whom the ancients called Helvetii and Sequani, we now commonly call Burgundians and Swiss. Certain cities, too, have lost their primitive names, for who with his finger will point out on the River Rhine the cities Canodorum, Augusta Rauricum, Elcebus and Berthomagus mentioned by Ptolemy?

---

[320]See Gaetano Ferro et al., *Columbian Iconography* (see note 108 in Chap. 1 above), p. 496.
[321]Baldacci, *Columbian Atlas of the Great Discovery* (see note 194), p. 121.
[322]See http://www.kislakfoundation.org/download/Hessler-Sheet_9_of_1516_Carta_Marina_Translation.pdf.
[323]For indications that a print run of 1000 was not unusual in Italy in the early sixteenth century see Berta Maracchi Biagiarelli, "Il privilegio di stampatore ducale nella Firenze Medicea," *Archivio storico italiano* 123 (1965), pp. 304–370, at 348; Uwe Neddermeyer suggests that a print run of about a thousand was normal around the turn of the sixteenth century in "Möglichkeiten und Grenzen einer quantitativen Bestimmung der Buchproduktion im Spätmittelalter," *Gazette du livre médiéval* 28 (1996), pp. 23–32, but this claim is disputed by Joseph A. Dane, "Twenty Million Incunables Can't Be Wrong," in Joseph A. Dane, ed., *The Myth of Print Culture: Essays on Evidence, Textuality, and Bibliographical Method* (Toronto: University of Toronto Press, 2003), pp. 32–56 and 198–203, at 43–44.
[324]Jonathan Green, Frank McIntyre, and Paul Needham, "The Shape of Incunable Survival and Statistical Estimation of Lost Editions," *Papers of the Bibliographical Society of America* 105.2 (2011), pp. 141–175.
[325]See the list of examples on pp. 44–45 above.
[326]This passage is translated into English by Henry N. Stevens, *The First Delineation of the New World and the First Use of the Name America on a Printed Map* (London: H. Stevens, Son & Stiles, 1928), p. 40, and part is quoted in Robert W. Karrow Jr., "Intellectual Foundations of the Cartographic Revolution," Ph.D. Dissertation, Loyola University of Chicago, 1999, p. 185.

Waldseemüller mentions using some local maps (*particulares tabulas, chorographias*) in the composition of the *Carta marina*, and though we do not have any direct evidence about which local maps they might have been, R. A. Skelton has listed some of the works, including maps, that Waldseemüller used in making his 1513 Ptolemy.[327] It is worth pointing out that Waldseemüller's list of the sources he used is certainly not complete, though we would not really expect it to be so. Just to mention two examples, in the introduction above it was shown that Waldseemüller was using as a source for his depictions of the lands a large and elaborately decorated nautical chart, similar in nature to the Catalan Atlas of 1375, Mecia de Viladestes's chart of 1413, or the Catalan-Estense map of c. 1460, but he only mentions using nautical charts for his depiction of the ocean; and in the commentary on Legend 3.27 we saw that Waldseemüller was using the brief thirteenth-century account of a journey to the Holy Land by one Thetmar or Theitmar.

[327]R. A. Skelton, "Bibliographical Note," in *Ptolemy, Geographia, Strassburg, 1513* (Amsterdam: Theatrum Orbis Terrarum, 1966), p. xvii.

2.11 Sheet 10. Southern South America and the South Atlantic (Plate 2.10)

The coastal place names in South America come from Caverio's nautical chart of c. 1503; Caverio has a Portuguese flag at the southern end of the continent, but Waldseemüller replaced this with a Spanish flag. The name "America," which Waldseemüller so famously debuted in South America on his 1507 map, does not appear on the *Carta marina*. At the bottom of this sheet is a scale of Italian miles for measuring distances on the map: compare this with the scale of German miles on sheet 2. As Waldseemüller specifies in the *Cosmographiae introductio*, the German mile was four times as long as the Italian mile.[328]

10.1
BRASILLA SIVE TERRA PAPAGALLI Antropophagorum genus hic est

Brazil or the Land of Parrots. The people here are man-eaters.

The name Brazil (in one of its many variant spellings) seems first to have been applied to the New World in print in 1504; a "R. de Brasil" appears on the Ruysch map of 1508, and there is a "rio de brazil" and an "y de brassil" on the *Tabula Terre Nove* in the 1513 Ptolemy, but the first map on which the name is used for a region in South America is Gregor Reisch's world map in the 1513 edition of his *Margarita philosophica*.[329] It is interesting that Reisch's map is largely based on those in the 1513 Ptolemy: he evidently took the idea that the name Brazil applied to the whole region from another source. Incidentally Fries discusses Brazil in Chap. 82 of the *Uslegung*.[330]

The first cartographic appearance of New World parrots is on the Cantino chart of c. 1502; they also appear on the Caverio chart, and one of the few artistic decorations on Waldseemüller's 1507 map is a parrot in South America that is perched on a banner that reads *Rubei psitaci*, i.e. red parrots.[331] Given that the *Carta marina* is iconographically richer than both the Caverio chart and his own 1507 map, it is interesting that Waldseemüller chose to use the name Terra Papagalli, but not to represent the birds. This choice seems to reflect the cartographer's greater interest in Asia than in the New World: both legends and decorations in the New World are quite sparse compared to those in Asia. We cannot be certain where Waldseemüller obtained the name Terra Papagalli, but it does appear in the *Paesi*, Chap. 125, in a letter from Giovanni Matteo Cretico dated June 27, 1501.[332]

For discussion of the long tradition of reports of cannibals in the New World, see the commentary on Legend 5.5 above.

10.2
PASSIM INCOLITVR HEC REGIO QVE PLERISQVE ALTER TERRARVM ORBIS EXISTIMATVR. FEMINE AC MARES VEL NVDI PRORSVS VEL INTEXTIS RADICIBVS AVIVM PENNIS VARII COLORIS ORNATI LABIISQVE PERFORATIS INCEDVNT. APVD MVLTOS VIVITVR IN COMMVNI. NVLLA RELIGIONE. BELLA FREQVENTISSIME GERVNT, HVMANA CAPTIVORVM CARNE VESCVNTVR. AERE ADEO CLEMENTI VTVNTVR VT SVPRA ANNVM 150. PLVRES VIVANT. RARO ERGOTANT ET SI SE PERTVRBATVROS SENSERINT RADICIBVS HERBARVM CITO CVRANTVR.

---

[328]See Fischer and von Wieser, *The 'Cosmographiae introductio' of Martin Waldseemüller* (see note 7 in Chap. 1 above), pp. xxxvi (Latin, towards the end of Chap. 9) and 77 (English).

[329]See "Beschreibung der Meerfahrt von Lissabon nach Calacutt, vom Jahre 1504," *Jahresbericht des Historischen Kreisvereins im Regierungsbezirke von Schwaben und Neuburg* 26 (1860), pp. 160–162, at 161; and Walter B. Scaife, "Brazil, as a Geographical Appellation," *Modern Language Notes* 5.4 (1890), pp. 105–107. On Reich's map see Rodney W. Shirley, *The Mapping of the World: Early Printed World Maps 1472–1700* (London: Holland Press, 1983), pp. 40–42.

[330]Fries's passage on Brazil is translated into modern German by Petrzilka, *Die Karten des Laurent Fries* (see note 202 in Chap. 1 above), p. 150.

[331]For discussion of images of New World parrots on maps see Gaetano Ferro et al., *Columbian Iconography* (see note 108 in Chap. 1 above), pp. 528–535; and Renate Pieper, "Amerikanische Papageien als Symbol der Neuen Welt," in her *Die Vermittlung einer neuen Welt: Amerika im Nachrichtennetz des Habsburgischen Imperiums 1493–1598* (Mainz: P. von Zabern, 2000), pp. 245–271; a shorter version of her study was published in Spanish as "Papagayos americanos, mediadores culturales entre dos mundos," in Eddy Stols, Werner Thomas, and Johan Verberckmoes, eds., *Naturalia, mirabilia & monstrosa en los imperios Ibéricos (siglos xv–xix)* (Leuven: Leuven University Press, 2005), pp. 123–134.

[332]The passage in the *Paesi*, Chap. 125, that includes the name "Land of the Parrots" runs: *di sopra dal capo di Bonasperanza uerso garbin hanno scoperto una terra nova. la chiamano de li Papaga: per essergene di longeza di brazo .i. & mezo di uarii colori: de li quali ne hauemo uisto doi: iudicando questa terra esser terra ferma perche scorseno per costa piu di do.M. miglia ne mai trouono fine…*; "Above the Cape of Good Hope to the west they have discovered a new land. They call it that of the parrots, because some are found there which are an arm and a half in length, of various colours. We saw two of these. They judged that this was mainland because they ran along the coast more than two thousand miles but did not find the end." The letter of Cretico is translated into English by Greenlee, *The Voyage of Pedro Álvares Cabral* (see note 70 in Chap. 1 above), pp. 119–123, with this passage on p. 120.

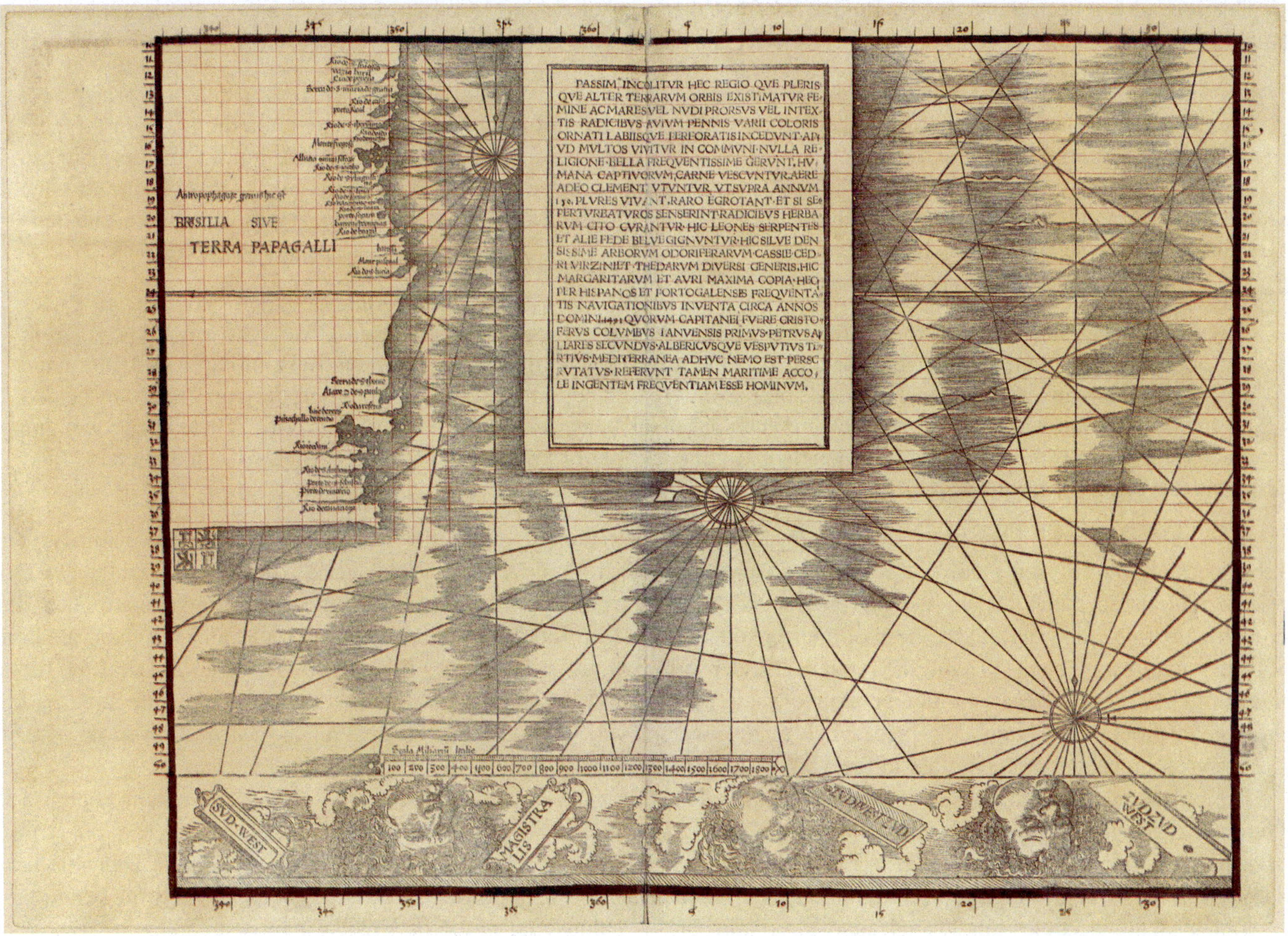

Plate 2.10 Sheet 10 of the *Carta marina*: Southern South America and the South Atlantic. Courtesy of the Library of Congress

HIC LEONES SERPENTES ET ALIE FEDE BELVE GIGNVNTVR. HIC SILVE DENSISSIME ARBORVM ODORIFERARVM, CASSIE, CEDRI, VIRZINI ET THEDARVM DIVERSI GENERIS. HIC MARGARITARVM ET AVRI MAXIMA COPIA. HEC PER HISPANOS ET PORTOGALENSES FREQVENTATIS NAVIGATIONIBVS INVENTA CIRCA ANNOS DOMINI 1492 QVORVM CAPITANEI FVERE CRISTOFERVS COLVMBVS IANVENSIS PRIMVS, PETRVS ALIARES SECVNDVS, ALBERICVSQVE VESPVTIVS TERTIVS. MEDITERRANEA ADHVC NEMO EST PERSCRVTATVS, REFERVNT TAMEN MARITIME ACCOLE INGENTEM FREQVENTIAM ESSE HOMINVM.

This land, which many believe to be a whole other world, is inhabited throughout. The women and men walk around either completely naked or decorated with roots interwoven with feathers of various colors, and they have their lips perforated. Many of them live communally, and have no religion. They wage war very frequently, and they eat the flesh of their prisoners. The climate is so mild that many of them live beyond 150 years of age. They seldom fall ill, and if they perceive they are about to become unwell, they are quickly cured by the roots of herbs. Lions, snakes and other wild beasts live here, and there are very dense forests of odiferous trees, such as cassia, cedar, brazilwood, and pines of various types. There are great quantities of pearls and gold. This was discovered in about 1492 through numerous voyages by the Spaniards and Portuguese, under the following captains: first Christopher Columbus, then Pedro Álvares [Cabral], and third, Alberico [Amerigo] Vespucci. So far no one has explored the hinterland, but the inhabitants of the coasts say that it is enormous and full of people.

This legend has been translated by Baldacci,[333] but the translation here is mine. More than half of the legend seems to have been inspired (like Legend 5.3 above) by a legend on Ruysch's world map of 1507–08[334]:

Passim incolitur haec regio, quae a plerisque alter terrarum orbis existimatur. Feminae maresque vel nudi prorsus vel intextis radicibus aviumque pennis varii coloris ornati incedunt. Vivitur multis in commune nulla religione, nullo rege. Bella inter se continenter gerunt, humanaque captivorum carne vescuntur. Aero adeo clementi utuntur ut supra annum 150 vivant. Raro aegrotant, tunque radicibus tantum curantur herbarum. Leones hic giguntyr, serpentesque et aliae foedae belluae sylvae. Insunt montes fluminaque. Margaritarum atque auri maxima copia. Avehuntur hinc a Lusitanis ligna brasi alias verzini et cassiae.

This region is everywhere inhabited; it is considered another world by most people. Women and men go either completely naked or dressed in woven fibres and birds' feathers of various colors. Many live together with no religion and no king. They constantly wage war amongst themselves. They feed on the human flesh of captives. They have such a mild climate that they live beyond their 150th year. They are rarely ill, and when they are, they are healed simply by the roots of herbs. Lions live here and snakes and other dreadful beasts of the forest. There are mountains and rivers. There is a great quantity of pearls and gold. Timber of brasil [used for red dye], also called "verzin," and cassia are exported from here by the Portuguese.

As with Legend 5.3 I do not think that Waldseemüller's reliance on Ruysch here has been noted previously. The substances of both Ruysch's and Waldseemüller's legends come from the account of Vespucci's third voyage in the *Paesi* or *Itinerarium*: the idea that the newly discovered lands are a new world, from Chap. 114; the fact that the people are nude, live communally, have no king, eat the flesh of their prisoners, live to be 150, heal themselves with roots, and that there are lions and serpents, from Chap. 117; and the reference to dense forests, gold, and pearls from Chap. 118. I do not find the list of trees in the *Paesi* or *Itinerarium*, but a couple of them come from Ruysch. It is worth remarking that while the latter part of the legend proclaims Columbus's precedence as discoverer over Vespucci, and in general the *Carta marina* recognizes Columbus's precedence (in contrast to the 1507 map), most of the legend in fact comes from Vespucci.

The phrase at the beginning of the legend to the effect that most people consider it another world recalls a phrase in Pomponius Mela 3.70 about the island of Taprobana: *Taprobane aut grandis admodum insula, aut prima pars orbis alterius... dicitur*, "Taprobana is said to be either a very large island, or the first part of another world."[335] This similarity shows that the difficulty of determining the nature of distant lands was a problem that explorers had faced for centuries.

Johann Schöner copies this legend almost verbatim on his manuscript globe of 1520, where it straddles the Tropic of Capricorn in the interior of the South American continent. However, Schöner takes the last sentence of the legend and places it north of the equator at an angle, even though he had room to write it at the end of the legend. This change was perhaps motivated by a desire to have the sentence about the interior of the continent closer to its center.

---

[333]See Baldacci, *Columbian Atlas of the Great Discovery* (see note 194), p. 124.

[334]The transcription and translation of Ruysch's legend are from John Thorley, *Documents in Medieval Latin* (Ann Arbor: University of Michigan Press, 1998), pp. 134 and 196; the legend is also transcribed by Mauro Bini et al., *Alla scoperta del mondo: l'arte della cartografia da Tolomeo a Mercatore* (Modena: Il bulino, 2001), p. 173. For references on the Ruysch map see note 172.

[335]See Veit Rosenberger, "Taprobane: Trauminsel oder der Beginn einer neuen Welt?" *Laverna* 7 (1996), pp. 1–16. Also see Pliny 6.81: *Taprobanen alterum orbem terrarum esse diu existimatum est Antichthonum appellatione.*

2.12 Sheet 11. Southern Africa and the Southwestern Indian Ocean (Plate 2.11)

At the top of the sheet in Africa are the *Montes Lune* or Mountains of the Moon, the source of the Nile according to Ptolemy, whose tributaries we see flowing northward. The southern part of the continent is labeled NOVE COGNITE AFRICE PARTIS EXTENSIO—"extension of the newly discovered part of Africa"—meaning the part that was unknown to Ptolemy. South of the southern tip of the continent is a large image of King Manuel of Portugal riding a sea monster to express his nation's mastery of the oceans, particularly of the sea route around the southern tip of Africa to Asia. Curiously, the alternative name given for the island of Madagascar is the Island of St. George, rather than the Island of St. Laurence, as it usually is.

11.1
Sub istis montibus sunt basilisci Merguli [for *et reguli*] serpentes ita ut totum istud latus sit quasi desertum propter eos.

Beneath these mountains are basilisks and serpents called *reguli*, so that this whole area is all but deserted because of them.

As indicated in the transcription, "Merguli" must be an error for "et reguli": the *mergulus* is the bird called the diver in English, which does not live under mountains and would not cause an area to be deserted, but the *regulus* and *basilicus* do make sense together, as Isidore, *Etymologiae* 12.4.6–9 and Bartholomaeus Anglicus, *De proprietatibus rerum* 18.15 say that they are the same creature.[336] I am not familiar with a text that says that these serpents live under mountains; nor have I found a convincing source for Waldseemüller's illustrations of the serpents.

11.2
CRISTIANISSIMI EMANUELIS REGIS PORTOGALIE VICTORIA

The victory of Emanuel, the very Christian King of Portugal

The victory of King Manuel is Portugal's mastery of the sea route to India around the Cape of Good Hope, pioneered by Vasco da Gama, which is symbolized here by an image of Manuel riding a sea monster near the southern tip of Africa: his control of the sea monster expresses Portugal's confidence in navigating the ocean. In the introduction I discussed this image of Manuel as an expression of the title *Senhor da conquista e da navegação e comércio de Etiópia, Arábia, Pérsia e da Índia*, "Lord of the conquest, and navigation, and commerce of Ethiopia, Arabia, Persia and India"—the adoption of which title is reported in the *Paesi* and *Itinerarium*, Chap. 62.[337] I also suggested that the image was inspired by that of Neptune riding a sea monster on Jacopo de' Barbari's view of Venice. Waldseemüller's image should be considered in relation to Portuguese imagery of Manuel: the armillary sphere and a grouping of spheres around the earth, both expressing world control, were one of his most common symbols.[338] Also, the image of Manuel riding a sea monster may be compared with that of Manuel at the beginning of Book 2 of the *Ordenações Manuelinas*, that is, *Liuro primeiro [-quinto] das Ordenações. Nouame[n]te corrigido na segu[n]da e[m]pressam* (Lisbon: Ioham Pedro Bonhomini, 1514): in the foreground, the image shows Manuel's power through his close relationship with the clergy, and in the background there is a chorographic image of the world showing the economic activities and trade over which Manuel presides, and which support his power.[339]

---

[336]The passage in Isidore is translated into English in *The Etymologies of Isidore of Seville*, trans. Stephen A. Barney et al. (Cambridge, UK and New York: Cambridge University Press, 2006), p. 255. For descriptions of the basilisk and the *regulus* in so-called Second Family bestiaries, where their chapters are adjacent, see Willene B. Clark, *A Medieval Book of Beasts: The Second-Family Bestiary: Commentary, Art, Text and Translation* (Woodbridge: Boydell, 2006), p. 195.

[337]The passage about Manuel's title appears in the so-called Second Letter of Girolamo Sernigi, and is translated into English in Ravenstein, *A Journal of the First Voyage* (see note 57 in Chap. 1 above), p. 141.

[338]See Ana Maria Alves, *Iconologia do poder real no período manuelino: à procura de uma linguagem perdida* (Lisbon: Impr. Nacional-Casa da Moeda, 1985), pp. 117–139, with many examples illustrated in Figs. 1–12, 14, 25–67, and 72–74.

[339]The image is reproduced in António Alberto Banha de Andrade, *Mundos novos do mundo: panorama da difusão, pela Europa, de notícias dos descobrimentos geográficos portugueses* (Lisbon: Junta de Investigações do Ultramar, 1972), between pp. 224 and 225, and my interpretation of the scene is paraphrased from his. Also the *Ordenações* are available in PDF from the Biblioteca Nacional de Portugal at http://purl.pt/index/geral/PT/index.html. For discussion of the development of Manuel's court see Susannah Charlton Humble, "From Royal Household to Royal Court: A Comparison of the Development of the Courts of Henry VII of England and D. Manuel of Portugal," Ph.D. Dissertation, Johns Hopkins University, 2003.

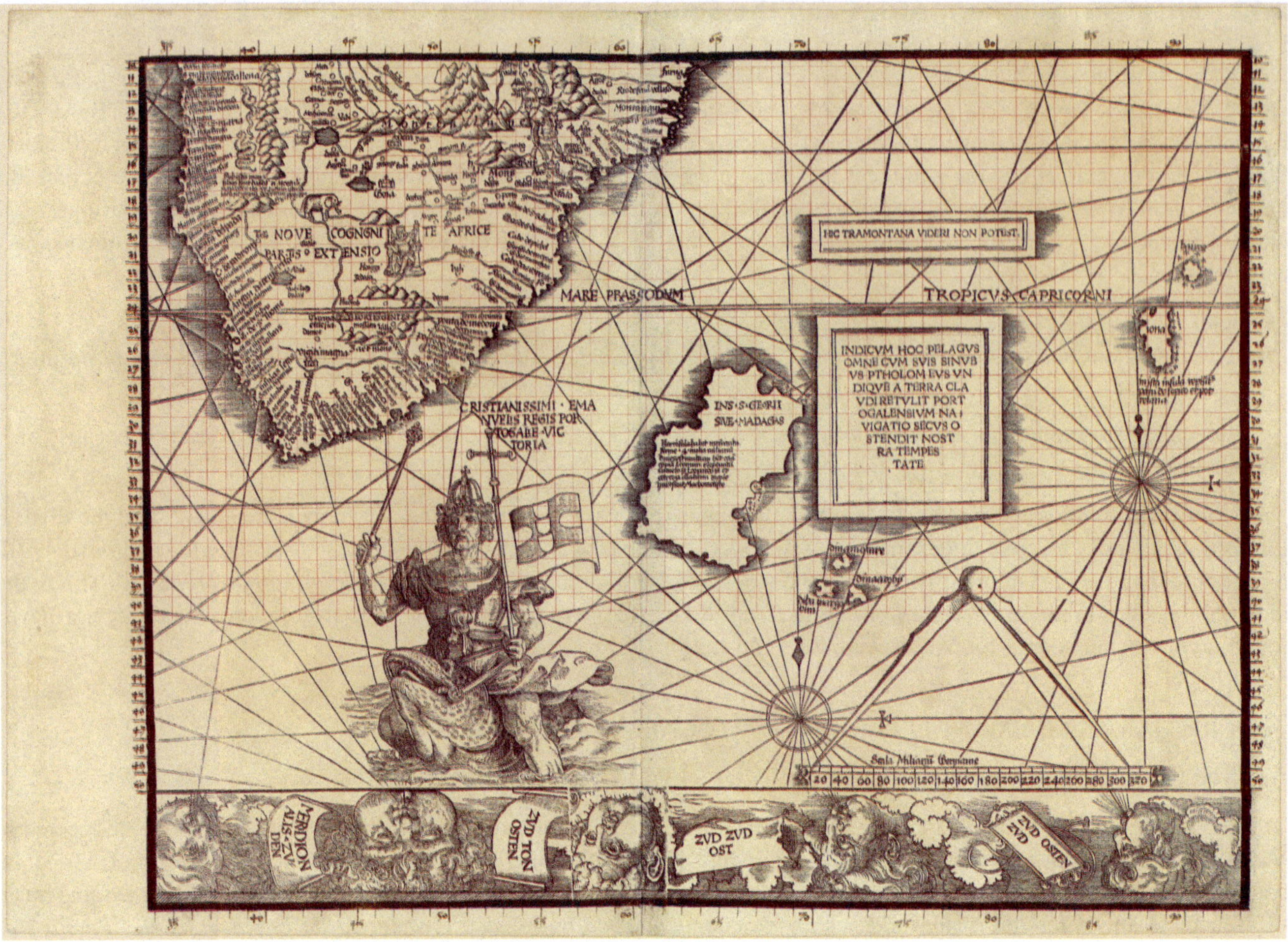

Plate 2.11 Sheet 11 of the *Carta marina*: Southern Africa and the Southwestern Indian Ocean. Courtesy of the Library of Congress

11.3
INS. S. GEORII SIVE MADAGAS Hec insula habet in circuitu ferme .4. milia miliarium dives est multum habet eciam copiam leonum elephantum camelorum Leopardorum et ceterorum animalium incole huius sunt Machometiste

The island of St. George or Madagascar. This island is about 4000 miles in circumference, and is very rich. It has an abundance of lions, elephants, camels, leopards and all other animals. The inhabitants of this island are followers of Mohammed.

Madagascar was described by Marco Polo, and curiously although the legends on both the 1507 map and the *Carta marina* derive from Marco Polo, they are different: the legend on the 1507, which falls into two parts, is longer and follows Polo a bit more closely,[340] except in the matter of the circumference of the island, where the 1507 map indicates 2000 miles, while Polo and the 1516 map give 4000 miles.[341] Evidently Waldseemüller could not find a more recent source that gave a

[340]The two legends on Madagascar on the 1507 map run: *Silua sandali Hec est maior Et ditior insula totius mundi continet.n. ambitus eius in circuitu miliaria 2000 habitatores sunt Saracen[ni] macometiste non habentes regem*, "Forests of sandalwood. This is the largest and richest island in the whole world. Its circumference is 2000 miles, and its inhabitants are Islamic Saracens who do not have a king"; and *Habet haec insula nemora sandalorum & omnia genera specierum etiam elephantes leones linces leopardos ccruos* [*sic*, for *ceruos*] *damos et aues multarum specierum*, "This island has groves of sandalwood and all types of spices, as well as elephants, lions, lynxes, leopards, deer, bucks, and many species of birds."
[341]See Marco Polo, *Marka Pavlova z Benátek, Milion* (see note 54), pp. 184–185; and *The Book of Ser Marco Polo* (see note 54), Book 3, Chap. 33, vol. 2, pp. 345–354.

detailed description of the island; Varthema, for example, just mentions it briefly.[342] It is puzzling that Waldseemüller refers to Madagascar as the island of St. George, which is a mistake for St. Laurence, the name by which it was known following its rediscovery, if we may use that term, by the Portuguese early in the sixteenth century.[343] For a description of trade on Madagascar a bit more than a hundred years after the making of the *Carta marina* see Roberts, *The Merchants Mappe* (1638), Chap. 35, p. 111: he says that the inhabitants would not let traders land on the island.

11.4
HIC TRAMONTANA VIDERI NON POTEST

Here the North Star cannot be seen.

A few of the authors that Waldseemüller consulted mention not being able to see the North Star from areas of the Indian Ocean: Marco Polo says this of Java Minor,[344] Odoric says it of Sumatra,[345] and Varthema of the island of Monoch, probably the Moluccas, and also of the sea near Java,[346] but I do not know of any source that Waldseemüller used that mentions this near Madagascar. The legend is at almost the same latitude (just above the Tropic of Capricorn) as Legend 12.1, which talks about the northern stars not being visible near Taprobana, whose southern tip is just to the north of that legend. So perhaps this legend (11.4) is simply the result of Waldseemüller copying information westward from that legend (12.1)—westward to a point that would be relevant to ships journeying between Europe and Asia by way of the Cape of Good Hope. Another possibility is that the legend is to be connected with Legend 6.3 about the compass needle turning south: both legends may come from a source describing a voyage around Africa, but if so, I do not know which voyage.

11.5
INDICUM HOC PELAGUS OMNE CUM SUIS SINUBUS PTHOLOMEUS UNDIQUE A TERRA CLAUDI RETULIT. PORTOGALENSIUM NAVIGATIO SECUS OSTENDIT NOSTRA TEMPESTATE

Ptolemy said that this Indian sea with all its bays was totally enclosed by land, but recent Portuguese voyages demonstrate otherwise.

This legend is transcribed and translated by Baldacci, but I give a somewhat different translation here.[347] Waldseemüller's legend is very similar to one on Ruysch's map[348]:

Indicum hoc pelagus quod omne cum suis sinubus undique claudi a terra ptolemeus retulit: partem oceani esse lusitanorum navigationes ostenderunt hoc tempore.

Ptolemy said that this Indian sea with all its bays was totally enclosed by land, but recent Portuguese voyages have demonstrated that it is part of the ocean.

It is clear that Waldseemüller took his legend from Ruysch (as he did with Legends 5.3 and 10.2), or else the two cartographers were working from the same source. As in the cases of Legends 5.3 and 10.2, to my knowledge the connection here between Waldseemüller's legend and Ruysch's has not been pointed out before.

[342]See Varthema, *The Travels of Ludovico di Varthema* (see note 103), p. 296.
[343]On the discovery of Madagascar by the Portuguese and historical cartography of the island see Albert Kammerer, "La découverte de Madagascar par les Portugais et la cartographie de l'île (1500–1667)," *Boletim da Sociedade de Geografia de Lisboa* 67.9–10 (1949), pp. 517–632; and also Gabriel Gravier, *La cartographie de Madagascar* (Rouen: E. Cagniard, and Paris: A. Challamel, 1896).
[344]See Marco Polo, *Marka Pavlova z Benátek, Milion* (see note 54), pp. 160 and 162; and *The Book of Ser Marco Polo* (see note 54), Book 3, Chap. 9, vol. 2, p. 284; and Book 3, Chap. 10, vol. 2, p. 292.
[345]Yule, *Cathay and the Way Thither* (see note 61), vol. 2, pp. 146–147 (English), 299 (Latin), and 344 (Italian).
[346]See Varthema, *The Travels of Ludovico di Varthema* (see note 103), pp. 246 and 249.
[347]See Baldacci, *Columbian Atlas of the Great Discovery* (see note 194), p. 124.
[348]Unfortunately this legend is not discussed by John Boyd Thacher, *The Continent of America* (see note 172), in his chapter on the legends on the Ruysch map, pp. 212–215. For references on the Ruysch map see note 172.

Fra Mauro on his mappamundi of c. 1455 has a legend pointing out that Ptolemy was incorrect about the Indian Ocean being enclosed by land,[349] and in Gregor Reisch's *Margarita philosophica,* first published in 1503, there is a Ptolemaic world map with the usual Ptolemaic land bridge joining southern Africa to southern Asia, but across that land bridge is written *Hic non terra sed mare est; in quo mire magnitudinis Insulae sed Ptolomeo fuerunt incognite*, "Here is not land but sea, in which there are islands of remarkable size unknown to Ptolemy."[350]

11.6
iona in ista insula reperiuntur panni de serico et porcelana

Java. In this island are found silk cloth and porcelain.

This is one of the relatively few legends on the *Carta marina* that is similar to one on the 1507 map, where the legend reads *Iona In ista insula reperiuntur panni de serico texti et porcellana vel bombex*, "Java. In this island are found silk cloth and porcelain or shells." These legends on both of Waldseemüller's maps derive from that on Caverio's map: *Y. Iana. Em ista inssulla a multo benioim et seda et porcelanas*, "The island of Java. In this island there is much benzoin resin and silk and porcelain."[351]

[349]See Falchetta, *Fra Mauro's World Map* (see note 143 in Chap. 1 above), pp. 192–193, *53: "Some authors write that the Sea of India is enclosed like a pond and does not communicate with the ocean. However, Solinus claims that it is itself part of the ocean and that it is navigable in the southern and south-western parts. And I myself say that some ships have sailed it along that route. This is confirmed by Pliny when he says that in his day two ships loaded with spices coming from the Sea of Arabia sailed around these regions to Spain and unloaded their cargo at Gibraltar (he gives the reason for this choice of route, but I omit it here). Fazio [degli Uberti] says the same; and those who have taken this route, men of great prudence, agree with these writers."

[350]Gregor Reisch, *Margarita philosophica* (Freiburg: J. Schott, 1503), with the map between signatures ovii^{v} and oviii^{r}. For discussion of the legend on the map see Francesc Relaño, *The Shaping of Africa: Cosmographic Discourse and Cartographic Science in Late Medieval and Early Modern Europe* (Aldershot, England, and Burlington, VT: Ashgate, 2002), pp. 189–190; and for discussion of the Ptolemaic land bridge generally see Wilcomb E. Washburn, "A Proposed Explanation of the Closed Indian Ocean on Some Ptolemaic maps of the 12th–15th Centuries," *Revista da Universidade de Coimbra* 32 (1986), pp. 431–441.

[351]The legend on the Caverio chart is transcribed, but with some errors, by Armando Cortesão, *Cartografia e cartógrafos portugueses dos séculos XV e XVI* (Lisbon: Seara Nova, 1935), vol. 1, p. 155.

2.13 Sheet 12. The Southern Indian Ocean (Plate 2.12)

This is one of the parts of the *Carta marina* that is the most different with respect to the 1507 map. Gone are the islands that come from Ptolemy; instead, Waldseemüller offers descriptions of the islands' inhabitants and natural riches based on the latest sources available to him, such as the travel narrative of Ludovico di Varthema, who journeyed in the East from 1502 to 1508. This sheet is dominated by a long legend on the spice trade in Calicut (now Kozhikode), India: the legend indicates the system of weights and the currency used in Calicut, and lists the sources and prices of many different spices, woods, and precious stones in that market.

> 12.1
> Taprobanam insulam sub equatoris circulo constitutam Ptolomeus directe retulit. Solinus autem secus ostendit, ubi septentriones nequaquam conspici et virgilie nunquam apparere illic posse. Lunamque ab octava in sextam decimam tantum supra terram videri, quod ac etiam Portugaliensium nauigatio hac positione clarissime aprobat.
>
> Ptolemy unambiguously indicates that Taprobana is located on the equator. But Solinus shows otherwise, as the northern stars cannot be seen there, and the Pleiades can never appear there. The moon is only above the horizon there from the eighth to the sixteenth day of the month, and the Portuguese voyages in this area unambiguously confirm this.

Ptolemy locates Taprobana on the equator in his *Geography* 7.4, and the passage from Solinus is in *Polyhistor* 53.[352] As discussed in the introduction, on the *Carta marina* Waldseemüller uses many fewer ancient authorities generally, and much less of Ptolemy in particular, whom he had so eagerly embraced on the 1507 map, and this legend is a good example of his turn against Ptolemy—though on the other hand he is citing another ancient authority, Strabo. This legend seems to have been composed by Waldseemüller rather than being a summary of something in one of his sources. For more on Waldseemüller's thoughts about Taprobana see Legend 8.10.

We can get a good idea of the difficulty cartographers faced in trying to assimilate conflicting sources in a statement by Robert Hues in his *Tractatus de globis et eorum usu* of 1594, who contradicts the statement by Solinus referred to by Waldseemüller, and implicitly the confirmatory Portuguese sources that Waldseemüller mentioned: "These Stars are reported by Pliny and Solinus to be never seene at all in the Isle Taprobana; but this is ridiculous, and fit to bee reported by none but such as Pliny and Solinus. For those that inhabite that Isle have them almost over their heads."[353]

> 12.2
> Juxta samotram et iauam sunt insularum (uti fertur) numero .8000. quarum alique habitate et alique deserte
>
> Near Sumatra and Java it is said that there are 8,000 islands, some inhabited and others not.

This legend comes from Varthema: "The captain of the ship said that around the island of Giava, and around the island of Sumatra, there were more than eight thousand islands."[354] This legend on the *Carta marina* in effect replaces a legend from Ptolemy 7.4 on the 1507 map that runs: *Ante taprobanam chortes sunt insularum quas dicunt esse .1378. numero quarum tamen traduntur haec sunt*, "Near Taprobana there are many islands—they say there are 1,378; these are the ones whose names are known." Thus, this legend about the 8,000 islands is another part of Waldseemüller's effort to replace information from Ptolemy with more recent data. At the same time, it is interesting to note some of the more modern sources that

[352]Solinus, *Polyhistor* 53: *nulla in navigando siderum observatio: utpote ubi septemtriones nequaquam videntur vergiliaeque nunquam apparent. lunam ab octava in sextam decimam tantum supra terram vident,* "Observations are not taken of the stars in navigating: for the northern stars cannot be seen, and the Pleiades never appear. They see the moon above the earth only from the eighth day to the sixteenth day."

[353]Robert Hues, *Tractatus de globis et eorum usu: A Treatise Descriptive of the Globes Constructed by Emery Molyneux and Published in 1592 (i.e. 1594)* (London: Printed for the Hakluyt Society, 1889), Chap. 4, p. 56.

[354]See Varthema, *The Travels of Ludovico di Varthema* (see note 103), p. 259.

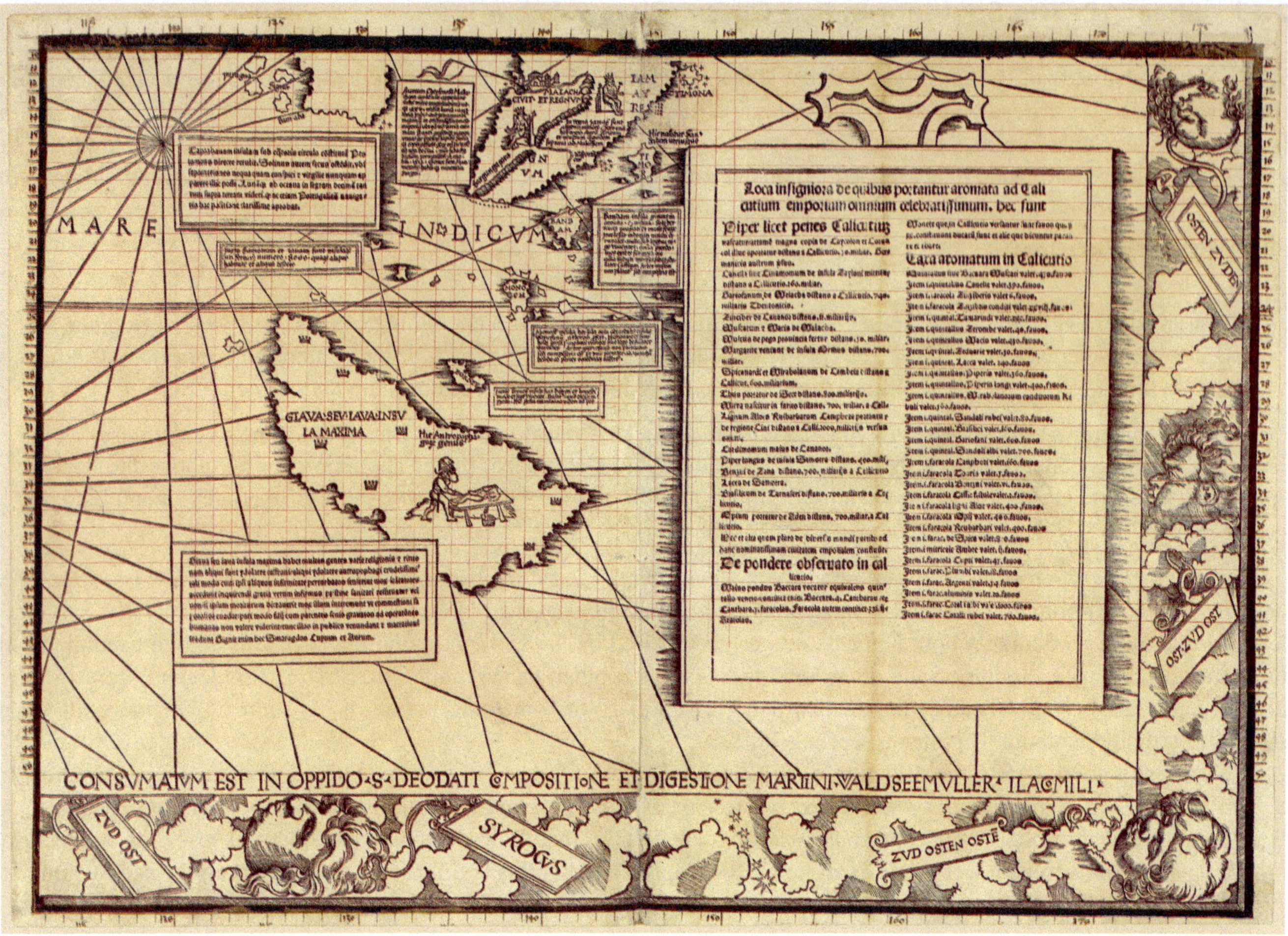

Plate 2.12 Sheet 12 of the *Carta marina*: The Southern Indian Ocean. Courtesy of the Library of Congress

Waldseemüller did not use for the number of islands in the Indian Ocean. In particular, he did not use Marco Polo's figure of 12,700, with which he was probably familiar,[355] or the figure of 7,743 on Ruysch's map, which he also certainly knew.[356] His decision not to follow Polo should be taken as additional evidence of his lack of confidence in that author.

12.3
Hic Antropophagorum genus

Here is a race of man-eaters.

The anthropophagy on Java is mentioned by Varthema, and it is worth noting that Marco Polo does not mention anthropophagy on Java Major.[357] As I remarked in the introduction, the scene of cannibalistic butchery here is copied by the maker of the Vallard Atlas of c. 1547: the scene on Fries's edition of the *Carta marina* is different, and it is clear that the maker of the Vallard Atlas was copying from Waldseemüller's map rather than Fries's in this instance, so this is good evidence for the diffusion of Waldseemüller's *Carta marina*.

---

[355]For Marco Polo's indication that there are 12,700 islands in the Indian Ocean see Marco Polo, *Marka Pavlova z Benátek, Milion* (see note 54), p. 188; and *The Book of Ser Marco Polo* (see note 54), Book 3, Chap. 34, vol. 2, p. 424. Fra Mauro on his mappamundi of c. 1455 indicates that there are 12,600, no doubt a slip for Marco Polo's number: see Falchetta, *Fra Mauro's World Map* (see note 143 in Chap. 1 above), pp. 190–191, *51. For a survey of the number of islands thought to be in the Indian Ocean by various medieval authors see Christine Gadrat, *Une image de l'Orient au XIV siècle les 'Mirabilia descripta' de Jordan Catala de Sévérac* (Paris: École des Chartes, 2005), p. 137.

[356]The legend on Ruysch's map is translated by Thacher, *The Continent of America* (see note 172), p. 215.

[357]See Varthema, *The Travels of Ludovico di Varthema* (see note 103), pp. 255–256; for the account of Java in Marco Polo see Marco Polo, *Marka Pavlova z Benátek, Milion* (see note 54), p. 159; and *The Book of Ser Marco Polo* (see note 54), Book 3, Chap. 6, vol. 2, pp. 272–275.

12.4
Giaua seu iaua insula maxima habet multas gentes varie religionis et ritus nam aliqui sunt ydolatre caffrani, aliqui ydolatre antropophagi crudelissimi tali modo cum ipsi aliquos infirmitate perturbatos senserint mox incantatores accedunt inquirendi gratia vtrum infirmus pristine sanitati restituatur vel non. si ipsum moriturum dictauerit mox illum interemunt ut commestioni saporosior evadat, pari modo filii cum parentes senio grauatos ad operationes humanas non valere viderint, tunc illos in publico venundant et mactationi tradunt. Gignit enim hec [insula] Smaragdos Cuprum et Aurum.

Giava or Java is a very large island that has many different races of different religions and rites. For some are Caffrani idolaters, while others are man-eating idolaters, extremely cruel, so that when one of them feels himself falling ill, the enchanters come and ask whether or not the sick person can be restored to good health. If he says that he is going to die, they quickly kill him so that he will be tastier when he is eaten, and similarly when sons see that their parents are so weakened with age that they cannot perform basic tasks, they sell them in public, and give them over to slaughter. This island produces emeralds, copper, and gold.

This legend comes from Varthema.[358] The legends on the island on the 1507 map are less sensationalistic, focusing on the spices the island produces and the religion of the people.[359] The legends on the island on the 1507 map come from Marco Polo, so this legend on the *Carta marina* is another case where Waldseemüller preferred the more recent information in Varthema to that in Marco Polo.[360]

12.5
Auream Chersonesum Malacham acole nunc appellant urbs mire magnitudinis utque. xxv. milia larium acque tocius Indie aut potius mundi maximum et celebratissimum est emporium ubi non modo varia aromata sed auri argentique margaritarum ac pretiosorum lapidum sericique copia affluit. Rex tributarius est regi de Cini, nunc subactus dictioni portugallensium, Anno domini .1512. Gentes sunt Macometani, habentque mineralia stagni

The Golden Chersonese, which the inhabitants now call Malacca. The city is very large, with 25,000 houses, and it is the largest and most famous emporium in India or indeed the world, where not only various spices but also gold, silver, pearls, precious stones, and silk flow in abundance. The king pays tribute to the king of China, but now, in the year 1512, has been brought under the dominion of the Portuguese. The people are followers of Mohammed, and they mine tin.

Much of this legend comes from a source that Waldseemüller does not name in the long text block on sheet 9, namely a letter from King Manuel to Pope Leo X dated June 1513, in which the king informs the pontiff about Afonso de Albuquerque's conquest of Malacca in July of 1511.[361] The passage in the letter from which Waldseemüller was drawing runs as follows[362]:

[358]See Varthema, *The Travels of Ludovico di Varthema* (see note 103), p. 251 on the various religions on Java, p. 252 on the emeralds; p. 253 on the gold and copper; and pp. 255–256 on the killing and eating of people.

[359]The legends on Java on Waldseemüller's 1507 map, which derive from Marco Polo (see note 357) read: *Silua nucum muscatum*; *Omnes habitatorre[s] sunt ydolatre istius insule*; *Hec inueniuntur hic piperis nucum muscatorum spici galange garioffali ceterorum aroma[tor]m copia*; *silva piper*; *Hec insula in circuitu suo habet mensuram miliariorum trium milium*; *rex insule nemini tributarius*; "Forest of nutmeg; All of the inhabitants of this island are idolaters; These things are found here: pepper, nutmeg, spikenard, galingale, cloves, and an abundance of other kinds of spices; Forest of pepper; This island is three thousand miles in circumference; The king of the island pays tribute to no one.".

[360]Incidentally there is discussion of Java in Fries's *Uslegung*, Chap. 64, which is translated into modern German by Petrzilka, *Die Karten des Laurent Fries* (see note 202 in Chap. 1 above), p. 143. For a description of trade at Java about a bit more than a century later see Roberts, *The Merchants Mappe* (see note 275), Chap. 104, pp. 203–204.

[361]The letter was published in several editions, the first being *Epistola potentissimi ac inuictissimi Emanuelis Regis Portugaliae & Algarbiorum &c. De victoriis habitis in India & Malacha: ad S. in Christo Patrem & D[omi]n[u]m nostrum D[omi]n[u]m Leonem X. Pont. Maximum* (Rome: Impressa per Iacobum Mazochium, 1513). The copy in the Bayerische Staatsbibliothek is available in PDF via www.digital-collections.de; the text was also printed in the collection of travel narratives *Nouus orbis regionum ac insularum ueteribus incognitarum* (Basel: Apud Io. Heruagium, 1532), pp. 184–187; the text of the letter is transcribed in William Roscoe, *The Life and Pontificate of Leo the Tenth*, revised by Thomas Roscoe (London: Chatto and Windus, 1876), vol. 1, pp. 521–524; the text is also supplied by Salvatore de Cutiis, *Une ambassade portugaise à Rome au XVIe siècle: Mémoire lu au IVe Congrès Scientifique International des Catholiques à Fribourg (1897)* (Naples: Michèle d'Auria, 1899), pp. 4–8. For discussion of the letter see Banha de Andrade, *Mundos novos do mundo* (see note 339), vol. 2, pp. 652–660; Luís de Matos, *L'expansion portugaise dans la littérature latine de la Renaissance* (Lisbon: Fundação Calouste Gulbenkian, Serviço de Educação, 1991), pp. 345–354; and Jean Aubin and Luis Filipe F. R. Thomaz, "Un opuscule latin sur la prise de Malacca par les Portugais, imprimé en Italie en 1514," *Archipel* 74 (2007), pp. 107–138, with an English abstract on p. 268.

[362]This passage is on signatures Ai^v–Aii^r of the pamphlet, and is transcribed by Roscoe, *The Life and Pontificate of Leo the Tenth*, pp. 521–522.

...Alphonsus de Albicherque protho-capitaneus noster, ut jacturam, quam superioribus annis nostri fecerent, injuriamque ulcisceretur, auream Chersonesum, Malacham accolae appellant, contendit, ea est inter Sinum magnum et Gangeticum sita, Urbs mirae magnitudinis, utque vigintiquinque millium et amplius larium censeatur, terra ipsa fecundissima, ac nobilissimarum quas fert India mertium feracissima, celebratissimum ob id Emporium, ubi non modo varia aromata et omnigeni odores, sed Auri quoque, argenti, margaritarum ac preciosorum lapillorum magna copia affluit. Hanc Rex Maurus gubernabat, eatenus vires suas Maumetica Secta protendente caetera Gentiles tenent.

...Afonso de Albuquerque, our high captain, in order to avenge the loss and injury that our men sustained in the previous years, set his sights on the Golden Chersonesus, which the inhabitants call Malacca. This is located between the Great Bay and the Ganges Bay, a city of remarkable size, with 25,000 or more houses, and very fertile soil, that abundantly produces the excellent goods that India generates. Because of this, the city is a famous emporium, where not only various spices and all sorts of scents, but also gold, silver, pearls and precious stones flow in great abundance. The city was governed by a Moorish King, the Muslim sect extending its strength thus far, while pagans hold the rest.

Waldseemüller gives the year of the conquest of Malacca as 1512 rather than 1511, but the pamphlet is not very clear about the date. Two of the details in Waldseemüller's legend, namely the indication that the king of Malacca used to pay tribute to the king of China, and the reference to the production of tin, come from Varthema.[363] Although Waldseemüller says in his legend that Malacca was the largest and most famous emporium in India,[364] other texts on his map, particularly Legends 7.18 and 12.11, point to Calicut.

12.6
In regno jamay sunt argenti minere, fert etiam hec regio Aurum Sericum et mustum, que quidem super terram ad Malacham hins [for *hinc*] portantur

In the kingdom of Jamay there are mines of silver, and in fact this region produces gold, silk, and musk, which are carried from here overland to Malacca.

Jamay is a province in Laos, but the place name does not appear in any of the sources that Waldseemüller lists on sheet 9 of the *Carta marina*. I have not been able to determine the source of this legend. It probably comes from the same source as the following legend about the island of Timor.

12.7
TIMOR Hic nascitur Sandalum utriusque

Timor. Here there grows sandalwood of both [types].

I have not been able to determine the source of this legend, though it is tempting to think that it comes from the same source as Legend 12.6 on Jamay. Thomas Suárez suggests that it comes from some excerpts from Tomé Pires's *Suma Oriental* of 1515 that somehow reached Waldseemüller.[365] Pires describes the Timor islands in this work, but

[363]See Varthema, *The Travels of Ludovico di Varthema* (see note 103), pp. 224–225.
[364]On the trade in Malacca see Laurence Noonan, "The Portuguese in Malacca: A Study of the First Major European Impact on East Asia," *Studia* 23 (1968), pp. 33–104; Luís Filipe F. R. Thomaz, "Malaca e as suas comunidades mercantis na viragem do século XVI," in his *De Ceuta a Timor* (Linda a Velha, Portugal: DIFEL, 1994), pp. 513–535; on the later fortunes of Malacca as a trading center see Roberts, *The Merchants Mappe* (see note 275), Chap. 97, pp. 199–201.
[365]Thomas Suárez, *Early Mapping of Southeast Asia* (Singapore: Periplus, 1999), p. 113.

Waldseemüller's depiction does not agree well with Pires's text, which says that there are two islands called Timor from which sandalwood comes.[366] Another remotely possible source is a map from the atlas of Francisco Rodrigues, who traveled in the Indian Ocean in 1511–1512, and made his atlas of maps of the region c. 1513.[367] The map on f. 37 of this atlas shows Timor, with the legend *A Jlha de timor homde naçe o ssamdollo*, "The Island of Timor where the sandalwood grows," which is quite similar to Waldseemüller's legend.[368] It is not at all clear, though, how Waldseemüller could have obtained a map from Rodrigues's atlas. Duarte Barbosa mentions that sandalwood comes from Timor, but as his book was not completed until 1516, it seems very unlikely that Waldseemüller could have used it.[369] The idea that Waldseemüller might somehow have used Barbosa's book is perhaps rendered still less likely by the fact that Barbosa says nothing about Jamay. None of the literature on Timor, its sandalwood trade, or the Portuguese presence there sheds light on Waldseemüller's likely source.[370]

12.8

Bandam insula girans in circuitu. c. miliaria. Sola hec nucis muscati et macis ferax. Incole sunt indomiti inculti et rurales. nulla sub lege ac rege viuentes. omnia preter domos exiles sunt apud eos quoniam induti camisiis suntque ydolatre caffrani Arbor muscati non plantatur sed campestris est

The island of Banda, which has a circumference of 100 miles. Only this island produces nutmeg and mace. The inhabitants are ungoverned, wild, and rustic; they have no law and no king. Everything among them, except their

---

[366]Tomé Pires, *The Suma oriental of Tomé Pires, an Account of the East, from the Red Sea to Japan, written in Malacca and India in 1512–1515*, ed. Armando Cortesão (London: The Hakluyt Society, 1944), vol. 1, pp. 203–204: "Between the islands of Bima and Solor there is a wide channel along which they go to the sandalwood islands. All the islands from Java onwards are called Timor, for timor means 'east' in the language of the country, as if they were saying the islands of the east. As they are the most important, these two from which the sandalwood comes are called the islands of Timor. The islands of Timor have heathen kings. There is a great deal of white sandalwood in these two. It is very cheap because there is no other wood in the forests. The Malay merchants say that God made Timor for sandalwood and Banda for mace and the Moluccas for cloves, and that this merchandise is not known anywhere else in the world except in these places; and I asked and enquired very diligently whether they had this merchandise anywhere else and they said not....".

[367]Rodrigues's atlas is in Paris, Bibliothèque de l'Assemblée Nationale, MS 1248; for discussion see Heinrich Winter, "Francisco Rodrigues' Atlas of ca. 1513," *Imago Mundi* 6 (1949), pp. 20–26; Armando Cortesão and Avelino Teixeira da Mota, *Portugaliae monumenta cartographica* (Lisbon, 1960–62), vol. 1, pp. 79–84 and plates 34–36. The text of the atlas has been edited in Tomé Pires and Francisco Rodrigues, *The Suma Oriental of Tomé Pires, an Account of the East, from the Red Sea to Japan, written in Malacca and India in 1512–1515, and The Book of Francisco Rodrigues, Rutter of a Voyage in the Red Sea, Nautical Rules, Almanack and Maps, Written and Drawn in the East before 1515, translated from the Portuguese MS in the Bibliothèque de la Chambre des Députés, Paris, and edited by Armando Cortesão* (London: The Hakluyt Society, 1944). The atlas has been reproduced in facsimile as Francisco Rodrigues, *O livro de Francisco Rodrigues: o primeiro atlas do mundo moderno*, ed. José Manuel Garcia (Porto: Editora da Universidade do Porto, 2008).

[368]This map is reproduced in Manuel Francisco de Barros e Sousa Santarém, *Atlas composé de mappemondes* (Paris: E. Thunot, 1849), as map no. 20 in the set of maps from Rodrigues; in the reprint edition, *Atlas de Santarém: Facsimile of the Final Edition, with Explanatory Texts by Helen Wallis and A. H. Sijmons* (Amsterdam: Rudolf Muller, 1985), it is reproduced in plate 70, no 20. The map is also reproduced in Tomé Pires and Francisco Rodrigues, *The Suma oriental of Tomé Pires* (see note 595), in plate 27, between pp. 208 and 209, with the legend transcribed on p. 203, and a brief description of the map in Appendix 2, p. 523; in Cortesão and Teixeira da Mota, *Portugaliae monumenta cartographica* (Lisbon, 1960–62), vol. 1, plate 35, no. 10; and in the facsimile edition, Francisco Rodrigues, *O livro de Francisco Rodrigues* (see note 367), with commentary on pp. 98–99.

[369]See Duarte Barbosa, *A Description of the Coasts of East Africa and Malabar in the Beginning of the Sixteenth Century* (London: Printed for the Hakluyt society, 1866), p. 199. On the transmission of the book see José Manuel Herrero Massari, "La transmisión textual del 'Libro de Duarte Barbosa'," *Revista de filología románica* 11–12 (1994–1995), pp. 391–402, who notes (p. 391) that in the prologue to the first edition of Barbosa's text, in vol. 1 of Giovanni Battista Ramusio's *Delle navigationi et viaggi* (Venice, 1550), Barbosa writes: *nel presente anno 1516 io diedi fine a scrivere il presente libro.*

[370]For discussion of Timor and the sandalwood trade see John Villiers, "As derradeiras do mundo: The Dominican Mission and the Sandalwood Trade in the Lesser Sunda Islands in the Sixteenth and Seventeenth Centuries," in Luís de Albuquerque and Inácio Guerreiro, eds., *II Seminário Internacional de História Indo-Portuguesa: Actas* (Lisbon: Instituto de Investigação Científica Tropical, Centro de Estudos de História e Cartografia Antiga, 1985), pp. 572–600; also printed in his *East of Malacca: Three Essays on the Portuguese in the Indonesian Archipelago in the Sixteenth and Early Seventeenth Centuries* (Bangkok: Calouste Gulbenkian Foundation, 1985); Rui Manuel Loureiro, "Os portugueses em Timor: relance histórico," in Rui Manuel Loureiro, ed., *Onde nasce o sândalo: os portugueses em Timor nos séculos XVI e XVII* (Lisbon: Grupo de Trabalho do Ministério da Educação para as Comemorações dos Descobrimentos Portugueses, 1995), pp. 27–44, esp. 31–34, "Primeiras notícias de Timor"; Roderich Ptak, "Some References to Timor in Old Chinese Records," *Ming Studies* 17 (1983), pp. 37–48; reprinted in his *China's Seaborne Trade with South and Southeast Asia, 1200–1750* (Aldershot and Brookfield, VT: Ashgate, 1999), article 1; and Roderich Ptak, "The Transportation of Sandalwood from Timor to China and Macao, ca. 1350–1600," *Revista de Cultura* 1 (1987), pp. 36–45; reprinted in Roderich Ptak, ed., *Portuguese Asia: Aspects in History and Economic History (Sixteenth and Seventeenth Centuries)* (Stuttgart: Franz Steiner, 1989), pp. 89–109.

houses, is very poor, for they are clad in [nothing but] shirts. They are Caffrani idolaters. The nutmeg tree is not cultivated but grows wild.

This legend comes from Varthema.[371] If Waldseemüller took his information about Timor and Jamay from a map by Rodrigues, it is strange that this map seems not to have influenced other nearby parts of the *Carta marina* as well.

12.9
Monoch insula hec sola cum adiacentibus insulis Gariofanum arborem profert. Habitatores sunt homines grossi simplices rurales sine lege bestialiter viuentes. Arbor gariofanus non plantatur sed campestris est. Et hinc portatur ad Malacham deinde ad partes occidentales transfertur.

The Moluccas. This island, together with those nearby, are the only ones that produce the clove tree. The inhabitants are fat, simple, and rustic, living like beasts without any law. The clove tree is not cultivated, but grows wild. From here it is carried to Malacca and from there to the West.

Some of this legend comes from Varthema, but not all of it: Varthema does not say that cloves grow only on these islands, nor does he say that the clove tree grows wild, nor that cloves are taken from the Moluccas to Malacca.[372] The chapters about the spices in Calicut in the *Paesi* and *Itinerarium* (chapters 82 and 83, see Legend 12.11) say that cloves come from Meluza (the Moluccas) to Calicut, not to Malacca. I have not been able to determine Waldseemüller's source for these pieces of information. Duarte Barbosa says that cloves were brought from the Moluccas to Malacca, and also that the clove trees grow wild, but Barbosa did not finish writing his book until 1516, which would mean it would not have been possible for Waldseemüller to have used his book for the *Carta marina*.[373] None of the literature on the Moluccas or the Portuguese presence there sheds light on Waldseemüller's likely source.[374]

12.10
Incole Borney insule sunt discreti et honesti racione et lege viuentes. Nascitur apud eos Camphora. Hic stella tramontana videre non potest.

The inhabitants of Borneo are prudent and honest, living according to laws and reason. Camphor grows there. Here the North Star cannot be seen.

This legend comes from Varthema[375]; the detail about the North Star not being visible is from the chapter on Monoch, i.e. the Moluccas, which immediately precedes his chapter on Borney (Borneo).[376] Waldseemüller also mentions the North Star being invisible in Legends 11.4 and 12.1.

[371]See Varthema, *The Travels of Ludovico di Varthema* (see note 103), pp. 243–244. For discussion of the sixteenth-century trade involving the Banda islands see John Villiers, "Trade and Society in the Banda Islands in the Sixteenth Century," *Modern Asian Studies* 15.4 (1981), pp. 723–750; revised version published as John Villiers, "Da verde noz tomando seu tributo: The Portuguese in the Banda Islands in the Sixteenth Century," in his *East of Malacca: Three Essays on the Portuguese in the Indonesian Archipelago in the Sixteenth and Early Seventeenth Centuries* (Bangkok: Calouste Gulbenkian Foundation, 1985), pp. 1–30.
[372]See Varthema, *The Travels of Ludovico di Varthema* (see note 103), pp. 245–246.
[373]See Duarte Barbosa, *A Description of the Coasts of East Africa and Malabar in the Beginning of the Sixteenth Century* (London: Printed for the Hakluyt society, 1866), pp. 201–202. On the date of the book see note 369.
[374]See E. C. Abendanon and E. Heawood, "Missing Links in the Development of the Ancient Portuguese Cartography of the Netherlands East Indian Archipelago," *The Geographical Journal* 54.6 (1919), pp. 347–355; and Armando Cortesão, "As mais antigas cartografia e descrição das Moluccas," in A. Teixeira da Mota, ed., *A viagem de Fernão de Magalhães e a questão das Moluccas: Actas do II Colóquio luso-espanhol de história ultramarina* (Lisbon: Junta de Investigações científicas do Ultramar, 1975), pp. 49–75—but Cortesão's claim that Asia on the *Carta marina* is "riddled with reminiscences of Ptolemy" (p. 64) is false, and betrays a very superficial study of the map. Also see António da Silva Rego, "As Molucas em princípios do século XVI," in the same volume, pp. 75–89. On the history of cloves see R. A. Donkin, "Byzantium and the Asiatic Antecedents of *caryophyllon*," in his *Between East and West: The Moluccas and the Traffic in Spices up to the Arrival of Europeans* (Philadelphia, PA: American Philosophical Society, 2003), pp. 111–116; and C. R. de Silva, "The Portuguese and the Trade in Cloves in Asia during the Sixteenth Century," in M. A. Hassan and N. H. S. N. Abd Rahman, eds., *The Eighth Conference: International Association of Historians of Asia. Selected Papers* (Bangi, Selangor Darul Ehsan, Malaysia: Organising Committee, Eighth International Conference IAHA, and History Dept., University Kebangsaan Malaysia, 1988), pp. 251–260; reprinted in M. N. Pearson, ed., *Spices in the Indian Ocean World* (Aldershot, Hampshire, UK; Brookfield, VT: Variorum, 1996), pp. 259–268.
[375]See Varthema, *The Travels of Ludovico di Varthema* (see note 103), pp. 246–248.
[376]Manuel Teixeira, "Early Portuguese & Spanish Contacts with Borneo," in Chang Kuei-yung et al., eds., *International Association of Historians of Asia: Second Biennial Conference Proceedings* (Taipei: s.n., 1962), pp. 485–526; reprinted in *Boletim da Sociedade de Geografia de Lisboa* 82 (1964), pp. 299–335, and excerpts in John Bastin and Robin W. Winks, eds., *Malaysia: Selected Historical Readings* (Nendeln, Liechtenstein: KTO Press, 1979), pp. 44–48.

12.11

Loca insigniora de quibus portantur aromata ad Calicutium emporium omnium celebratissimum. hec sunt Piper licet penes Calicutium nascatur, attamen magna copia de Caycolon et Coruncol[377] illuc aportatur distans a Callicutio .50. miliar. Germanicis austrum versus.[378]
Canella siue Cinamomun de insula Zayloni mittitur distans a Callicutio .260. miliar.
Gariofanum de Melacha distans a Callicutio. 740. miliaris Theutonicis.
Zinciber de Cananor distans .12. miliariis.
Muscatum et Macia de Malacha.[379]
Muscus de pego prouincia fertur distans .50. miliar.[380]
Margarite veniunt de insula Ormus distans .700. miliar.
Spicanardi et Mirabolanum de Cambeia distans a Callicut .600. miliarium.[381]
Thus portatur de Seer distans .800. miliariis.
Mirra nascitur in farico distans .700. miliar. a Call.[382]
Lignum Aloes Ruibarbarum Camphora portantur de regione Cini distans a Calli .2000. miliariis versus ortum.[383]
Cardimomum maius de Cananor.[384]
Piper longus de insula Samotra distans .400. mili.[385]
Benzui de Zana distans .700. miliariis a Callicutio[386]
Lacca de Samotra.[387]
Brasilicum de Tarnaseri distans .700. miliaris a Callicutio.[388]
Opium portatur de Aden distans .700. miliar. a Callicutio.
Hec et alia quam plura de diuersis mundi partibus ad hanc nominatissimam ciuitatem emporialem confluunt.
De pondere obseruato in callicutio.

Maius pondus Baccara vocatur equiualens quintallo veneto, continet enim Baccara .4. Cantharas et Canthara .5. faracolas. Faracola autem continet.xxiiii.Aratolas.[389]

Monete que in Callicutio versantur sunt fauos quorum xx. constituunt ducatum. sunt et alie que dicuntur parante et chare.
Taxa aromatum in Calicutio
Quintallus siue Bacaara [*sic*] Muscati valet. 450 fauos
Item i. quintallus Canelle valet. 390. fauos.
Item i. faracola Zinziberis valet 6. fauos
Item i. faracola Zinzibris conditi valet .xxviii. fauos
Item i. quintal. Tamarindi valet .xxx. fauos.

---

[377]Caycolon is perhaps to be understood as an alternative for Coruncol. The *Paesi novamente ritrovati* gives Chorunchel, which Greenlee suggests represents Cranganore: see Greenlee, "The Anonymous Narrative" (see note 67 in Chap. 1 above), p. 93. The *Itinerarium* does not give either of these place names, and this is excellent evidence that Waldseemüller was not using the *Itinerarium* here.

[378]The *Paesi* indicates the unit of measure for all of the distances in this list as *leghe*, or leagues.

[379]The *Paesi* indicates that nutmeg and mace come from *melucha*, which Greenlee, "The Anonymous Narrative" (see note 67 in Chap. 1 above), p. 93, correctly interprets as Molucca, but Waldseemüller interprets it as Malacca. Waldseemüller omits the distance of "Malacha" (i.e. melucha) from Calicut, which is supplied in the *Paesi*.

[380]The *Paesi* gives 500, and the Trevisan manuscript gives 600—for the latter number see Eric Dursteler, "Reverberations of the Voyages of Discovery" (see note 54 in Chap. 1 above), p. 60.

[381]Waldseemüller omits what is the next line in the *Paesi*, which says that cassia-fistula grows in Calicut.

[382]On the trade in myrrh see Bradley Z. Hull, "Frankincense, Myrrh, and Spices: The Oldest Global Supply Chain?" *Journal of Macromarketing* 28.3 (2008), pp. 275–288.

[383]Waldseemüller omits the next line of the *Paesi*, which says that *zeromba* or zerumbet grows in Calicut.

[384]Waldseemüller omits the *Paesi*'s indication that Cananor is twelve leagues beyond Calicut.

[385]Waldseemüller has added the detail that Sumatra is 400 miles beyond Calicut from a few lines lower down in the *Paesi*.

[386]Waldseemüller omits the next two lines in the *Paesi*, which say that tamarind and zedoary grow in Calicut.

[387]Waldseemüller omits the *Paesi*'s indication that Sumatra is 400 leagues beyond Calicut, no doubt because he included this information two lines above.

[388]The *Paesi* indicates the distance as 500 leagues (which would be 500 miles according to the conversion factor Waldseemüller is using).

[389]Waldseemüller's explanation of the system of weights is very different than it is in the *Paesi* and the *Itinerarium*. The *Paesi* has *baar*, or *bacar*, or *bacaro* where Waldseemüller has *baccara*, and the *Itinerarium* gives *bacar*. Also, neither of the two printed texts says that the *baar* or *bacar* is equivalent to the Venetian *quintal*, as Waldseemüller does, and Waldseemüller fails to indicate that the *aratola* is a Portuguese weight, which is specified in the *Paesi* and the *Itinerarium*, but the Trevisan manuscript does indicate that "A baar is 640 lb according to Venetian usage"—see Dursteler, "Reverberations of the Voyages of Discovery in Venice" (see note 54 in Chap. 1 above), p. 61.

Item i. quintallus Zerombe valet .40. fauos.
item i. quintallus Macis valet .430. fauos.
Item i. quintal Zeduarie valet .30. fauos.
Item i. quintal Lacca valet .240. fauos[390]
Item i. quintallus. Piperis valet .360. fauos.
Item i. quintallus. Piperis longi valet .400. fauos.
Item i. quintallus. Mirabolanorum conditorum Rebuli valet .560 fauos.
Item i. quintal. Sandali rubei valet .80. fauos.
Item i. quintal. Brasilici valet .160. fauos.
Item i. quintal. Gariofani valet .600. fauos.[391]
Item i. quintal. Sandali albi valet .700. fauos
Item i. faracola Canphori valet .160. fauos
Item i. faracola Thuris valet .5. fauos.
Item i. faracola Bentzui valet .vi. fauos.
Item i. faracola Cassie fistule valet .2. fauos.
Item i. faracola ligni Aloe valet .400. fauos.
Item i. faracola Opii valet .400. fauos.
Item i. faracola Reubarbari valet .400. fauos.
Item i. farac. de Spica valet .820. fauos[392]
Item.i mitricale Ambre valet .ii. fauos.[393]
Item i. faracola Cupri valet .45. fauos[394]
Item i. farac. Plumbi valet .18. fauos
Item i. farac. Argenti valet .54. fauos
Item i. farac. aluminis valet .20. fauos
Item.i. farac. Corali albi valet .1000. fauos
Item i. farac Coralli rubei valet .700. fauos.

The more important places from which spices are brought to Calicut, the most famous market city of all. These include pepper, which grows at Calicut, but nonetheless a great quantity is brought from Caycolon and Coruncol [Cranganore][395] to Calicut, a distance of 50 German miles[396] to the south [from Calicut]. Canella or cinnamon is sent from the island of Ceylon, which is 260 miles distant from Calicut.
Cloves come from Molucca, 740 German miles from Calicut.
Ginger comes from Cananor, which is 12 miles from Calicut.
Nutmeg and mace come from Molucca.[397]
Musk comes from a province called Pegu, 50 miles from Calicut.[398]
Pearls come from Hormuz, 700 miles from Calicut.
Spikenard and myrobalans come from Combaia, 600 miles from Calicut.[399]

---

[390]The *Paesi* gives 260, as does the Trevisan manuscript—see Dursteler, "Reverberations of the Voyages of Discovery" (see note 54 in Chap. 1 above), p. 59. Also, Waldseemüller omits the next line in the *Paesi*, *Itinerarium*, and Trevisan manuscript; in the *Paesi*, it reads "Item uno bacar di macis ual.ccccxxx. fauos," "A bacar of mace is worth 430 favos.".

[391]The order of the items now differs somewhat from the order in the *Paesi* and the *Itinerarium*.

[392]The *Paesi* gives 800, as does the Trevisan manuscript—for the latter see Dursteler, "Reverberations of the Voyages of Discovery" (see note 54 in Chap. 1 above), p. 60.

[393]The *Paesi*, *Itinerarium*, and Trevisan manuscript give the same price for a mitricale of amber, so it seems that Waldseemüller's pen slipped here.

[394]In the *Paesi*, *Itinerarium*, and Trevisan manuscript, these last items in Waldseemüller's list are in a separate list that indicates the prices of various goods that are carried from Lisbon to Calicut, but Waldseemüller does not distinguish between the two groups, and omits the so-called bastard coral from the latter list.

[395]Caycolon is perhaps to be understood as an alternative for Coruncol. The *Paesi* gives Chorunchel, which Greenlee suggests represents Cranganore: see Greenlee, "The Anonymous Narrative," in *The Voyage of Pedro Álvares Cabral* (see note 67 in Chap. 1 above), pp. 53–94, at 93. The *Itinerarium* does not give either of these place names, and this is excellent evidence that Waldseemüller was not using the *Itinerarium* here.

[396]The *Paesi* indicates the unit of measure for all of the distances in this list as *leghe*, or leagues.

[397]The *Paesi* indicates that nutmeg and mace come from *melucha*, which Greenlee, "The Anonymous Narrative" (see note 67 in Chap. 1 above), p. 93, correctly interprets as Molucca, but Waldseemüller interprets it as Malacca. Waldseemüller omits the distance of "Malacha" (i.e. melucha) from Calicut, which is supplied in the *Paesi*.

[398]The *Paesi* gives 500, and the Trevisan manuscript gives 600—for the latter number see Dursteler, "Reverberations of the Voyages of Discovery" (see note 54 in Chap. 1 above), p. 60.

[399]Waldseemüller omits what is the next line in the *Paesi*, which says that cassia-fistula grows in Calicut.

Incense is brought from Seer, which is 800 miles from Calicut.
Myrrh grows in Fartak, which is 700 miles from Calicut.[400]
Aloe-wood and rhubarb and camphor come from China, which is 2000 miles to the east from Calicut.[401]
Very large cardamons come from Cananore.[402]
Long pepper grows in Sumatra, which is 400 miles away.[403]
Benzoin comes from Siam, 700 miles from Calicut.[404]
Lac comes from Sumatra.[405]
Brazil-wood comes from Tenasserim, which is 700 miles from Calicut.[406]
Opium comes from Aden, which is 700 miles from Calicut.
These and many other spices come to the famous market-city from various parts of the world.

Of the system of weights used in Calicut. The largest weight is the *baccara*, which is equal to a Venetian *quintal*. A baccara contains 4 *cantharas*, and a *canthara* 5 *faracolas*. A *faracola* contains 24 *aratolas*.[407] The coins used in Calicut are the *favos*, of which 20 equal one ducat. There are also others which are called *parane* and *chare*.

The prices of spices in Calicut:
A *quintal* or *bacara* of nutmeg is worth 450 *favos*.
A *quintal* of cinnamon is worth 390 *favos*.
A *faracola* of ginger is worth 6 *favos*.
A *faracola* of ginger preserved in sugar is worth 28 *favos*.
A *quintal* of tamarind is worth 30 *favos*.
A *quintal* of zerumbet is worth 40 *favos*.
A *quintal* of mace is worth 430 *favos*.
A *quintal* of zeduary is worth 30 *favos*.
A *quintal* of lac is worth 240 *favos*.[408]
A *quintal* of pepper is worth 360 *favos*.
A *quintal* of long pepper is worth 400 *favos*.
A *quintal* of preserved sebuli myrobalans is worth 560 *favos*.
A *quintal* of red sandalwood is worth 80 *favos*.
A *quintal* of brazil-wood is worth 160 *favos*.
A *quintal* of cloves is worth 600 *favos*.[409]
A *quintal* of white sandalwood is worth 700 *favos*.
A *faracola* of camphor is worth 160 *favos*.
A *faracola* of incense is worth 5 *favos*.
A *faracola* of benzoin is worth 6 *favos*.
A *faracola* of cassia-fistula is worth 2 *favos*.
A *faracola* of aloe-wood is worth 400 *favos*.
A *faracola* of opium is worth 400 *favos*.
A *faracola* of rhubarb is worth 400 *favos*.

---

[400]On the trade in myrrh see Bradley Z. Hull, "Frankincense, Myrrh, and Spices The Oldest Global Supply Chain?" *Journal of Macromarketing* 28.3 (2008), pp. 275–288.
[401]Waldseemüller omits the next line of the *Paesi*, which says that *zeromba* or zerumbet grows in Calicut.
[402]Waldseemüller omits the *Paesi*'s indication that Cananor is 12 leagues beyond Calicut.
[403]Waldseemüller has added the detail that Sumatra is 400 miles beyond Calicut from a few lines lower down in the *Paesi*.
[404]Waldseemüller omits the next two lines in the *Paesi*, which say that tamarind and zedoary grow in Calicut.
[405]Waldseemüller omits the *Paesi*'s indication that Sumatra is 400 leagues beyond Calicut, no doubt because he included this information two lines above.
[406]The *Paesi* indicates the distance as 500 leagues (which would be 500 miles according to the conversion factor Waldseemüller is using).
[407]Waldseemüller's explanation of the system of weights is very different than it is in the *Paesi* and the *Itinerarium*. The *Paesi* has *baar*, or *bacar*, or *bacaro* where Waldseemüller has *baccara*, and the *Itinerarium* gives *bacar*. Also, neither of the two printed texts says that the *baar* or *bacar* is equivalent to the Venetian *quintal*, as Waldseemüller does, and Waldseemüller fails to indicate that the *aratola* is a Portuguese weight, which is specified in the *Paesi* and the *Itinerarium*, but the Trevisan manuscript does indicate that "A *baar* is 640 lb according to Venetian usage"—see Dursteler, "Reverberations of the Voyages of Discovery" (see note 54 in Chap. 1 above), p. 61.
[408]The *Paesi* gives 260, as does the Trevisan manuscript—see Dursteler, "Reverberations of the Voyages of Discovery" (see note 54 in Chap. 1 above), p. 59. Also, Waldseemüller omits the next line in the *Paesi*, *Itinerarium*, and Trevisan manuscript; in the *Paesi*, it reads "Item uno bacar di macis ual.ccccxxx. fauos," "A *bacar* of mace is worth 430 *favos*.".
[409]The order of the items now differs somewhat from the order in the *Paesi* and the *Itinerarium*.

A *faracola* of spikenard is worth 820 *favos*.[410]
A *mitricale* of amber is worth 2 *favos*.[411]
A *faracola* of copper is worth 45 *favos*.[412]
A *faracola* of lead is worth 18 *favos*.
A *faracola* of silver is worth 54 *favos*.
A *faracola* of alum is worth 20 *favos*.
A *faracola* of white coral is worth 1000 *favos*.
A *faracola* of red coral is worth 700 *favos*.

This long legend about the sources and prices of spices in Calicut comes from chapters 82 and 83 of the *Paesi nouamente retrovati*, which are part of the so-called Anonymous Narrative of Cabral's voyage.[413] As the legend is quite long and detailed, it offers a much better opportunity than Waldseemüller's other legends to determine which version of the *Paesi* he was using, and it is immediately clear that he was not using the *Itinerarium Portugallensium*, the Latin translation of the *Paesi novamente ritrovati*, for the *Itinerarium* does not include some of the place names that Waldseemüller gives at the beginning of the legend (see the notes above), which do appear in the *Paesi*. And the German translation of the *Paesi*, the *Newe unbekanthe landte* of 1508, does not include these chapters, so we may be quite confident that Waldseemüller was using one of the Italian editions of the work.

Waldseemüller has changed various aspects of the way this information is presented in the *Paesi*. The *Paesi* lists the prices of the items first (in Chap. 82), and then the sources (Chap. 83), but Waldseemüller reverses this order. Also, at the end of the *Paesi*'s list of prices of spices in Calicut, there is a separate list of the prices that some European goods fetch in Calicut, but Waldseemüller has simply combined the two lists into one. Also, after the first paragraph of the list of sources, Waldseemüller omits the indications of spices that grow in the Calicut area, perhaps because he already mentioned those in Legend 7.18. There are other differences between Waldseemüller's presentation of this data and that in the *Paesi*, particularly in his presentation of the system of weights used in Calicut (see the notes above), but these differences are not great enough to support a hypothesis that Waldseemüller was using a manuscript of the *Paesi* rather than the printed edition.

It seems likely, as I suggested in the introduction (see p. 18), that Waldseemüller drew inspiration for using this material from the *Paesi* from the long legend about the route that spices took from the Indian Ocean to Europe on Martin Behaim's globe of 1492.[414] Waldseemüller's inclusion of the prices of spices on his map is a significant innovation, and raises interesting questions about the purpose of the *Carta marina*: this information might be useful to a ship captain sailing for the Indian Ocean, and the map is based on a nautical chart model, that is, on the type of map used by ship captains. Yet it seems unlikely that Waldseemüller intended the map to be taken to sea, particularly given the distance of Saint-Dié (or Strasbourg,

410The *Paesi* gives 800, as does the Trevisan manuscript—for the latter see Dursteler, "Reverberations of the Voyages of Discovery" (see note 54 in Chap. 1 above), p. 60.

411The *Paesi*, *Itinerarium*, and Trevisan manuscript give the same price for a *mitricale* of amber, so it seems that Waldseemüller's pen slipped here.

412In the *Paesi*, *Itinerarium*, and Trevisan manuscript, these last items in Waldseemüller's list are in a separate list that indicates the prices of various goods that are carried from Lisbon to Calicut, but Waldseemüller does not distinguish between the two groups, and omits so-called bastard coral from the latter list.

413These chapters are translated into English by Greenlee, *The Voyage of Pedro Álvares Cabral* (see note 67 in Chap. 1 above), pp. 53–94, at 91–94. There is a Portuguese translation of the Anonymous Narrative in T. O. Marcondes de Sousa, "Relação do Pilôto Anônimo," *Revista do Instituto Histórico e Geográfico de São Paulo* 45 (1945/1950), pp. 82–108, with the chapters in question translated on pp. 106–108. The somewhat different version in the Trevisan Manuscript in the Library of Congress is translated by Dursteler, "Reverberations of the Voyages of Discovery" (see note 54 in Chap. 1 above), pp. 59–61; who notes on p. 52 of his article that the information about the prices and sources of the spices for sale in Calicut is also included in another manuscript of the *Paesi*, namely Venice, Biblioteca Marciana, IT VI 277 (5806), ff. 56r–75r.

414There is some price information about luxury goods in the ancient world in the Edict of Diocletian, which was issued in the year 301. See Elsa Rose Graser, "The Significance of Two New Fragments of the Edict of Diocletian," *Transactions and Proceedings of the American Philological Association* 71 (1940), pp. 157–174, esp. 166; and Elsa Rose Graser, "The Edict of Diocletian on Maximum Prices," in Tenney Frank, ed., *An Economic Survey of Ancient Rome* (Baltimore: Johns Hopkins University Press, 1933–40), vol. 5, pp. 305–421, esp. 417–421.

where the map was probably printed) from the sea. It seems likely that Waldseemüller intended merely to satisfy the curiosity of the viewers of his map about the price of these expensive commodities at their source.

Information similar to that in the *Paesi* and on the *Carta marina* was compiled by Duarte Barbosa, a Portuguese officer and writer, in about 1516, but his work was not published until 1550, in volume 1 of Ramusio's *Navigationi e viaggi*.[415]

The prices of spices in the sixteenth century is an active area of research, but to my knowledge neither the data from the *Paesi* nor that on Waldseemüller's *Carta marina* has been incorporated into those discussions.[416] Similarly, this material has not been used in discussions of the systems of weights in the Indian Ocean area.[417] Incidentally the indication of a price for opium in the *Paesi* and on the *Carta marina* are some of the earliest indications of Renaissance European involvement in the opium trade.[418]

The importance of Calicut on the *Carta marina* is abundantly clear: two long legends (7.18 and 12.11) are devoted to the city, and it is thus effectively the most important city in the world. Adding to this prominence more subtly is the image of King Manuel riding the sea monster off the southern tip of Africa, proclaiming Portugal's dominance of the route to India: this accomplishment would not be so important if it were not for the spices (no insignificant part of them from Calicut) flowing along that route back to Portugal.[419]

12.12
CONSUMATUM EST IN OPPIDO DEODATI COMPOSITIONE ET DIGESTIONE MARTINI WALDSEE-MULLER ILACOMILI

[This map] was completed in the town of Saint-Dié through the composition and arrangement of Martin Waldseemüller, Ilacomylus.

Waldseemüller's indicates that the *Carta marina* is his creation; there is no such indication on his 1507 world map or in the 1513 edition of Ptolemy's *Geography*. "Ilacomylus" is the Latinized version of his last name, meaning "Miller of the lake in the woods."

---

[415]See Duarte Barbosa, "The Divers Kinds of Spices, Where They Grow, What They Are Worth at Calicut, and Whither They Are Carried"; and "Weights of Portugal and of India, and the Correspondence Between These and Those of Portugal," in *The Book of Duarte Barbosa: An Account of the Countries Bordering on the Indian Ocean and their Inhabitants, Written by Duarte Barbosa and Completed about the Year 1518 A.D.*, trans. Mansel Longworth Dames (London: Printed for the Hakluyt Society, 1918–1921), vol. 2, pp. 227–231 and 232, respectively. On Ramusio's publication of the work see George B. Parks, "The Contents and Sources of Ramusio's *Navigationi*," *Bulletin of the New York Public Library* 59.6 (1955), pp. 279–313.

[416]For discussion of the prices of spices in the sixteenth century see Donald F. Lach, "Appendix: Pepper Prices in the Sixteenth Century," in his *Asia in the Making of Europe*, vol. 1, *The Century of Discovery* (Chicago: University of Chicago Press, 1965), Book 1, pp. 143–147; Vitorino Magalhães-Godinho, "Le repli vénitien et égyptien et la route du cap," in *Éventail de l'histoire vivante: Hommage à Lucien Febvre* (Paris: Colin, 1953), vol. 2, pp. 283–300; reprinted in M. N. Pearson, ed., *Spices in the Indian Ocean World* (Aldershot, Hampshire, UK, and Brookfield, VT: Variorum, 1996), pp. 93–110; and Kevin H. O'Rourke and Jeffrey G. Williamson, "Did Vasco da Gama Matter to European Markets?" *Economic History Review* 62.3 (2009), pp. 655–684.

[417]See Gabriel Ferrand, "Les poids, measures et monnaies des Mers du Sud aux XVI et XVII siècles," *Journal Asiatique* (1920), pp. 5–312 (also published separately, Paris: Imprimerie Nationale, 1921).

[418]See Geneviève Bouchon, "Notes on the Opium Trade in Southern Asia during the Pre-Colonial Period," in Roderich Ptak and Dietmar Rothermund, eds., *Emporia, Commodities, and Entrepreneurs in Asian Maritime Trade, c. 1400–1750* (Stuttgart: Steiner Verlag, 1991), pp. 95–106; reprinted in her *Inde découverte, Inde retrouvée, 1498–1630: études d'histoire indo-portugaise* (Lisbon: Fundação Calouste Gulbenhian; and Paris: Centre Culturel Calouste Gulbenkian and Commission Nationale pour les Commémorations des Découvertes Portugaises, 1999), pp. 203–214.

[419]For discussion of the route from Portugal to India see T. Bentley Duncan, "Navigation Between Portugal and Asia in the Sixteenth and Seventeenth Centuries," in Cyriac K. Pullapilly and Edwin J. Van Kley, eds., *Asia and the West: Encounters and Exchanges from the Age of Explorations* (Notre Dame: Cross Cultural Publications, 1986), pp. 3–25; reprinted in Om Prakash, ed., *European Commercial Expansion in Early Modern Asia* (Aldershot, England; Brookfield, VT: Variorum, 1997); data from this article is also used in Jan de Vries, "Connecting Europe and Asia: A Quantitative Analysis of the Cape-Route Trade, 1497–1795," in Dennis O. Flynn, Arturo Giraldez, Richard von Glahn, eds., *Global Connections and Monetary History, 1470–1800* (Burlington, VT: Ashgate, 2003), pp. 35–106. For discussion of the significance of spices in European culture see Stefan Halikowski Smith, "The Mystification of Spices in the Western Tradition," *European Review of History* 8.2 (2001), pp. 119–136.

The manufacturer's authorised representative in the EU is Springer Nature Customer Service Centre GmbH, Europaplatz 3, 69115 Heidelberg, Germany. If you have any concerns regarding our products, please contact ProductSafety@springernature.com

Printed and bound by CPI Group (UK) Ltd, Croydon, CR0 4YY
30/09/2026
02225479-0004